Symposium on

Carbon-Fluorine Compounds

Chemistry, Biochemistry & Biological Activities, London, 1971

A Ciba Foundation Symposium

1972

Elsevier · Excerpta Medica · North-Holland
Associated Scientific Publishers · Amsterdam · London · New York

XS 1512

© *Copyright 1972 Ciba Foundation*

ISBN Excerpta Medica 90 219 4002 7
ISBN American Elsevier 0-444-10373-2

Library of Congress Catalog Card Number 72-76005

Published in 1972 by Associated Scientific Publishers, P.O. Box 3489, Amsterdam, and 52 Vanderbilt Avenue, New York, N.Y. 10017.
Suggested series entry for library catalogues: Ciba Foundation Symposia.

Printed in The Netherlands by Mouton & Co., The Hague

Carbon-Fluorine Compounds

Chemistry, Biochemistry & Biological Activities

The Ciba Foundation for the promotion of international cooperation in
medical and chemical research is a scientific and educational charity established by
CIBA Limited – now CIBA-GEIGY Limited – of Basle. The Foundation operates
independently in London under English trust law.

Ciba Foundation Symposia are published in collaboration with
Associated Scientific Publishers (Elsevier Scientific Publishing Company, Excerpta Medica,
North-Holland Publishing Company) in Amsterdam.

Associated Scientific Publishers, P.O. Box 3489, Amsterdam

Contents

Contributors

Symposium on Carbon-Fluorine Compounds: Chemistry, Biochemistry and Biological Activities
held at the Ciba Foundation, London, 13th-15th September 1971

Chairman: SIR RUDOLPH PETERS Department of Biochemistry, University of Cambridge, Tennis Court Road, Cambridge CB2 1QW

W. D. ARMSTRONG Department of Biochemistry, University of Minnesota, 227 Millard Hall, Minneapolis, Minnesota 55455, USA

J. E. G. BARNETT Department of Physiology and Biochemistry, The University of Southampton, Highfield, Southampton SO9 5NH

E. D. BERGMANN Department of Organic Chemistry, The Hebrew University of Jerusalem, Jerusalem, Israel

P. BUFFA Istituto di Patologia Generale, Università degli Studi di Modena, Via Campi 287, 41100 Modena, Italy

R. B. CAIN Biological Laboratory, The University of Kent at Canterbury, Beverley Farm, Canterbury, Kent

R. A. DWEK Department of Biochemistry, University of Oxford, South Parks Road, Oxford OX1 3QU

A. B. FOSTER Department of Chemistry, Chester Beatty Research Institute, Royal Cancer Hospital, Fulham Road, London SW3

L. FOWDEN Department of Botany and Microbiology, University College London, Gower Street, London WC1

E. M. GÁL Division of Neurobiochemistry, Department of Psychiatry, The University of Iowa, 500 Newton Road, Iowa City, Iowa 52240, USA

P. GOLDMAN Department of Health, Education and Welfare, Public Health Service, Building 10, Room 8D17, National Institutes of Health, Bethesda, Maryland 20014, USA

L. D. HALL Department of Biochemistry, The University of Sheffield, Sheffield S10 2TN

D. B. HARPER   Department of Biological Sciences, University of Kent at
    Canterbury, Canterbury, Kent
C. HEIDELBERGER   McArdle Laboratory for Cancer Research, Medical Center,
    University of Wisconsin, Madison, Wisconsin 53706, USA
P. W. KENT   Department of Biochemistry, University of Oxford,
    South Parks Road, Oxford OX1 3QU
E. KUN   Department of Pharmacology, University of California School of
    Medicine, San Francisco, California 94122, USA
G. W. MILLER   Huxley College of Environmental Studies, West Washington
    State College, Bellingham, Washington 98225, USA
B. C. SAUNDERS   Department of Organic and Inorganic Chemistry,
    University of Cambridge, Cambridge CB2 1QW
A. G. SHARPE   Department of Chemistry, University of Cambridge,
    Lensfield Road, Cambridge CB5 8BL
M. SHORTHOUSE   Department of Biochemistry, University of Cambridge,
    Tennis Court Road, Cambridge CB2 1QW
J. L. SHUPE   Department of Veterinary Science, Utah State University,
    Logan, Utah 84321, USA
D. R. TAVES   Department of Pharmacology and Toxicology, The University
    of Rochester School of Medicine and Dentistry, 260 Crittenden
    Boulevard, Rochester, New York 14620, USA
N. F. TAYLOR   Biochemistry Group, School of Biological Sciences,
    The University of Bath, Claverton Down, Bath BA2 7AY
P. F. V. WARD   Biochemistry Department, Agricultural Research Council
    Institute of Animal Physiology, Babraham, Cambridge
A. WETTSTEIN   CIBA-GEIGY Limited, CH-4002 Basle, Switzerland

Editors: KATHERINE ELLIOTT and JOAN BIRCH (The editors gratefully
    acknowledge the invaluable assistance of Dr P. W. Kent in the
    preparation of this book)

Introduction

SIR RUDOLPH PETERS

Department of Biochemistry, University of Cambridge

Some two years ago interest in the carbon-fluorine bond seemed to be increasing and Dr P. W. Kent and I thought that a meeting with an interdisciplinary approach to these problems might bring the present studies to a focus and provide a spring-board for fresh attacks on these problems.

Interest in the monofluorocarbon bond goes back to the curiosity of an organic chemist in Belgium in the last century. It was later picked up in Germany

FIG. 1. *Dichapetalum cymosum* (gifblaar). (Photograph kindly presented by Dr Dyer.)

FIG. 2. Seed of *Dichapetalum toxicarium*. The kernels contain fluoro-oleic acid and some other acid (from Peters *et al*. 1960).

as a lead to possible insecticides; and compounds with the carbon-fluorine bond became known in the United Kingdom early in the 1940's when it was imperative to look at the toxicity of these compounds, to be forewarned in case they were used against the allies in World War II. The main work was done in Cambridge by Dr Saunders and his group, and we shall hear from him about this (pp. 9–27). Until 1943 fluoroacetate was just an interesting chemical compound, possibly toxic, with a very stable carbon-fluorine bond, resistant even to hot concentrated sulphuric acid. But then the interest was suddenly widened by the discovery by J. S. C. Marais (1943) that fluoroacetate was the toxic principle of a well-known South African plant, *Dichapetalum cymosum* or gifblaar (Fig. 1). This plant grows in the Transvaal region and causes considerable losses of cattle, especially in the spring, when they eat the fresh shoots coming up among the rocks. These plants are spread by rhizomes and are very difficult to eradicate. The toxicity varies from a maximum in the South African spring to a minimum in their autumn. Many other varieties of *Dichapetalum* are now known in Africa—most of them less toxic. An interesting one is *D. toxicarium* from W.

FIG. 3. *Acacia georginae* (gidyea) trees growing in the vicinity of Alice Springs. (Photograph kindly presented by Mr L. R. Murray.)

Africa; the plant contains traces of fluoroacetate but the main toxicity is in the seeds (Fig. 2) and is due to fluoro-oleic acid (18 carbon atoms) and small amounts of shorter-chained fluoro acids with even numbers of carbon atoms. In these compounds the fluoride ion is in the ω-position. Nearly 70 years ago, Renner (1904) described how the 'local' doctors used the seeds of this plant to produce loss of motor activity and sensation, and presumably death.

Up to 1961 this was the extent of our plant knowledge, though it led to much interesting research and to the extensive use of fluoroacetate (called Compound 1080) to poison rodents and rabbits, especially in Australia and New Zealand. It was also found to be a good systemic insecticide for aphids in concentrations which did not kill the plants.

In 1961, while in Australia, I was invited to visit Drs J. H. Whittem and J. D. McConnell and Mr L. R. Murray at Alice Springs, where they thought a small acacia tree, *Acacia georginae* or gidyea (Fig. 3), was killing cattle, and that its toxicity was due to fluoroacetate (Murray, McConnell and Whittem 1961). I helped to prove this by sending samples of a poisoned guinea-pig back to England for analysis. Simultaneously, Oelrichs and McEwan (1961), working

FIG. 4. *Oxylobium parviflorum* Benth. (box poison).

in the Brisbane area, came to the same conclusion. The discovery not only widened the field, but made the finding of some cure even more necessary, as losses of cattle due to the gidyea could amount to £50 000 on one of the larger farms.

Since then the number of plants known to contain fluoroacetate has risen to about 36. Many have been found in Western Australia, where a campaign of

FIG. 5. *Gastrolobium bilobum* R.Br. (heart leaf poison).

eradication is in progress. The plants are varieties of *Gastrolobium* and *Oxylobium* (family, Leguminosae). The most toxic reported so far (Aplin 1969) are box poison (*Oxylobium parviflorum* Benth.) (Fig. 4) and heart leaf poison (*Gastrolobium bilobum* R. Br.) (Fig. 5). At their peak these plants may contain 12 500 p.p.m. dry weight fluoroacetate, and 28g wet leaves may be enough to kill a sheep of 50 kg. Toxic plants have also been found in South America; in

Fɪɢ. 6. *Palicourea marcgravii* (Rosaceae). [A drawing given by A. Saint-Hilaire (1825) *Mem. Mus. Hist. Nat., Paris* **12**, Planche 11.]

the *Palicourea marcgravii* (Fig. 6), the toxic agent is again fluoroacetate (Oliveira 1963). However, this plant belongs to another botanical family—the Rosaceae. It seems most likely that other plants will be found which can accumulate toxic amounts of carbon-fluorine compounds.

Apart from the obvious scientific interest in carbon-fluorine compounds, they are extremely toxic and human and animal accidents inevitably occur. An extreme example took place in the South of England a few years ago when fluoroacetamide spilt from a factory into fields, causing the death of many cows and putting the fields out of action for two years. So may I stress the importance of trying to find a cure for the poisoning, and hope that these days of discussion will lead to new discoveries.

References

APLIN, T. E. H. (1969) *J. Agr. West Aust.* **10**, No. 6 & 8.

MARAIS, J. S. C. (1943) *Onderstepoort J. Vet. Sci. Anim. Ind.* **18**, 203–206.

MURRAY, L. R., J. D. McCONNELL and J. H. WHITTEM (1961) *Aust. J. Sci.* **24**, 41–42.

OELRICHS, P. B. and T. McEWAN (1961) *Nature (Lond.)* **190**, 808.

OLIVEIRA, M. M. DE (1963) *Experientia* **19**, 586.

PETERS, R. A., R. J. HALL, P. F. V. WARD and N. SHEPPARD (1960) *Biochem. J.* **77**, 17–23.

RENNER, W. (1904) *Br. Med. J.* **1**, 1314.

SAINT-HILAIRE, A. (1825) *Mem. Mus. Hist. Nat., Paris* **12**, planche 11.

Chemical characteristics
of the carbon-fluorine bond

B. C. SAUNDERS

University Chemical Laboratory, Cambridge

Fluoroacetic acid, FCH_2COOH, was first prepared by Swarts (1896), but surprisingly he made no mention of any toxic effects. Certain Polish researchers (Gryskiewicz-Trochimowski, Sporzynski and Wnuk 1947) worked on fluoro-acetates before the Second World War but their results were not published until 1947.

During World War II Saunders* and co-workers at Cambridge, in collaboration with Cambridge biochemists, physiologists, pharmacologists and pathologists, carried out extensive work on a variety of organic compounds containing the carbon-fluorine (C−F) bond (McCombie and Saunders 1946; Saunders and Stacey 1948a; Saunders 1957).

Schrader, in Germany, also worked on C−F compounds (particularly for use as insecticides) during the war; his results, however, were not known until later.

Our work on C−F compounds is quite distinct from researches on phosphorus-fluorine compounds—the nerve gases (McCombie and Saunders 1946; Saunders and Stacey 1948b; Saunders 1957) carried out in Cambridge during the same period.

Reference must be made to the outstanding biochemical work Sir Rudolph Peters and his colleagues carried out during and after the war, and also to the work of Pattison in Canada, Chenoweth and Kharasch in America, and to the contributions made by Gál, Kent and Bergmann.

---

* For security reasons work on C-F compounds was not published during World War II, but our reports (which were made available to American workers) were regularly submitted to the Ministry of Supply from 1941.

FLUOROACETATES AND RELATED COMPOUNDS

Methyl fluoroacetate

The first compound to be investigated in detail was methyl fluoroacetate, FCH_2COOCH_3 (MFA). It was readily prepared by heating methyl chloroacetate and potassium fluoride (not sodium fluoride) in a rotating autoclave (Saunders and Stacey 1948a). This method formed the basis for the production of MFA and related substances on a large scale.

MFA is a mobile liquid, b.p. 104°C, m.p. −32°C, with an extremely faint odour.

During a 10-minute exposure to a lethal concentration of the vapour, small animals did not appear to be affected. After exposure no obvious symptoms developed for 30–60 minutes (depending on the concentration). The symptoms shown depended on the species, but all animals suffered convulsions from which a partial recovery was sometimes made. However, a recurrence of the convulsions would cause death.

The LC_{50} for rabbits, rats and guinea-pigs was about 0.1 mg/l. Mice were more resistant. In rabbits the LD_{50} by intravenous injection was 0.25 mg/kg body weight; in mice it was 7–10 mg/kg body weight for subcutaneous injections. MFA was also toxic by absorption through the skin, but less so than by other routes. When MFA was placed on the clipped backs of rabbits the LD_{50} was about 20 mg/kg body weight. Free evaporation of the drops was permitted, and the animals showed the usual symptoms.

By contrast a massive concentration of methyl chloroacetate vapour (1:1000; 4.85 mg/l) did not kill any animals.

MFA is practically odourless. When exposed to a concentration of 1:1 000 000 in a 10 m³ chamber, we were unable to detect the compound. Even at a concentration of 1:100 000 (30 seconds for reasons of safety) the compound possessed only a faint fruit-like odour indistinguishable from that of many harmless esters not containing fluorine.

The fluorine atom in methyl fluoroacetate (and in many other compounds containing the ·CH_2F group) is firmly bound. When MFA was boiled with 20% alcoholic potassium hydroxide for 5 minutes no free fluoride was formed. Even after some 20 hours there was only about a 50% conversion into potassium fluoride. The C—F bond is not broken by boiling 100% sulphuric acid.

This stability is in marked contrast to chloroacetic acid, bromoacetic acid and iodoacetic acid which have chemically very active halogens and are powerful lacrimators. Fluoroacetic acid and its esters have no lacrimatory properties.

Fluoroethanol

It was necessary to prepare fluoroethanol, FCH_2CH_2OH (FEA), to enable its toxicity to be investigated and for use as a starting-point in the synthesis of other fluorine-containing compounds (Saunders, Stacey and Wilding 1949). Fluoroethanol, b.p. 103–105 °C, is a stable mobile liquid, completely miscible with water and practically odourless. At comparable concentrations it produced convulsions in animals similar to those caused by MFA. The fluorine atom (β to OH) in FEA is much more reactive than in MFA (see p. 22, 23). Some 300 carbon-fluorine compounds were synthesized from FEA and MFA and from the results of gross toxicity tests it was possible to deduce certain relationships between chemical constitution and toxic action.

It was of interest to determine whether other esters of fluoroacetic acid would be more or less toxic than the methyl ester. In the phosphorofluoridate series, for example, we found that esters of secondary alcohols were far more potent than those of primary alcohols; for instance, di-isopropyl phosphoro-fluoridate (I) was a compound of great activity (Saunders and Stacey 1948*b*). Accordingly ethyl, *n*-propyl and isopropyl fluoroacetates were prepared. The toxicity figures for these esters were very similar to those for methyl fluoro-acetate.

$$(CHMe_2 \cdot O)_2 POF \qquad CHMeF \cdot CO_2Me \qquad CMe_2F \cdot CO_2Me$$

$$\text{(I)} \qquad\qquad \text{(II)} \qquad\qquad \text{(III)}$$

It was next important to determine the effect of altering the groups adjacent to the fluorine atom. Thus methyl α-fluoropropionate (II) and methyl α-fluoro-isobutyrate (III) were prepared. Both these compounds were non-toxic. This result indicated that the $FCH_2CO \cdot$ group is necessary to produce a 'fluoro-acetate-like' toxicity. This was confirmed by the following results.

Fluoroacetyl chloride, FCH_2COCl	Toxicity similar to that of methyl fluoroacetate.
Chloroacetyl fluoride, $ClCH_2COF$	Non-toxic. Labile fluorine.
Fluoroacetyl fluoride, FCH_2COF	Toxicity similar to that of methyl fluoroacetate. One fluorine stable.

These findings were in accordance with expectation, and it was becoming obvious that the toxicity was bound up with the $FCH_2CO \cdot$ group, whereas the $-C{\overset{\displaystyle O}{\underset{\displaystyle F}{}}}$ group was ineffective. Further confirmation of this point was provided by the observation that ethyl fluoroformate, $FCOOC_2H_5$ (which contains the

$-C\overset{\displaystyle O}{\underset{\displaystyle F}{\Big\langle}}$ group and has an easily hydrolysable fluorine), was non-toxic.

2-*Fluoroethyl fluoroacetate*

As fluoroethanol produced a toxic effect comparable with that of fluoroacetic acid, it seemed worth while synthesizing a compound in which the 'active' parts of these molecules were combined, in the hope of obtaining a substance of increased potency. We prepared (Saunders and Stacey 1949) 2-fluoroethyl fluoroacetate, $FCH_2COOCH_2CH_2F$. This compound was found to possess greatly enhanced toxic properties, and a 10-minute exposure to 0.092 mg/l killed 70% of a batch of rabbits, guinea-pigs and rats. The LC_{50} for rabbits by inhalation was 0.05 mg/l; that for MFA is 0.1 mg/l. In short, the compound was about twice as toxic as methyl fluoroacetate (weight for weight). In addition it possessed an extremely faint odour.

Certain other 2-fluoroethyl esters also showed increased toxicities and so it seems that these esters have some kind of toxic action *per se*, independent of hydrolysis products.

As might be expected, fluoroacetamide, FCH_2CONH_2, a white crystalline compound (with an unreactive fluorine atom), exhibits a toxicity comparable with that of methyl fluoroacetate or fluoroacetic acid. Fluoroacetamide would readily hydrolyse to fluoroacetic acid in the mammalian body. Similarly the following substituted amides (Buckle, Heap and Saunders 1949) were found to be toxic, the essential feature for this being the $FCH_2CO\cdot$ group: $FCH_2CO-NHCH_3$; $FCH_2CON(NO)CH_3$; $FCH_2CONHCH_2CH_2OH$; $FCH_2CONH-CH_2CH_2Cl$; $FCH_2CON(CH_2CH_2Cl)_2$.

The compounds (IV) and (V) were readily prepared (Saunders 1949) and were found to be relatively non-toxic. This is consistent with the observation that if the fluorine compound is readily hydrolysable or oxidizable to fluoroacetic acid, then the compound is toxic. If, on the other hand, the compound is not readily convertible to fluoroacetic acid then it will not be toxic.

Compounds (VI) and (VII), as might be expected, showed considerable 'fluoroacetate' toxicity as both these compounds can be hydrolysed to fluoro-ethanol and thence oxidized to fluoroacetic acid:

$$\overset{+}{N}H_3CH_2COOCH_2CH_2F \quad Cl^- \quad (VI)$$

$$Me_3\overset{+}{N}CH_2COOCH_2CH_2F \quad Cl^- \quad (VII)$$

Sesqui-H, $ClCH_2CH_2SCH_2CH_2SCH_2CH_2Cl$, is known to be a powerful vesicant, but the fluorine analogue (VIII) of sesqui-H is neither a vesicant nor does it cause fluoroacetate poisoning, for this compound

$$FCH_2CH_2SCH_2CH_2SCH_2CH_2F$$

$$(VIII)$$

is not likely to be metabolized in the animal body to a compound containing the $FCH_2CO\cdot$ group. Further, the fluorine atom (unlike the chlorine in the corresponding chloro compound) is chemically unreactive and so is unlikely to produce vesication.

ω-*Fluorocarboxylic acids*

We demonstrated (Saunders 1947; Buckle, Pattison and Saunders 1949) a striking alternation in the toxic properties of ω-fluorocarboxylic acid as follows:

FCH_2COOR	toxic	$F(CH_2)_5COOR$	toxic
FCH_2CH_2COOR	non-toxic	$F(CH_2)_7COOR$	toxic
$F(CH_2)_3COOR$	toxic	$F(CH_2)_{10}COOR$	non-toxic
$F(CH_2)_4COOR$	non-toxic	$F(CH_2)_{11}COOR$	toxic

Knoop in 1904 suggested that fatty acids were oxidized in the animal body by the loss of two carbon atoms at a time owing to oxidation occurring at the carbon atom which was in the β position with respect to the carboxyl group (1).

$$R\overset{\beta}{C}H_2CH_2COOH \longrightarrow R.COOH + CH_3COOH \quad (1)$$

It will be readily seen in the $F(CH_2)_nCOOH$ series above that when n is an odd number this process of β-oxidation would yield the toxic fluoroacetic acid, whereas when n is even, the compound will be oxidized only as far as the non-toxic β-fluoropropionic acid, FCH_2CH_2COOH (or perhaps to the unstable and non-toxic fluoroformic acid). Our results are in complete accord with this hypothesis and provide a proof of a kind not hitherto achieved of the process of β-oxidation in the living animal body.

If the theory of alternating toxicities is right, then if the β-position in the chain is 'blocked' so that oxidation cannot take place, then the compound should be devoid of toxic properties. For this purpose we synthesized ethyl β,β-dimethyl-γ-fluorobutyrate (IX),

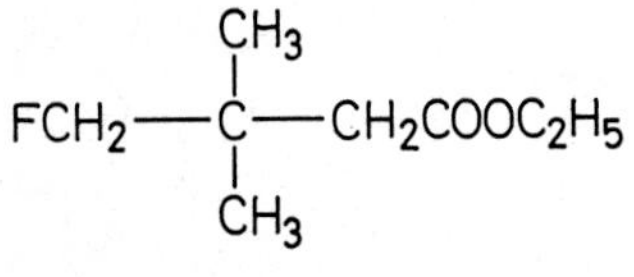

$$FCH_2-\overset{\displaystyle CH_3}{\underset{\displaystyle CH_3}{\overset{|}{\underset{|}{C}}}}-CH_2COOC_2H_5$$

(IX)

and showed that it was indeed non-toxic. Another way of testing the theory was to build a carbon-ring on the β-carbon atom so that the animal body could not oxidize the compound to fluoroacetic acid. Accordingly we synthesized the compound (X) and showed that it was non-toxic (Pattison and Saunders 1949), whereas the parent compound (XI) is extremely toxic.

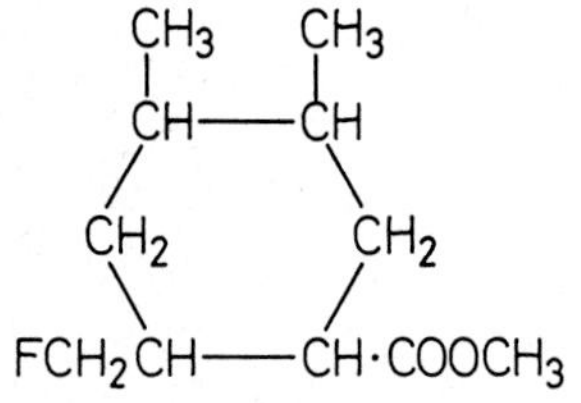

(X) $FCH_2CH_2CH_2COOCH_3$

(XI)

p-Fluorophenylacetic acid (XII), which has the carbon skeleton of the highly toxic 5-fluorohexanoate (XIII),

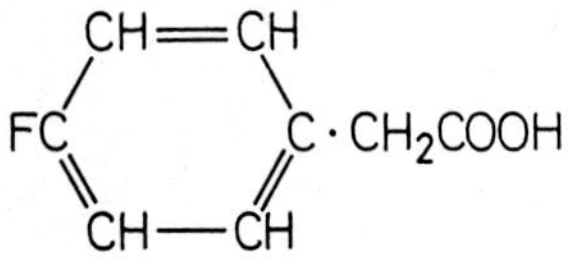

(XII) 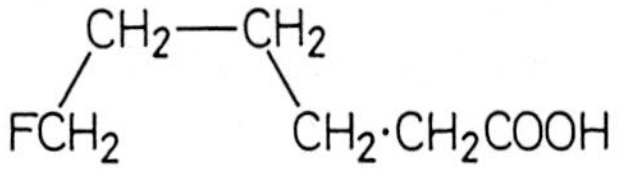   (XIII)

was found to be non-toxic, and, of course, is unlikely to be oxidized to fluoro-acetic acid. Pattison (1959) has investigated several other homologous series and has found comparable alternations in toxicities.

CAUSE OF TOXIC ACTION OF CARBON-FLUORINE COMPOUNDS

Unlike the fluoroacetates, phosphorofluoridates have a quick 'knock-out' action and are toxic by virtue of their powerful anticholinesterase activity. As Sir Rudolph Peters will explain (p. 55), he and his colleagues have clearly demonstrated that poisoning by fluoroacetate is brought about by blocking the tricarboxylic acid cycle *in vivo*. They have proved that fluoroacetate is converted into fluorocitrate (2) which blocks the enzyme aconitase. This enzyme (within the mitochondria) is concerned with the conversion of *cis*-aconitate to either citric acid or isocitric acid. Accumulation of citric acid should therefore be a characteristic feature of fluoroacetate poisoning and indeed is.

$$FCH_2COOH \longrightarrow \begin{array}{l} F \cdot CH \cdot COOH \\ HO \cdot C \cdot COOH \\ CH_2COOH \end{array} \quad (2)$$

Such a mechanism would appear to be dependent upon (*a*) the stability of the C—F bond in the fluoroacetate; and (*b*) the small size of the fluorine atom. γ-Fluorobutyrate is converted by β-oxidation to fluoroacetate and thence to fluorocitrate. It has been noted that the fluorine atom in β-fluoropropionic acid is very labile and is easily removed from the molecule (Mirosevic-Sorgo and Saunders 1959). Hence β-fluoropropionic acid would be expected to be non-toxic, which it is.

Difluoroacetic acid (XIV) and trifluoroacetic acid (XV) are non-toxic although the C—F bonds in these compounds are extremely stable. Here, however, because the fluorine atoms are slightly larger than the hydrogen atoms, the molecules can no longer mimic acetic acid.

$$\begin{array}{l} F \\ \diagdown \\ CH \cdot COOH \\ \diagup \\ F \end{array} \qquad \begin{array}{l} F \\ \diagdown \\ F{-\!\!-}C \cdot COOH \\ \diagup \\ F \end{array}$$

$$(XIV) \qquad\qquad (XV)$$

A CHEMICAL SYNTHESIS OF MONOFLUOROCITRIC ACID

Apart from the biochemical syntheses of this compound so ably demonstrated by Peters (1954), the first chemical synthesis was achieved by Rivett (1953), but his compound contained several impurities. We have recorded (Brown and

Saunders 1962) an independent synthesis of diethyl monofluorocitrate which appears to give a purer product.

The new synthesis consists in condensing malonic acid with ethyl ethoxalyl fluoroacetate (XVI) in pyridine. Diethyl monofluorocitrate (XVII) is thus

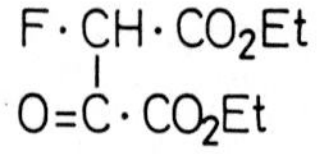

$$F \cdot CH \cdot CO_2Et$$
$$O = C \cdot CO_2Et$$

(XVI)

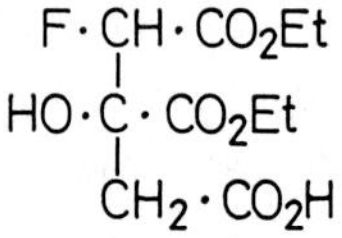

$$F \cdot CH \cdot CO_2Et$$
$$HO \cdot C \cdot CO_2Et$$
$$CH_2 \cdot CO_2H$$

(XVII)

obtained as an oil and acid hydrolysis gives the required free monofluorocitric acid. The latter, being very hygroscopic, is isolated as the barium salt in 50% overall yield. Blank, Mager and Bergmann (1955) have shown that 91–92% of ethyl ethoxalyl fluoroacetate exists in the keto form, and we have confirmed this by n.m.r. examination (Brown and Saunders 1962).

Preliminary biochemical tests with aconitase have shown that the barium fluorocitrate has an inhibitory power of the same order as that of the most toxic component of a sample prepared according to Rivett's method.

It was important to be able to prepare a stable crystalline derivative of α-fluorocitric acid. One method was to convert diethyl α-fluorocitrate by means of diazomethane into diethyl methyl α-fluorocitrate (XVIII), which crystallized on standing.

We draw attention to the possibility of α-fluorocitric acid taking up a conformation in which there is a maximum amount of hydrogen bonding and in which three six-membered rings are formed (XIX). This restricted-rotational form may thereby have increased stability.

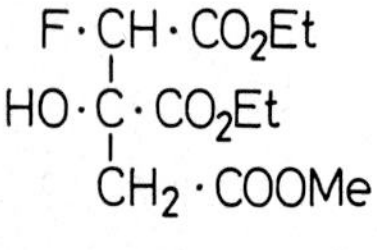

$$F \cdot CH \cdot CO_2Et$$
$$HO \cdot C \cdot CO_2Et$$
$$CH_2 \cdot COOMe$$

(XVIII)

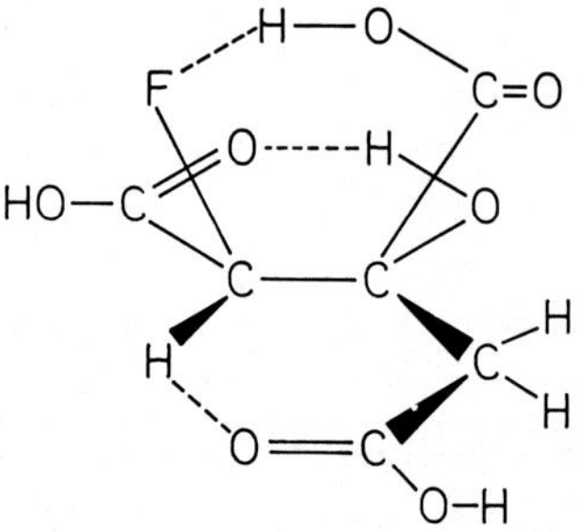

(XIX)

THE REACTIONS BETWEEN BENZYLAMINE AND α-FLUOROCITRIC ACID AND SOME OTHER FLUORINE COMPOUNDS

The tribenzylamide of citric acid is readily prepared as a well-defined highly crystalline solid with a sharp melting-point (Mann and Saunders 1971). This compound can be obtained from the free acid or from one of its esters with benzylamine and a catalyst (3).

$$\begin{array}{ccc} CH_2COOH & & CH_2 \cdot CONHCH_2C_6H_5 \\ | & 3C_6H_5CH_2NH_2 & | \\ C(OH)COOH & \xrightarrow{\hspace{2cm}} & C(OH)CONHCH_2C_6H_5 \quad (3) \\ | & & | \\ CH_2COOH & & CH_2 \cdot CONHCH_2C_6H_5 \end{array}$$

It was hoped that α-fluorocitric acid would similarly give a characteristic tribenzylamide. With this in mind the action of benzylamine on two model compounds was investigated in detail (Brown and Saunders 1971, unpublished).

In the reaction between ethyl chloroacetate and benzylamine, ethyl benzyl-aminoacetate hydrochloride (XX) is formed first (4), and can be hydrolysed to the corresponding substituted acetic acid. Further reaction of (XX) with benzylamine gives benzylaminoacetobenzylamide hydrochloride (XXI) (5).

$$Cl \cdot CH_2COOC_2H_5 \xrightarrow{C_6H_5CH_2NH_2} (C_6H_5CH_2NH_2CH_2COOC_2H_5)^+Cl^- \quad (4)$$
$$(XX)$$

$$\xrightarrow{C_6H_5CH_2NH_2} (C_6H_5CH_2NH_2CH_2CONHCH_2C_6H_5)^+Cl^- \quad (5)$$
$$(XXI)$$

The chlorine is therefore more easily removed than the ethoxide group $(-OC_2H_5)$ of the ester.

When an equimolecular mixture of ethyl fluoroacetate and benzylamine was left to stand at room temperature for 24 hours, the benzylamide of fluoroacetic acid (XXII) was obtained in 75% yield (6).

$$FCH_2COOC_2H_5 \xrightarrow{C_6H_5CH_2NH_2} FCH_2CONHCH_2C_6H_5 \quad (6)$$
$$(XXII)$$

This benzylamide of fluoroacetic acid was then heated under reflux with more benzylamine for 6 hours, after which 25% of the fluoroacetobenzylamide remained unchanged and 65% of the benzylaminoacetobenzylamide hydrofluoride (XXIII) was obtained.

$$(C_6H_5CH_2NH_2CH_2CONHCH_2C_6H_5)^+F^-$$
$$(XXIII)$$

This means that the fluorine (unlike chlorine) is less easily replaced than the ethoxide ($-OC_2H_5$) of the ester. In fact the fluorine is remarkably resistant to prolonged treatment with benzylamine at high temperatures.

ATTEMPTED FORMATION OF TRI-*N*-BENZYLAMIDE OF α-FLUOROCITRIC ACID

The reactions described above gave reasonable hope that a tribenzylamide of α-fluorocitric acid would be readily obtainable by reaction with benzylamine and a catalyst. Accordingly diethyl α-fluorocitrate, benzylamine and ammonium chloride were warmed on a water-bath. Rather surprisingly the dibenzylamide of oxalic acid was obtained in 30% yield together with a compound not containing fluorine and so far not identified (7).

$$\begin{array}{c} \text{FCH·COOEt} \\ | \\ \text{C(OH)COOEt} \\ | \\ \text{CH}_2\text{COOEt} \end{array} \xrightarrow{\text{C}_6\text{H}_5\text{CH}_2\text{NH}_2} \begin{array}{c} \text{CO·NHCH}_2\text{C}_6\text{H}_5 \\ | \\ \text{CO·NHCH}_2\text{C}_6\text{H}_5 \end{array} \quad (7)$$

This isolation of a derivative of oxalic acid prompted us to investigate the action of strong bases on diethyl and triethyl α-fluorocitrate and triethyl α,α'-difluorocitrate. On warming each of these esters with 30% sodium hydroxide solution, sodium oxalate was obtained in over 80% yield. Citric acid does not behave analogously.

Although the above decompositions to oxalate take place under vigorous conditions, one is tempted to speculate whether appropriate enzymes under mild biological conditions might convert α-fluorocitrate in part to toxic oxalic acid.

THE REACTION BETWEEN DIETHYLOXALYL α-FLUOROACETATE AND PYRIDINE

In one of our preparations of diethyl α-fluorocitrate, the diethyloxalyl α-fluoroacetate (XXIV) was added to pyridine and pyridinium malonate solution over 5 minutes (Brown and Saunders 1971, unpublished). In addition to the expected diethyl α-fluorocitrate, a neutral white crystalline solid, m.p.

108 °C, was obtained (8). This compound contained no fluorine, and after recrystallization from ethanol, analysis showed it to be $C_{17}H_{19}O_6N$. It was stable and odourless. The reaction was then repeated without the malonic acid and the same crystalline compound was again obtained.

Infrared and n.m.r. spectra and general reactions showed that the compound was triethyl indolizine 1,2,3-tricarboxylate (XXV).

$$FCH \cdot CO \cdot COOC_2H_5 \longrightarrow \text{(XXV structure)} \quad (8)$$

$$\underset{COOC_2H_5}{|}$$

(XXIV)   (XXV)

For this strange reaction, the following mechanism (9) is suggested:

(9)

ENZYMIC CLEAVAGE OF A CARBON-FLUORINE BOND

The South African plant *Dichapetalum cymosum* (gifblaar) contains potassium fluoroacetate and hence is poisonous to cattle. As explained above, the chemical synthesis of fluoroacetate requires a high temperature. It is not known how the plant builds up fluoroacetic acid, but it is certain that it involves inorganic fluoride. The enzymes involved in this synthesis are not known, nor indeed until recently were the enzymes concerned with the reverse process, that is the breaking of the C—F bond. When the plant dies fluoroacetate does not appear to accumulate in the soil to any great extent, but seems to revert to fluoride. Dr Goldman (1972) has now broken the C—F bond in fluoroacetate by enzymes.

In *p*-fluoroaniline the fluorine atom is very firmly bound, but we have been able to break it by a single pure enzyme (Hughes and Saunders 1954).

In the course of our investigations on peroxidase-catalysed oxidations we effected an enzymic cleavage of the C—F bond as follows. In acetate buffer (pH 4.5) and at room temperature *p*-fluoroaniline was readily oxidized by hydrogen peroxide and peroxidase to give mainly the red crystalline 2-amino-5-*p*-fluoroanilinobenzoquinone di-*p*-fluoroanil (XXVI) (10).

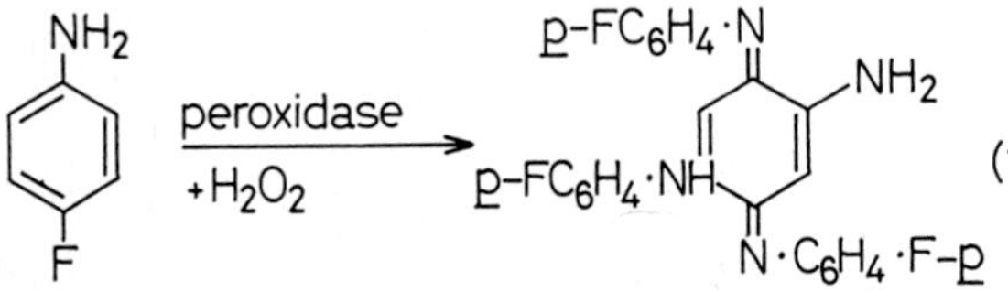

(10)

(XXVI)

The formation of (XXVI) requires the elimination of one fluorine atom per four molecules of amine. The fluorine was expelled as F^-, and since the enzyme reaction is retarded by F^-, the process is self-poisoning. The reaction stopped at 30% completion and F^- was detected as a product.

We believe that the oxidation of *p*-fluoroaniline was the first recorded case of an enzymic cleavage of a C—F bond.

ADDITIONAL OBSERVATIONS OF THE CLEAVAGE OF THE CARBON-FLUORINE BOND BY ALKALI AND OTHER REAGENTS

It is frequently stated, in general terms, that the carbon and fluorine atoms in fluoroacetates and related compounds are very firmly bound. By comparison with the corresponding chloro, bromo and iodo compounds this is true; nevertheless there are a number of instances where the C—F link is easily ruptured, particularly by nucleophilic reagents. Attention is drawn to the action of sodium hydroxide on 'fluoroacetates' in connection with analytical methods for determining fluorine content.

When methyl fluoroacetate or fluoroacetic acid is heated under reflux with 10% aqueous sodium hydroxide solution for one hour no F^- is produced. Rather surprisingly little serious attention has been paid to the action of boiling concentrated (i.e. 30%) aqueous alkali on the stability of the C—F link in these compounds. We therefore investigated the reaction in detail by heating the fluoro compound under reflux with a large excess of 30% sodium hydroxide solution for one hour. The free fluoride was determined by precipitation as lead chlorofluoride followed by a Volhard volumetric estimation (Chapman, Heap and Saunders 1948). Almost all the fluorine was removed under these conditions. In some instances the fluorine was eliminated within 15 minutes.

Trifluoroacetic acid when heated under reflux for one hour with 30% sodium hydroxide solution gave no fluoride ion.

Hoffmann (1949) observed the reactivity of the halogen atoms in a series of 1-bromo- and 1-chloro-ω-fluoroalkanes using M-sodium ethoxide in ethanol and M-potassium hydroxide solution in 70% ethanol. He concluded that, in general, there was great resistance to nucleophilic attack when the C—F bond was not influenced by neighbouring groups. The highest percentage of fluorine hydrolysed was 11% compared with 82% of bromine and 52% of chlorine in a wide range of experiments under standard conditions. Also, as the fluorine atom becomes further separated from the other halogen atom in the molecule, so resistance to hydrolysis increases.

In the compound FCH_2CH_2Br, the bromine atom is more reactive than the fluorine atom (e.g. 11) (Saunders, Stacey and Wilding 1949).

$$FCH_2CH_2Br \xrightarrow{\text{KSCN}} FCH_2CH_2SCN \qquad (11)$$

Nevertheless the bromine atom is found to be less reactive than might be expected. No doubt the carbon-bromine (C—Br) bond is strengthened by the strong inductive effect of the adjacent fluorine atom. The electrons in the C—Br bond are held more closely to the carbon atom than in CH_3CH_2Br, thus making their loss from the negatively charged bromine atom more difficult (12).

$$F \twoheadleftarrow \overset{\delta++}{C} \text{———} \overset{\delta+}{C} \text{———} Br \qquad (12)$$

Because of slow hydrolysis FCH_2CH_2Br and FCH_2CH_2Cl exhibit relatively low toxicities.

It is evident from the above that any reaction at the C—F bond would have to be carried out with some strong nucleophilic reagent to have any chance of success. The obvious choice was the Grignard reagent. It is known that phenyl magnesium bromide converts methyl chloroacetate into 1,1-diphenyl-2-chloro-ethanol in small yield. We have shown that when methyl fluoroacetate reacted under reflux, without cooling, with three moles of phenyl magnesium bromide, 1,2,2-triphenylethanol* was produced (13) (Mirosevic-Sorgo and Saunders 1959). This facile rupture of the C—F bond was unexpected, and separate experiments showed that at least 77.5% of the fluorine was eliminated as fluoride ion.

$$FCH_2COOMe \xrightarrow{\text{3PhMgBr}} Ph_2CH \cdot CH(OH)Ph \qquad (13)$$

When the reaction was carried out at $0\,°C$, 1,1-diphenyl-2-fluoroethanol was obtained. Three moles of phenyl magnesium bromide reacted with one mole of fluoroacetyl chloride, without cooling, to give 1,2,2-triphenylethanol.

Although Saunders, Stacey and Wilding (1949) demonstrated the difference in reactivity of the fluorine and bromine atoms in 1-fluoro-2-bromoethane, ethanolic potassium cyanide will remove both the halogen atoms from 1-fluoro-2-bromoethane, giving ethylene dicyanide in 50% yield (Mirosevic-Sorgo and Saunders 1959).

Since alcoholic potassium cyanide solution does not convert methyl fluoro-acetate to methyl cyanoacetate (but yields only potassium fluoroacetate), and in view of Hoffmann's experiments cited above, the following mechanism (14) is suggested for the action of potassium cyanide on 1-fluoro-2-bromoethane:

$$CH_2F \cdot CH_2 \cdot Br \xrightarrow{\text{KCN}} CH_2F \cdot CH_2 \cdot CN \xrightarrow{\text{-HF}} CH_2 = CH \cdot CN \xrightarrow{\text{HCN}} \qquad (14)$$
$$CH_2CN \cdot CH_2 \cdot CN$$

This emphasizes that the C—F bond is weakened when the carbon is β to a positive carbon atom.

The ease with which the β-fluorine atom can be lost is demonstrated by the difficulty in preparing β-fluoropropionic acid by any substitution reaction. Also, whereas boiling 10% sodium hydroxide solution did not remove any fluorine from methyl fluoroacetate in one hour, it removed all the fluorine

---

* One might expect that 1,1,2-triphenylethanol would be formed, but this is not so. A mixed m.p. showed that our product (1,2,2-triphenylethanol) and the compound obtained by reducing 1-bromo-1,1-diphenylacetophenone were identical.

from β-fluoropropionic acid under the same conditions. Even on allowing β-fluoropropionic acid to stand at room temperature for 24 hours with 10% aqueous sodium hydroxide, removal of 94.6% of the fluorine took place.

It was considered that, in the next homologue γ-fluorobutyric acid, the fluorine was too far removed from the carboxyl group to undergo facile elimination as hydrogen fluoride.

Methyl γ-fluorobutyrate was allowed to react with three equivalents of phenyl magnesium bromide in the cold, followed by heating under reflux. A crystalline solid, m.p. 65–66 °C, was obtained. The compound was fluorine-free, and the analysis and reactions agreed with that of 2,2-diphenyltetrahydrofuran. The following pathway (15) for the reaction is suggested:

$$F(CH_2)_3COOCH_3 \xrightarrow{2PhMgBr} F(CH_2)_3\overset{\overset{\displaystyle Ph}{|}}{\underset{\underset{\displaystyle Ph}{|}}{C}}\text{-OH} \xrightarrow{PhMgBr} \begin{array}{c}\text{(furan ring)}\end{array}\!\!\overset{Ph}{\underset{Ph}{}} \quad (15)$$

Under the same conditions phenyl magnesium bromide and methyl γ-chloro-butyrate failed to produce the diphenyltetrahydrofuran.

It has often been stated that in fluoroethanol the fluorine atom is unaffected by all but the most drastic treatment: this requires considerable modification. We found (Bronnert and Saunders 1960) that when fluoroethanol was allowed to stand at room temperature for 12 hours with 10% sodium hydroxide solution, 80% of the fluorine appeared as fluoride. In these circumstances the C−F bond in MFA was not affected.

On heating 2-fluoroethanol with phenyl magnesium bromide in ether, the fluorine atom was eliminated and phenylethanol resulted. The reaction most probably follows the course (16):

$$FCH_2CH_2OH \xrightarrow{PhMgBr} FCH_2CH_2OMgBr \longrightarrow MgBrF + \underset{\underset{\displaystyle O}{\diagdown\!\diagup}}{CH_2\text{--}CH_2}$$

$$\xrightarrow{PhMgBr} PhCH_2CH_2OMgBr \qquad (16)$$

SOME REACTIONS OF ω-FLUOROACETOPHENONE

In the course of our investigations into the stability of the carbon-fluorine bond in 2-fluoroethanol, we considered the stability of ω-fluoroacetophenone (Bronnert and Saunders 1965). When a methanolic solution of ω-fluoroaceto-phenone was treated with a methanolic solution of sodium methoxide, a yellow colour appeared and deepened to orange and a white precipitate came down.

The latter was an organic solid (contaminated with sodium carbonate and sodium fluoride). It crystallized from methanol in long white acicular crystals and was proved by analysis, degradation and i.r., u.v., and n.m.r. spectra to be 2,3-epoxy-4-fluoro-1,3-diphenylbutan-1-one (XXVII). The filtrate from the reaction yielded a dark red solid with a low melting-point—2-fluoro-1,3-diphenylbutan-1-one-4-al (XXVIII). A fluoride determination showed that 65% of the fluorine in the original ω-fluoroacetophenone had been eliminated as F⁻.

The reactions postulated to account for the formation of (XXVII) and (XXVIII) are set out below (17).

$$C_6H_5CO\ CH_2F \longrightarrow C_6H_5COCHF-\overset{\overset{\displaystyle C_6H_5}{|}}{\underset{\underset{\displaystyle OH}{|}}{C}}-CH_2F \qquad (17)$$

$$C_6H_5COCH\overset{\displaystyle C_6H_5}{-\!\!-\!\!C} \qquad \overset{\displaystyle C_6H_5}{C}-\!\!-\!\!CH_2$$

(XXVII)   (XXVIII)

$$C_6H_5COCHF-\overset{\overset{\displaystyle C_6H_5}{|}}{C}H-CHO$$

The first step is an 'aldol' condensation giving a compound containing two fluorine atoms, each of which is α to the hydroxyl group. We suggested in our work on 2-fluoroethanol that compounds of this type readily lose hydrogen fluoride to form an epoxide. In the reaction between sodium methoxide and ω-fluoroacetophenone, the primary product can eliminate hydrogen fluoride in two ways to give two different compounds. The present work shows that both are formed, and that the proportions are comparable (17).

By heating (XXVII) with methanolic hydrogen chloride we have been able to open the epoxide ring and have thus obtained 2-chloro-4-fluoro-3-hydroxy-1,3-diphenylbutan-1-one (XXIX) and 2,3-dihydro-3-oxo-2,4-diphenylfuran (XXX) (18).

On treating (XXVIII) with concentrated hydrochloric acid, 3-fluoro-2,4-diphenylfuran (XXXI) was readily obtained. Compound (XXXI) was readily brominated by bromine in chloroform to give 5-bromo-3-fluoro-2,4-diphenyl furan (XXXII) (19).

Since (XXVIII) is effectively a 1,4-diketone, it should be possible to obtain other heterocycles containing fluorine in the 3-position, and if this reaction is general with ketones containing the FCH_2CO group, a wide range of heterocycles can be envisaged.

$$(XXVII) \xrightarrow{HCl} C_6H_5COCH-\underset{\underset{Cl}{|}}{C}\underset{\underset{OH}{|}}{\overset{C_6H_5}{\diagup}}_{CH_2F} \quad (XXIX)$$

$$\longrightarrow C_6H_5COCH-\underset{\underset{OH}{|}}{C}\underset{\underset{OH}{|}}{\overset{C_6H_5}{\diagup}}_{CH_2F}$$

$$\xrightarrow{-HF} C_6H_5COCH-\underset{\underset{OH}{|}}{C}\underset{O}{\overset{C_6H_5}{|}}-CH_2 \quad (18)$$

$$\longrightarrow C_6H_5COCH-\underset{\underset{OH}{|}}{CH}-\underset{\underset{C_6H_5}{|}}{CH}:O$$

$$\longrightarrow C_6H_5-\underset{\underset{OH}{|}}{C}=\underset{}{\overset{OH}{|}}C-\underset{\underset{OH}{|}}{C}=CH$$

$$\longrightarrow C_6H_5\text{-furan-}OH\text{,}C_6H_5 \quad (XXX)$$

$$(XXVIII) \longrightarrow C_6H_5\cdot\underset{\underset{OH}{|}}{C}=CF-\underset{\underset{OH}{|}}{C}=CH \longrightarrow C_6H_5\text{-furan-}F\text{,}C_6H_5 \quad (XXXI)$$

$$(19)$$

$$C_6H_5\text{-furan-}F\text{,}C_6H_5 \xrightarrow[CHCl_3]{Br_2} C_6H_5\text{-furan-}F, Br\text{,}C_6H_5 \quad (XXXII)$$

EPILOGUE

In the early stages of work on C—F compounds many of them were made as potential chemical warfare agents. Fortunately they were never employed for this purpose. Instead it has been an example of beating the proverbial swords into ploughshares. Much has been revealed on the relationship between chemical constitution and pharmacological activity. The role of many enzyme systems has become much clearer and many C—F compounds are of clinical interest.

SUMMARY

A brief historical background and a broad general survey are given. Reference is made to the very different kinds of toxicity associated with fluorine in one form or another, e.g. FCH_2COOR and $(RO)_2POF$. Compounds containing the CH_2F group are convulsant poisons with a delayed action provided that metabolism can give rise to fluoroacetic acid, FCH_2COOH, and thence to fluorocitric acid,

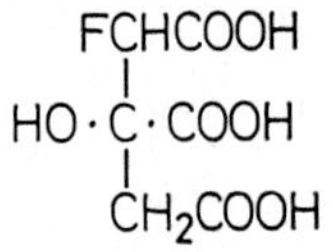

$$
\begin{array}{l}
\text{FCHCOOH} \\
\quad | \\
\text{HO·C·COOH} \\
\quad | \\
\text{CH}_2\text{COOH}
\end{array}
$$

Relationships between toxicity, chemical constitution and chemical characteristics of the C—F bond are discussed.

The striking alternation of physiological effects of the ω-fluorocarboxylic acids, $F(CH_2)_nCOOH$, is considered and the involvement of β-oxidation stressed.

The stability of the C—F bond towards a variety of reagents (including Grignard reagents and organic bases) are discussed and some reactions of fluorocitric acid described.

NOTE ON NOMENCLATURE

So much chemistry has been built up around accepted trivial names that the α, β, γ designations are used here for convenience. 'Trivial names are normally confined to the C_1–C_5 aliphatic acids and to the common fatty acids from C_{14} to C_{18}. Numerals (COOH = 1) are used except that, owing to a confusing earlier usage, Greek letters are used only with trivial names of C_2–C_5 acids ($\alpha = 2$, $\beta = 3$, $\gamma = 4$, $\delta = 5$)'. [Quoted from *Handbook for Chemical Society Authors*, special publication No. 14, pp. 108–109n (1960) The Chemical Society, London.]

References

BLANK, I., J. MAGER and E. D. BERGMANN (1955) *J. Chem. Soc.* 2190.
BRONNERT, D. L. E. and B. C. SAUNDERS (1960) *Tetrahedron* 10, 160.
BRONNERT, D. L. E. and B. C. SAUNDERS (1965) *Tetrahedron* 21, 3325.
BROWN, P. J. and B. C. SAUNDERS (1962) *Chem. & Ind.* 307.
BUCKLE, F. J., R. HEAP and B. C. SAUNDERS (1949) *J. Chem. Soc.* 912.
BUCKLE, F. J., F. L. M. PATTISON and B. C. SAUNDERS (1949) *J. Chem. Soc.* 1471.
CHAPMAN, N. B., R. HEAP and B. C. SAUNDERS (1948) *Analyst (Lond.)* 73, 434.
GOLDMAN, P. (1972) *This Volume*, pp. 335–349.
GRYSKIEWICZ-TROCHIMOWSKI, E., A. SPORZYNSKI and J. WNUK (1947) *Recl. Trav. Chim. Pays-Bas Belg.* 66, 419.
HOFFMANN, F. W. (1949) *J. Org. Chem.* 14, 105.
HUGHES, G. M. K. and B. C. SAUNDERS (1954) *Chem. & Ind.* 1265.
KNOOP, F. (1904) *Beitr. Chem. Physiol. Pathol.* 6, 150.
MCCOMBIE, H. and B. C. SAUNDERS (1946) *Nature (Lond.)* 158, 382.
MANN, F. G. and B. C. SAUNDERS (1971) *Practical Organic Chemistry*. London: Longmans.
MIROSEVIC-SORGO, P. and B. C. SAUNDERS (1959) *Tetrahedron* 5, 38.
PATTISON, F. L. M. (1959) *Toxic Aliphatic Compounds*. Amsterdam: Elsevier.
PATTISON, F. L. M. and B. C. SAUNDERS (1949) *J. Chem. Soc.* 2745.
PETERS, R. A. (1954) *Endeavour* 13, 147.
RIVETT, D. (1953) *J. Chem. Soc.* 3710.
SAUNDERS, B. C. (1947) *Nature (Lond.)* 160, 179.
SAUNDERS, B. C. (1949) *J. Chem. Soc.* 1279.
SAUNDERS, B. C. (1957) *Some Aspects of the Chemistry and Toxic Action of Organic Compounds containing Phosphorus and Fluorine*. London: Cambridge University Press.
SAUNDERS, B. C. and G. J. STACEY (1948a) *J. Chem. Soc.* 1773.
SAUNDERS, B. C. and G. J. STACEY (1948b) *J. Chem. Soc.* 695.
SAUNDERS, B. C. and G. J. STACEY (1949) *J. Chem. Soc.* 916.
SAUNDERS, B. C., G. J. STACEY and I. G. WILDING (1949) *J. Chem. Soc.* 773.
SWARTS, F. (1896) *Bull. Soc. Chim. Belg.* 15, 1134.

Discussion

Gál: Is it likely that the inductive effects in β-fluoropropionic acid lower the bond energy of C—F enough to allow exchange of F for OH to occur in cold alkaline conditions?

Saunders: The fluorobutyrates appear to have more stability. Could it be the loss of HF in much the same way as hydracrylic acid goes to acrylic acid so easily with the loss of water, whereas lactic acid does not behave in the same way? I am here drawing an analogy between OH and fluorine.

Sharpe: I suggest that the origin of the differences between, say, fluoroacetate and β-fluoropropionate is very unlikely to lie in a substantial difference in the C—F bond energies in these species. For reasons I shall mention later (pp. 33-49), reliable values for C—F bond energies in fluoro acids and esters are not available, but where data do exist for homologous series of organic compounds,

approximate constancy of bond energy terms is always found. It seems to me
that most of the variation in reactivity observed for C—F bonds lies in kinetic
rather than in thermodynamic behaviour, that is, in activation energies rather
than in free energies of reactions.

Heidelberger: We also have a series of compounds in which there is a striking
difference in the reactivities of the carbon-fluorine bond. 5-Fluorouracil (Fig. 1,
p. 126) has a very stable C—F bond. However, when a trifluoromethyl group is
put in the 5-position of uracil (Fig. 1, p. 126) this compound is converted completely to the corresponding carboxylic acid under very mild alkaline conditions. So this is an unusually labile C—F bond (Heidelberger, Parsons and
Remy 1964). We, and Santi and Sakai (1971), have come to the conclusion that
the mechanism of hydrolysis involves a nucleophilic attack on the C-6 carbon
atom, which then labilizes the C—F bond. The reason for believing this is that
the corresponding 6-aza compound (5-trifluoromethyl-6-azauracil), even though
it has unpaired electrons, is completely stable to alkaline hydrolysis (Dipple
and Heidelberger 1966).

Saunders: The carbonyl group probably activates the adjacent fluorines.

Heidelberger: No, I don't think so. In the 6-aza compound we also have
carbonyl groups, so the action must take place at the 6-position.

Kun: The crystallization of fluorocitrate as a dicyclohexylamine salt has been
very successful. The (—)-optical isomer was sufficiently crystalline to carry out
X-ray analysis on it, and in all probability the 1R:2R configuration corresponds
to the inhibitory isomer. I think the lack of toxicity of di- and trifluoroacetic
acid is related to the enolizability of the acetyl-CoA derivative. Synthetic di-
and trifluoroacetyl-CoA do not react with the condensing enzyme, as that
requires a certain amount of enolization which would not take place with more
than one fluorine atom of the α-carbon.

Goldman: Brady (1955) has shown that fluoroacetyl-CoA is considerably less
stable than acetyl-CoA at the pH at which the condensing enzyme is measured.
This instability, which is probably caused by the electronegativity of fluorine,
might be expected to increase in the CoA derivatives of di- or trifluoroacetate.
Could the failure of these compounds to react be due to this instability?

Kun: We do not have exact data on this. The stability is not much influenced
by the introduction of fluorine; it is the reactivity with the condensing enzyme
that is primarily diminished.

Gál: Kaufman (1961) and Millard and I (Gál and Millard 1971) have also
shown that 50% of the *p*-fluorophenylalanine is transformed into tyrosine
through the action of mono-oxygenases in the liver; the chloro- and bromo-
phenylalanines are hardly acted upon even though they produce inhibition and
are incorporated into protein.

Saunders: This bears out what I said: we should not think of fluorine as a halogen.

Gál: But the *m*-fluorophenylalanine may be metabolized to fluorocitric acid as postulated by Weissman and Koe (1967).

Peters: Dr Saunders, you mentioned a compound that had FCH_2 at each end and said that it was anomalous in its toxicity. Could you tell us more about it?

Saunders: FCH_2CH_2OH and FCH_2COOH have equal toxicity. If one puts them together in one molecule, $FCH_2COOCH_2CH_2F$, one can work out what the gross toxicity should be, because it hydrolyses to fluoroacetic acid plus fluoroethanol. But the toxicity by inhalation for mice of FCH_2COOCH_3 is 0.1 mg/l, and that of $FCH_2COOCH_2CH_2F$ should be much the same. However it is 0.05 mg/l. We also found that the 2-fluoroethyl esters of some of the higher ω-fluorocarboxylic acids had a greater toxicity than expected from a consideration solely of the hydrolysis products.

Taylor: The increased toxicity might be connected with transport to a particular site.

Saunders: Yes, that is the explanation we gave. We often find this in the phosphorofluoridate series: phosphorofluoroidic acid is non-toxic, but the esters (I) are toxic.

$$O=P\begin{array}{c}OR\\ \diagup\\ \diagdown\end{array}\!\!\begin{array}{c}OR\\ \\ F\end{array} \qquad O=P\begin{array}{c}OR\\ \diagup\\ \diagdown\end{array}\!\!\begin{array}{c}F\\ \\ F\end{array} \qquad O=P\begin{array}{c}OR\\ \diagup\\ \diagdown\end{array}\!\!\begin{array}{c}OH\\ \\ F\end{array}$$

(I) (II) (III)

I think that when there are two fluorine atoms in the phosphate molecule (II), one fluorine hydrolyses quickly to OH and the new molecule (III) does not then get through the plasma membrane. The anticholinesterase activity of di-isopropyl phosphorofluoridate (DFP) is $10 \times 10^{-10}M$ or $10 \times 10^{-11}M$, but that of the free acid or salts is only $10^{-2}M$.

Peters: It is curious that di- or triethylfluorocitrate are almost non-toxic when given by injection, but the monoethyl fluorocitrate is probably toxic.

Saunders: In general it makes little difference what R is in FCH_2COOR, as far as toxic function is concerned, but it does matter what R is in the α-position in FCH·COOH.

R

We made methyl fluoroacetate by the action of KF on methyl chloroacetate. NaF had little action. Can Dr Sharpe explain this?

Sharpe: The advantage of using KF rather than NaF for halogen exchange reactions is easily understandable if we remember than K^+ is a larger cation than Na^+, so the decrease in electrostatic interaction energy on conversion of KF into KCl is less than that on conversion of NaF into NaCl. CsF would be even better.

Saunders: I always suspected that.

Kun: Perhaps the toxicity of fluorocitric acid is related to its ability to co-ordinate with a metal present in a macromolecule. If it is esterified, then this possibility is lost. Dr Glusker has proposed a reasonable model for this co-ordination type of reaction (Carrel *et al.* 1970).

Wettstein: I do not understand the high toxicity of fluoroethanol, and especially not the equal toxicity of fluoroethanol and fluoroacetic acid which is allegedly caused by the transformation of the first to the second. Fluoro-ethanol is oxidized to fluoroacetic acid much more slowly than ethanol is to acetic acid. This is due to the effect of the unsuitable polarization of the hydro-gen bonds at the carbinol carbon atom. So why should these two molecules be equally toxic?

Saunders: There is little doubt that fluoroethanol is toxic.

Wettstein: What is the reason for its toxicity?

Saunders: I have always imagined that it is oxidized to fluoroacetate.

Wettstein: But much slower than ethanol?

Saunders: Yes.

Barnett: Might 2-fluoroethanol and 2-fluoroacetaldehyde be converted directly to fluoroacetyl-CoA by a system similar to the pyruvate decarboxylase complex, rather than by oxidation to fluoroacetate followed by conversion into fluoroacetyl-CoA by acetate thiokinase? The pyruvate decarboxylase complex catalyses the decarboxylation of pyruvic acid to the acetaldehyde-TPP complex, which is then converted directly into acetyl-CoA.

Saunders: Synthetic FCH_2-CHO has the toxicity one would expect from its oxidation to fluoroacetic acid. (Incidentally, the aldehyde is hydrated like chloral hydrate.)

Gál: There might be a difference between the fluoroacetaldehyde formed *in situ*, and injected fluoroacetaldehyde. Fluoroacetaldehyde if it is formed from fluoroethanol, may, like acetaldehyde, participate in a Schiff's condensation with biological receptor sites. Because of this readiness to form a Schiff's base, it may have toxic effects additional to those arising from its conversion to fluoroacetate.

Bergmann: (added in proof):

Regarding the stability of the C—F bond, we have recently observed two cases in which it appears that the proximity of an activated C—F and an N—H or C—H bond causes an unusual elimination of HF with concomitant ring formation.

(1) When *N*-bromomagnesyl-*N*-methylaniline reacted with ethyl α-fluoro-phenylacetate (I), instead of the expected amide, 1-methyl-3-phenyl-2-oxindole (II) was formed (I. Shahak, E. D. Bergmann and Y. Itach, unpublished results).

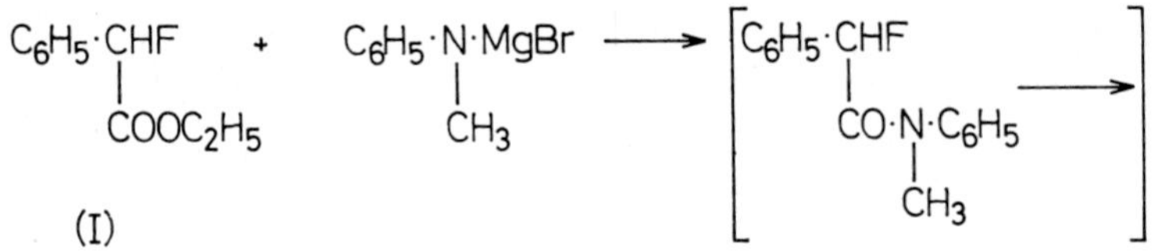

$$C_6H_5 \cdot \underset{\underset{COOC_2H_5}{|}}{CHF} \quad + \quad C_6H_5 \cdot \underset{\underset{CH_3}{|}}{N \cdot MgBr} \longrightarrow \left[C_6H_5 \cdot \underset{\underset{\underset{CH_3}{|}}{CO \cdot N \cdot C_6H_5}}{CHF} \right] \longrightarrow$$

(I)

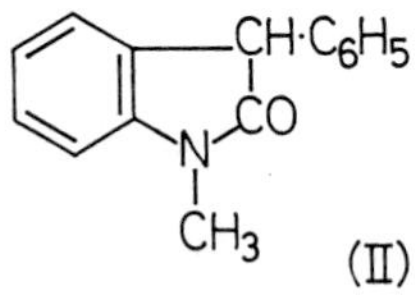

(II)

(2) In a study of the reaction of α, ω-dibromo-dicarboxylic acid di-(aryl-amides), Ar·NHCO·CH Br·(CH$_2$)$_n$·CHBr·CO·NH·Ar, with potassium fluoride, it was found (Shahak, Rozen and Bergmann 1971) that the adipic acid derivative ($n=2$) (III) gives only small yields of the desired α, ω-difluoroanalogues, and that the most characteristic product of the reaction is a new bicyclic ring system (IV).

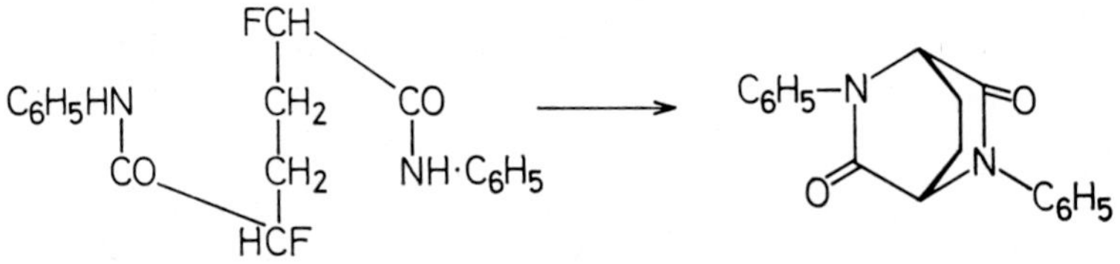

(III)   (IV)

Two other cases in which C—F bonds were broken—rather unexpectedly, as the fluorine atoms were vinylic ones—were the following:

(*a*) 3-Chloro-2-fluoro-1-phenylpropene gave with aromatic Grignard compounds 1,3-diarylallenes (V).

$$C_6H_5 \cdot CH = CF \cdot CH_2 \, Cl + ArMgBr \rightarrow C_6H_5 \cdot CH = CF \cdot CH_2 \cdot Ar(?) \rightarrow$$
$$C_6H_5CH=C=CHAr$$

(V)

In this way, 1,3-diphenyl-1-(*p*-chlorophenyl)-3-phenyl- and 1-(α-naphthyl)-

3-phenylallene were obtained (I. Shahak, E. D. Bergmann and J. Azran, unpublished results).

(b) Although diethyl 2-fluoro-3-phenylallylphosphonate anion (VI) normally gave with aromatic aldehydes in a Wittig-Horner reaction 1-phenyl-4-aryl-2-fluorobuta-1,3-dienes (VII), it yielded with benzophenone a fluorine-free compound $C_{22}H_{16}O$, for which the formula of 1-benzoyl-3,3-diphenylallene (VIII) was proved (I. Shahak, E. D. Bergmann and J. Azran, unpublished results). The mechanism is obviously quite complex.

$$C_6H_5 \cdot CH{=}CF \cdot CH \cdot P \overset{\overset{\displaystyle OEt}{\diagup}}{\underset{\underset{\displaystyle O}{\parallel}}{\diagdown}}_{OEt} \qquad\qquad C_6H_5 \cdot CH{=}CF \cdot CH{=}CH \cdot Ar$$

(VI)   (VII)

$$C_6H_5C \cdot CH{=}C{=}C(C_6H_5)_2 \qquad\qquad R \cdot CH{=}CF \cdot C\overset{\overset{\displaystyle CH_3}{\diagup}}{\underset{\underset{\displaystyle OH}{}}{\diagdown}}_{CH_3}$$

(VIII)   (IX)

$$R \cdot CH{=}CF \cdot C(CH_3){=}CH_2$$

(X)

These reactions are even more surprising since one finds that substituted α-fluoro-acrylates, $R \cdot CH = CF \cdot COOEt$, react normally with Grignard compounds, leading to compounds of type IX, for example, which can be dehydrated to the fluorinated dienes (X) (I. Shahak, E. D. Bergmann and Y. Itach, unpublished results).

References

Brady, R. O. (1955) *J. Biol. Chem.* **217**, 213–224.
Carrell, H. L., J. D. Glusker, J. J. Villafranca, A. S. Mildvam, R. J. Dummel and E. Kun (1970) *Science* **170**, 1412.
Dipple, A. and C. Heidelberger (1966) *J. Med. Chem.* **9**, 715.
Gál, E. M. and S. A. Millard (1971) *Biochim. & Biophys. Acta* **227**, 32–41.
Heidelberger, C., D. G. Parsons and D. C. Remy (1964) *J. Med. Chem.* **7**, 1.
Kaufman, S. (1961) *Biochim. & Biophys. Acta* **51**, 619–621.
Santi, D. V. and T. Sakai (1971) *Biochemistry* **10**, 3598.
Shahak, I., S. Rozen and E. D. Bergmann (1971) *J. Org. Chem.* **36**, 501.
Weissman, A. and B. K. Koe (1967) *J. Pharmacol. Exp. Ther.* **155**, 135–144.

The physical properties
of the carbon-fluorine bond

A. G. SHARPE

University Chemical Laboratory, Cambridge

This paper deals with the fundamental physicochemical properties of compounds containing the C—F (carbon-fluorine) bond. In it I shall try to give a critical summary of our quantitative knowledge of the properties of the C—F bond (with special reference to compounds containing one such bond) and an indication of its potential usefulness in the study of problems in organic chemistry and biochemistry. There is, unfortunately, often only a distant connection between what a physical chemist measures and what an organic chemist or a biochemist needs to know. I take the latter to include the answers to such questions as the following. How can some property of a compound A be calculated from other properties? Will A spontaneously, or under the influence of external factors, undergo chemical change (e.g. decomposition or polymerization) or will it react with B, C, or D? How complete and how fast will any such reactions be? Can their extents and rates be calculated from a knowledge of the structures and properties of the reactants and products? If so, what particular features of the structures and what properties are most important, and how can they be used to make at least qualitative predictions if quantitative treatment is impossible or too difficult?

In trying to show how these questions may be answered for compounds containing C—F bonds, I shall refer to modern physicochemical theories of equilibria and kinetics, and it may be useful, therefore, to begin by giving a brief outline of these subjects, further details of which are given in a number of standard texts (e.g. Gilliom 1969; Gould 1959; Gucker and Seifert 1967; Moore 1963).

CHEMICAL EQUILIBRIA AND KINETICS

Any physical or chemical change may be represented by means of a general equation

$$aA + bB + \ldots \rightarrow pP + qQ + \ldots \tag{1}$$

in which the physical states of all substances are specified. For all such changes there is an equilibrium constant, K, defined by

$$K = \frac{[P]^p [Q]^q \ldots}{[A]^a [B]^b \ldots} \tag{2}$$

where square brackets denote equilibrium concentrations (or, strictly, equilibrium activities). The value of K is the chemist's normal way of expressing how far a process goes. It is related to the standard free energy change of the process, $\Delta G°$ (the difference between the sum of the free energies of the substances on the left hand side of the equation and that of the substances on the right hand side when all are present at unit concentration), by the equation

$$-\Delta G° = RT \ln K$$
$$\text{or } -\Delta G° = 2.3 \, RT \log K \tag{3}$$

where R is the constant of the ideal gas equation $pV = RT$ (8.3 J or 2 cal $°C^{-1}$ mol^{-1}) and T is the absolute temperature. This equation may also be written in the form

$$K = e^{-\Delta G° / RT} \tag{4}$$

where e, 2.718, is the base of natural logarithms. The logarithmic nature of the relationship means that K is very sensitive to variation in $\Delta G°$, itself usually a small difference between large quantities. An error of 40 kJ (10 kcal) in $\Delta G°$ corresponds to an error of a factor of over a million in the value of K. Although it is sometimes possible to calculate standard free energies, and hence $\Delta G°$ and K, from first principles, the results in any but the very simplest systems are highly approximate; calculation of K from values of $\Delta G°$ obtained by independent experiments is, on the other hand, reliable and very important.

The rate of reaction (1) can always be represented by an expression of the form

$$\text{Rate} = k[A]^x [B]^y \ldots \tag{5}$$

where x and y, the orders of the reaction with respect to reactants A and B, may have any values (including zero); they need not be (and seldom are) the same as the stoichiometric coefficients a and b. It is important to emphasize that we can obtain x and y only by experiment; it is mainly from them that we infer the mechanism, though other considerations are also important. The constant k, which is temperature-dependent, is the rate constant of the reaction. Nearly all reactions involve molecular collisions, but it is easily shown that usually only a very small fraction of such collisions result in chemical change.

In the simplest theory of chemical kinetics (the so-called collision theory), the colliding molecules have to possess a certain minimum amount of energy (the activation energy of the reaction) and be suitably orientated (the steric factor-unity for very simple reactions) if a collision is to lead to reaction. In a later theory (transition state theory), which is much more useful for the discussion of reactions in solution, the reactants are considered to be in equilibrium with a low concentration of an activated complex (or transition state), and it is the position of this equilibrium that determines the rate constant of the reaction, all activated complexes decomposing to products at the same (or nearly the same) rate. In this case the relationship between k and ΔG°_{act}, the difference between the reactants and the activated complex, is given by

$$k = \frac{R}{N h} T e^{-\Delta G^{\circ}act/R T} \tag{6}$$

where N is Avogadro's number (6.06×10^{23} mol^{-1}) and h is Planck's constant (6.63×10^{-34} J s); at about 25°C, $R T/N h$ is about 6×10^{12}. The similarity of (4) and (6) is obvious, but there is a very important difference between ΔG° and ΔG°_{act}; whereas one can often obtain the value of ΔG° by methods that have nothing to do with the measurement of equilibrium constants, the only experimental route to ΔG°_{act} is by measurement of k. However, one can make postulates about the nature of the activated complex, try to estimate its proper-ties by extrapolation from those of stable species, and hence obtain an estimate of ΔG°_{act}. To this end both ΔG° and ΔG°_{act} must be considered further.

The standard free energy change of a process is itself a composite quantity; at a given temperature

$$\Delta G^{\circ} = \Delta H^{\circ} - T\Delta S^{\circ} \tag{7}$$

where ΔH° is the change in heat content for the process (and corresponds to the difference in the sum of the strengths of bonds in the reactants and that in the products) and ΔS° is the change in entropy (which is most usefully looked upon as the degree of randomness of the system). Among reactions which are accompanied by increases in entropy (and are thereby made more probable) are those in which solid or liquid reactants are converted into gaseous products, those in which oppositely charged particles in solution (which bring about orientation of solvent molecules) combine to form uncharged species, and isotopic exchange reactions. In the last of these, typified by

$$H_2O + D_2O = 2HOD$$

the value of ΔH° is very nearly zero, and the increase in entropy is the factor that makes ΔG° negative, corresponding to a value of K of approximately four (i.e. to the statistical distribution at equilibrium of two HOD, one H_2O, and one D_2O). Conversely, reactions in which gases are converted into solids or liquids, or charged species are formed in solution, are accompanied by decreases in entropy (and are thereby made less probable).

Combination of (7) with (4) and (6) respectively gives

$$K = e^{\Delta S°/R} e^{-\Delta H°/RT} \tag{8}$$

and

$$k = \frac{R}{Nh} T e^{\Delta S°_{act}/R} e^{-\Delta H°_{act}/RT} \tag{9}$$

where $\Delta S°_{act}$ and $\Delta H°_{act}$ are the entropy and heat content changes when the activated complex is formed from the reactants. For this process $\Delta H°_{act}$ is nearly always positive, but $\Delta S°_{act}$ may be positive or negative; if the activated complex is more ordered than the reactants (e.g. if there are rigid steric requirements to be met in forming it), $\Delta S°_{act}$ is negative and k is thereby made smaller; conversely, if the activated complex is less ordered than the reactants, $\Delta S°_{act}$ is positive and k is thereby made larger.

It will be apparent that if the thermochemical heats of formation of all the reactants and products in a reaction are known, then $\Delta H°$ for the reaction can easily be obtained. If one knows the entropies of all the reactants and products (they can be obtained by calorimetric methods or from spectroscopic data), one can also obtain $\Delta S°$, and can then evaluate $\Delta G°$ and hence K, the equilibrium constant. Conversely, by measuring K at different values of T one can obtain values of $\Delta G°$, $\Delta H°$, and $\Delta S°$. In this way values of $\Delta G°$ and $\Delta H°$ (relative to elements in their standard states) and of $S°$ can be assigned to a very large number of different substances; and from such values one can work out equilibrium constants of new reactions—and get the correct answers. This is chemical thermodynamics, the best-established and most respectable branch of physical chemistry.

Chemical kinetics is a much more difficult field to treat quantitatively. For a few very simple gas phase reactions (e.g. $D + H_2 \rightarrow HD + H$) it is possible to calculate reaction rates from first principles, but for reactions between more complex species taking place in solution most progress is made by comparing one reaction with another. As I mentioned earlier, mechanisms are usually inferred from orders of reaction; for the alkaline hydrolysis of alkyl bromides in aqueous ethanol, for example, the general reaction is

$$RBr + OH^- = ROH + Br^-$$

For $R = CH_3$, the rate of reaction is proportional to $[RBr][OH^-]$, but for $R = (CH_3)_3C$, it is proportional to $[RBr]$ only. From these observations it is inferred that the mechanisms of hydrolysis of methyl and tertiary butyl bromides are respectively

$$HO^- + CH_3Br \xrightarrow{slow} HOCH_3 + Br^-$$

and
$$(CH_3)_3CBr \xrightarrow{slow} (CH_3)_3C^+ + Br^-$$
$$(CH_3)_3C^+ + OH^- \xrightarrow{fast} (CH_3)_3COH$$

(Note, however, that these mechanisms make no mention of the possible role of the solvent, present in large excess throughout the reaction. Liquid water, for example, is 55M, and a dilute solution of a compound in an organic solvent which has not been specially dried may contain a molar concentration of water approaching that of the solute.) But increasing attention is now being paid to ΔH_{act}° and ΔS_{act}°, evaluated from measurements of k at different temperatures. Thus physical quantities such as bond energies and bond polarities are potentially of considerable interest in biological chemistry.

CLASSIFICATION OF PHYSICAL PROPERTIES

It is convenient to group physical properties of the C—F bond into two categories (though these are to some extent interdependent): those concerned with the strengths of bonds (bond energies, bond dissociation energies, bond lengths, bond stretching frequencies and force constants); and those concerned with their polarities (dipole moments, electronegativity differences, infrared absorption intensities, polarizabilities, and nuclear magnetic resonance chemical shifts). In connection with the second group it is also desirable to mention the qualitative effects widely involved in the systematization of organic chemistry, particularly the inductive and mesomeric effects; quantitative data are not yet so abundant that empirical correlations can be dismissed without comment. I shall, however, exclude treatment of hyperconjugation and so-called linear free energy relationships, partly because these are specialized topics but mainly because, so far as the C—F bond is concerned, much of the evidence about them comes from nuclear magnetic resonance spectroscopy, which forms the subject of a later paper. I shall also exclude much that is known about the physical properties of fluorocarbons (Reed 1964); although these substances are extremely important in the study of intermolecular attraction, they are extremely unreactive chemically and bear very little similarity to the compounds being discussed at this meeting.

Properties primarily concerned with bond strengths

Bond energy terms and bond dissociation energies. The sum of the bond energy terms is, by definition, the heat of formation of a gaseous molecule from atoms in the gaseous state and in their lowest electronic energy levels or ground states, for example carbon with the configuration $(2s)^2(2p)^2$ and fluorine with the configuration $(2s)^2(2p)^5$. For most organic compounds the sum of the bond energy terms is obtained from the heats of combustion of the compound and the heats of combustion and atomization of all elements in it. Fluorine-containing compounds, however, seldom burn quantitatively in oxygen, and the new method of fluorine combustion calorimetry has not yet been applied to them on any considerable scale. Reliable values for heats of formation are therefore available for only a small number of organic fluorine compounds, many of them chlorofluoro-hydrocarbons or -olefins important as refrigerants or polymer intermediates.

There is a further difficulty in assigning C—F bond energy terms. Where the molecule contains only one kind of bond (e.g. CF_4) the bond energy is unambiguously given by the appropriate fraction of the heat of formation from gaseous ground-state atoms. Where two or more kinds of bond are present (e.g. in CH_3F, FCH_2COOH), heats of formation can be used to provide bond energy data only if values are available for as many compounds as there are types of bond present, and if it is assumed that for a particular type of bond the bond energy is constant among the compounds chosen. The heat of formation of gaseous fluorobenzene is 194 kJ (46.4 kcal) greater than that of gaseous benzene, for example; but this information cannot be used to compare the C—H and C—F bond energies unless one makes the questionable assumption that the C—C and all other C—H bonds in the two compounds are equivalent,

TABLE 1

C—F bond energy terms

Compound	C—F bond energy		Notes
	kJ	kcal	
CF_4	485	116	
CHF_3	481	115	C—H bond energy assumed same as in CH_4
CH_2F_2	460	110	C—H bond energy assumed same as in CH_4
CH_3F	—	—	Heat of formation not measured. Estimates based on $n-C_3H_7F$ and related compounds suggest a value of 452 kJ (108 kcal)
CF_3Cl	481	115	C—Cl bond energy assumed same as in CCl_4
CF_2Cl_2	464	111	C—Cl bond energy assumed same as in CCl_4
$CFCl_3$	456	109	C—Cl bond energy assumed same as in CCl_4

and the bond energies of both C—H and C—C (aromatic) cannot be obtained from the value for benzene. Considerations of this kind account for the paucity of data in Table 1 (Patrick 1961; Lacher and Skinner 1968; Benson *et al.* 1969). (Reasonably reliable data are also available for fluoro- and chlorofluoro-olefins (Patrick 1961; Sheppard and Sharts 1969)). It seems, however, certain that C—F bond energies in fluoromethanes and chlorofluoromethanes increase with increase in fluorine content. No comparable effect is found for C—Cl and C—Br bonds in chloro- and bromo-methanes.

The energy required to bring about the gas phase reaction

$$CF_4 \rightarrow CF_3 + F$$

(for which $\Delta H°$ is $+510$ kJ or $+122$ kcal) is the C—F bond dissociation energy in this compound; it is not the same as the average C—F bond energy (485 kJ or 116 kcal), since the CF_3 radical produced is not simply three-quarters of a CF_4 molecule. The C—F bond dissociation energy in CF_4 is also larger than that in CH_3F (452 kJ or 108 kcal), again suggesting that the C—F bond strength increases with increase in fluorine content of the molecule. The C_6H_5—F bond dissociation energy (481 kJ or 115 kcal) has an intermediate value.

The C—F bond is the strongest single bond formed by carbon, but it is considerably weaker than the H—F bond (bond energy 570 kJ or 136 kcal) and mainly (though by no means entirely) for this reason is thermodynamically unstable to hydrolysis. Thus for the reaction

$$CF_4 \text{ (g)} + 2H_2O \text{ (liq)} \rightarrow CO_2 \text{ (aq)} + 4HF \text{ (aq)}$$

$\Delta H°$ is -180 kJ (-43 kcal) and $\Delta G°$ -222 kJ (-53 kcal); with so wide a margin of instability towards water, there is obviously no possibility of preparing carbon tetrafluoride in aqueous solution. However, for the reaction

$$CH_3F \text{ (g)} + H_2O \text{ (liq)} \rightarrow CH_3OH \text{ (aq)} + HF \text{ (aq)}$$

the best estimates that can be made at present are $\Delta H° = -38$ kJ (-9 kcal) and $\Delta G° = -13$ kJ (-3 kcal), nearly the same as for the analogous reaction of methyl chloride. In this instance $\Delta G°$ is so nearly zero as to indicate the possibility of forming the C—F bond in aqueous solution if the reaction is coupled to one which takes place spontaneously. The difference in $\Delta G°$ for the two hydrolytic reactions is in the opposite direction to what would have been predicted on the basis of the C—F bond strengths in CF_4 and CH_3F. The explanation resides in the nature of the carbon-containing hydrolysis product; if the hydrolysis of CH_3F had been written as

$$4CH_3F + 2H_2O \rightarrow 3CH_4 + CO_2 + 4HF$$

(in which the net reaction is replacement of four C—F and four O—H bonds by two C=O and four H—F bonds, as in the hydrolysis of CF_4) a different conclusion would have been reached. Thermodynamics therefore still needs to be supplemented by chemical experience or intuition if the choice of a reaction for study is involved.

The most important general reaction for the synthesis of organic fluorine compounds is halogen replacement of the type

$$\hspace{-1em} \text{>C—Cl} + \text{F}^- \rightarrow \text{>C—F} + \text{Cl}^-$$

If the C—F bond in aliphatic monofluoro-compounds is taken to be 113 kJ (27 kcal) stronger than the C—Cl bond in the corresponding chloro compounds, then whether a given source of fluoride ion will bring about halogen exchange depends largely upon whether the energy of interaction of F^- with the environment is less than 113 kJ (27 kcal) greater than that of Cl^- with the environment. This condition is best met if one uses the fluoride of a large cation in the anhydrous state [lattice energy is inversely proportional to $1/(r_+ + r_-)$] or a non-aqueous solvent of dielectric constant lower than that of water (to diminish the difference in solvation energies of F^- and Cl^-, since these depend on $\frac{1}{r}\left(1 - \frac{1}{\varepsilon}\right)$ where ε is the dielectric constant). Conversely, the thermodynamic reactivity of a C—F bond should be modifiable by a suitable choice of reagent or solvent; for the conversion of >C—F into >C—Cl, for example, chlorides of small cations (especially Li^+) are most effective (Sharpe 1967).

Bond lengths. Carbon-fluorine bond lengths are almost invariably determined by X-ray diffraction, electron diffraction, rotational spectroscopy, or rotational-vibrational spectroscopy. The first of these is an absolute method; for molecules containing more than a few atoms the others sometimes provide enough information for the evaluation of molecular parameters only if assumptions are made about some bond lengths and/or bond angles. None of the bond lengths quoted in Table 2 rests on assumptions concerning the remainder of the molecule.

It will be seen that the increase in the C—F bond energy in compounds containing more than one fluorine atom bonded to a carbon atom is accompanied by a decrease in bond length. This is another general manifestation of stronger bonding. Bonding to an unsaturated carbon atom also results in a shortening of the C—F bond. Like the bond energy, the length of an aromatic C—F bond remains somewhat problematical; older spectroscopic and electron diffraction data indicating lengths of 1.30–1.35 Å in fluorobenzene and *p*-bromo and *p*-chlorofluorobenzenes rest upon somewhat dubious assumptions about other bonds and bond angles in these molecules. The only published value available from X-ray diffraction is that of 1.368 in o-$C_6H_4F(COOH)$ (Krause and Dunken 1966) which, surprisingly, is nearly as long as that in CH_3F; however, doubt has been cast on the accuracy of this value (Ferguson and Islam 1966), and a recent three-dimensional X-ray study (G. Ferguson 1971, unpublished) indicates that the C—F bond length is really 1.330 ± 0.01 Å, which is much more in line with chemical expectations.

TABLE 2

Some C—F bond lengths

Compound	Bond length* (Å)	Reference
FCH_2CONH_2	1.406	Hughes and Small (1962)
CH_3F	1.384	Hencher and Bauer (1967)†
$FCH(COOH)_2$	1.364	Roelofson et al. (1971)
CH_2F_2	1.358	Hencher and Bauer (1967)
$F_3CCOONH_4$	1.346	Cruickshank, Jones and Walker (1964)
$H_2C=CHF$	1.344	Hencher and Bauer (1967)
HCOF	1.338	Hencher and Bauer (1967)
CF_4	1.335	Hencher and Bauer (1967)
$FHC=CHF$	1.335	Hencher and Bauer (1967)
CH_3CF_3	1.333	Hencher and Bauer (1967)
CHF_3	1.332	Hencher and Bauer (1967)
o-$C_6H_4F(COOH)$	1.330	Ferguson (1971, unpublished)††
$CH_2=CF_2$	1.323	Hencher and Bauer (1967)
COF_2	1.312	Hencher and Bauer (1967)
$CH \equiv CF$	1.279	Hencher and Bauer (1967)
$FC \equiv N$	1.262	Hencher and Bauer (1967)

* All estimated correct within ± 0.01 Å or less.
† Review paper in which data for several compounds not included in Table 2 are given.
†† Krause and Dunken (1966) in an earlier less complete study reported 1.368 Å.

The C—F bond is not much longer than the C—H bond (1.06–1.09 Å, the value increasing from saturated to acetylenic compounds), but it corresponds to the fluorine atom being large enough to impose some steric constraints on molecular structures (though it should be noted that steric effects cannot be isolated from electronic effects in actual molecules). In o-fluorobenzoic acid the carboxyl group is twisted 7° out of the plane of the benzene ring, which is less than in the chloro and bromo compounds (14° and 18° respectively). [Spectroscopic evidence on preferred conformations shows that the *gauche* rotational isomer of n-propyl fluoride is more stable than the *trans*-form by about 2 kJ (0.5 kcal), although for 1,2-difluoroethane the *gauche* and *trans*-forms have the same energy (Hirota 1962).]

Vibrational spectroscopy. When there is a single fluorine atom in the molecule of an organic compound a strong C—F stretching bond usually appears at 1100—1000 cm^{-1}. When more fluorine is introduced this frequency rises and splits into two, an asymmetric stretching vibration at 1250–1100 cm^{-1} and a symmetric stretching vibration at 1250–1050 cm^{-1}. Fully fluorinated hydrocarbons give intense and complex spectra, which are very useful for identification but impossible to interpret, in the 1350–1100 cm^{-1} region. A typical progression is given by CCl_3F (1102 cm^{-1}), CCl_2F_2 (1155 and 1095 cm^{-1}), and $CClF_3$ (1210 and 1102 cm^{-1}) (Brown and Morgan 1965; Conley 1966). It is tempting to try to establish correlations between stretching frequencies, bond lengths, and bond energies, but it must be borne in mind (i) that molecules vibrate as a whole, and the idea of characteristic bond frequencies is always an approximation and sometimes an untenable one, and (ii) that even for a series of very closely related molecules a simple quantitative relationship between the stretching frequency and the bond energy is not necessarily to be expected. These two quantities relate to a small displacement from equilibrium (1000 cm^{-1} = 12 kJ = 3 kcal approximately) and to bond breaking respectively, and there is no reason to expect potential energy/interatomic distance curves for all molecules to be of exactly the same form.

For comparison of the forces required to stretch different bonds account must be taken of atomic masses, and the ideal quantity for consideration becomes the force constant, which is a measure of the restoring force per unit displacement. For CH_3F and CH_3Cl, for example, the values are 5.6 and 3.4×10^2 N m^{-1} (5.6 and 3.4×10^5 dynes cm^{-1}) respectively (Linnett 1947). For more complicated molecules, however, the assignment of force constants is both difficult and arbitrary, and I do not think it has much to contribute to the understanding of the properties of compounds containing C—F bonds.

Properties primarily concerned with bond polarities

Dipole moments and electronegativities. The permanent dipole moments of the methyl halides, in which any contribution from the CH_3 group is assumed constant, are: CH_3F, 1.79; CH_3Cl, 1.87; CH_3Br, 1.80; CH_3I, 1.65 Debye units (D). If the bond lengths in these molecules are taken into account the respective values for charge separation/distance are 1.30, 1.05, 0.93, and 0.76. (These, however, are based on dipole moments of molecules, not of bonds, and it must be remembered that unshared pairs of electrons may contribute substantially to molecular dipole moments.) Chemists in general like to believe that the polar character of a bond can be expressed in terms of the difference in electronegativity of the atoms forming the bond, and many attempts have been made to define and measure this elusive quantity. Of these, much the most famous is that of Pauling (1940), who defined the difference in electronegativity of two atoms A and B, $x_A - x_B$, as

$$x_A - x_B = \sqrt{[E_{A-B} - (E_{A-A} + E_{B-B})/2]}/23.06$$

where E_{A-B} is the actual bond energy of an A—B bond, and E_{A-A} and E_{B-B} are the standard A—A and B—B bond energies (in kcal); the factor 23.06 merely converts kcal into electronvolts. The square root relationship is introduced empirically in the interests of self-consistency. This scale measures difference in electronegativity, so it is necessary to choose an arbitrary value for one element. Pauling chose 2.1 for hydrogen so as to be able to assign positive values to all elements; comparison of the C—F (in CF_4), H—F, C—C, F—F, C—H, and H—H bond energies then leads to electronegativities on this (thermochemical) scale of 4.1 and 2.5 for fluorine and carbon respectively. *If* the 'extra' energy of a bond between unlike atoms is considered to arise from ionic character, Δx is an indication of ionic character and the value of 1.6 for C—F may be compared with 0.4 and 0.5 (in the senses $H^+ - C^-$ and $C^+ - Cl^-$) for C—H and C—Cl (Pauling 1940). Many later authors have made minor modifications to Pauling's scale, but the basic features remain essentially unchanged. Pauling's scale, it should be noted, is based on molecular properties; more nearly absolute scales (such as that of Mulliken, who proposed taking the mean of the electron affinity and the ionization potential of gaseous atoms as a measure of electronegativity) have the disadvantage of being more remote from the context in which they are used.

Polarizability. This much-used noun has been in vogue in chemistry ever since the early days of the electronic theory of valency, when the covalent bond was considered to arise from 'polarization' of the electronic distribution in one ion (especially a large anion) by the 'polarizing power' of another ion (especially a small and highly-charged cation). It has commonly been assumed that the polarizing power of an ion is its charge/radius or charge/radius^2 ratio; but polarizability has a quantitative definition which leads to measurement in terms of quite different physical quantities. The polarizability of a molecule, α, is the ratio of the dipole moment induced in it by an electric field to the strength of the effective field (Davies 1965). It can be related, subject to the satisfaction of certain conditions during measurements, to the dielectric constant and, in turn, to the refractive index of the substance, n, which is the ratio of the velocity of electromagnetic waves (light) *in vacuo* to that in the substance. Molar refractions, R, are related to other quantities by the equations

$$R = \frac{(n^2 - 1)}{(n^2 + 2)} \frac{M}{d} = \frac{4\pi}{3} N \alpha$$

where M is the molecular weight, d the density, N Avogadro's number, and α the polarizability. Molar refractions can be split up into atomic, ionic, or bond contributions. Thus the fact that the bond refraction (in cm^3) of C—F (1.45) is much lower than that of C—Cl (6.51) means that the action of an electric field induces a smaller change of dipole moment for C—F than for C—Cl (it does not affect the fact that C—F has a bigger permanent dipole moment). In a similar way, F$^-$ has a lower ionic polarizability than Cl$^-$ (or, indeed, any other anion); the values for F$^-$ and Cl$^-$ are 0.96 and 3.60×10^{-24} cm^3 respectively. The concept of polarizability has some use in qualitative discussions of chemical phenomena, but obviously the relationship of a quantity whose dimensions are (length)^3 to heats and entropies of reactions can be only a very indirect one.

Other properties. Since the intensity of a vibration which is active in the infrared spectrum depends on the square of the change of dipole moment which results from the motion giving rise to the absorption, infrared intensities may in principle be used to provide information about polarity. The high intensities of C—F stretching vibrations, where they can be identified with certainty, is, for example, very strong evidence that they involve the stretching of a highly polar bond.

SOME ELECTRONIC EFFECTS IN ORGANIC CHEMISTRY

All the effects postulated to account for the redistribution of electron density in a molecule may operate in the equilibrium state or on the approach of a reagent (i.e. in the transition state). It seems reasonable to suppose that the effects are often much larger as a polar or charged reagent approaches, but, of course, only the equilibrium state, or a state near to it, is susceptible to direct examination, and the evidence presented below therefore relates to stable species.

The inductive effect

This is the name given to the transmission of charge along a chain of saturated carbon atoms, made manifest in, for example, the polarity of the C—F bond in alkyl fluorides and similar compounds. It is, for example, the commonly accepted basis of the increase in the dissociation constant when a hydrogen atom of acetic acid is replaced by a halogen, pK_a values for CH_3COOH, FCH_2COOH, $ClCH_2COOH$, $BrCH_2COOH$, and ICH_2COOH in aqueous solution being 4.76, 2.66, 2.86, 2.90, and 3.16 respectively. Just how the inductive effect operates in this instance is, however, a subtle matter. The heats of ionization of all five acids (which can be calculated from the temperature-dependence of pK_a) are very nearly zero, so the variation in pK_a is an entropy effect; it seems that the halogen atom operates by withdrawing charge from the $-COO^-$ group of the ion, thus causing it to produce less orientation of water molecules in its neighbourhood (i.e. leading to a less negative entropy of hydration for XCH_2COO^- than for CH_3COO^-).

The mesomeric effect

This involves electron redistribution by the use of π-orbitals in unsaturated systems. Its most important application in the chemistry of the C—F bond lies in the interpretation of the rates and positions of substitution in halogen-substituted aromatic hydrocarbons by positively charged reagents such as NO_2^+, the active entity in nitration by nitric acid–sulphuric acid mixtures. All the monohalogenobenzenes are less reactive than benzene, but the relative order of reactivity

$$C_6H_6 > C_6H_5I > C_6H_5F > C_6H_5Cl \sim C_6H_5Br$$

is not that predicted on the basis of the order of inductive effects F > Cl > Br > I. Unlike other groups which deactivate the ring towards NO_2^+, the halogens cause predominantly *ortho-para* substitution. This suggests that the electron-withdrawing properties of fluorine can be partly compensated by electron-releasing properties, commonly represented by writing the F—C bond as intermediate between $^-F—C^+$ and $^+F=C^-$; the feedback of electronic charge to the ring helps in the attainment of a transition state in which NO_2^+ is bonded to the ring before a hydrogen atom is displaced as H^+. Somewhat equivocal evidence for this concept is provided by the fact that the dipole moment of C_6H_5F is smaller than that of CH_3F by a larger amount (0.24D) than that by which the dipole moment of C_6H_5Cl is smaller than that of CH_3Cl (0.13D). If the C—F bond length in C_6H_5F is much shorter than in CH_3F, this would provide further evidence; as we have pointed out already, reliable data are not available for fluorobenzene, though the most recent value

(1.330 Å) for the C—F bond length in o-$C_6H_4F(COOH)$ is compatible with a bond order above unity. Bond energy data for fluoro-aromatic compounds are totally lacking. In sum, therefore, physical evidence for a mesomeric effect in fluorobenzene is scanty, and some of the finer details of the picture (e.g. the much lower ratio of *ortho*: *para* substitution products for fluorobenzene than for the other halogenobenzenes) are quite inexplicable in terms of independently examinable variables (Ingold 1969).

Various data for the strengths of carboxylic acids support the existence of an electronic effect in opposition to the inductive effect; for example fluorobenzoic acids are all somewhat weaker than the corresponding chlorobenzoic acids, and perfluoroacrylic acid is weaker than perchloroacrylic acid. For these systems, however, dissociation constants have not been measured at different temperatures, and hence we cannot separate heat and entropy effects.

PHYSICAL DATA FOR REACTIONS OF THE C—F BOND

Apart from studies of the kinetics of reactions of highly fluorinated hydro-carbons and olefins (which are mostly free radical gas-phase reactions at high temperatures) few quantitative data for reactions of the C—F bond are available. There have, however, been a number of investigations of the replacement of fluorine bonded to alkyl or aryl groups (Parker 1963). Such replacement usually takes place in solution. Under these conditions, even in a reaction which is first order with respect to the halide and zero order with respect to any species whose influence can be studied (i.e. not the solvent, always present in large excess), the relationship of the heat of activation to the bond dissociation energy is expected to be only an indirect one. In the formation of R^+ in solution from RX, desolvation of RX, the splitting of the bond to give R and X, the ionization energy of R and the electron affinity of X, and the solvation of the ions are all involved if we consider the ionization of RX in terms of independently measurable quantities according to the cycle

$$RX \text{ (solution)} \rightarrow RX \text{ (gas)} \rightarrow R \text{ (gas)} + X \text{ (gas)}$$
$$\downarrow \qquad\qquad\qquad\quad \downarrow \qquad\quad \downarrow$$
$$R^+ \text{ (solution)} + X^- \text{ (solution)} \leftarrow R^+ \text{ (gas)} + X^- \text{ (gas)}$$

For the second-order reaction

$$Y + RX \rightarrow YR^+ + X^-$$

two main possibilities arise. If the mechanism is represented by

$$Y + R-X \xrightarrow{\text{slow}} Y^+ - R^- - X$$
$$Y^+ - R^- - X \xrightarrow{\text{fast}} Y^+ - R + X^-$$

the greater electron-withdrawing power of fluorine should stabilize the forma-tion of $Y^+ - R^- - X$ more for $X = F$ than for $X = Cl$. If the mechanism involves a bimolecular rate-determining step in which the $C-X$ bond is broken and the $C-Y$ bond formed, the extent to which these operations are involved in the transition state should determine the relative rates of substitution for $X = F$ and $X = Cl$; the fluoride should be more reactive than the chloride if bond making has progressed further than bond breaking, and *vice versa*.

If we consider the reactions of a fluoride and of the corresponding chloride under the same experimental conditions, and call the ratio of the rate constant for $X = F$ to that for $X = Cl$ the F/Cl ratio, a value of F/Cl less than unity suggests a mechanism in which the rate-determining step involves considerable $C-X$ bond stretching; one greater than unity suggests a rate-determining step in which the making of the $Y-C$ bond has progressed further than the breaking of the $C-X$ bond; and a ratio close to unity suggests bond breaking and bond making are evenly balanced. For a large number of reactions of alkyl, benzyl, and triphenylmethyl halides with H_2O, OEt^-, or I^-, the F/Cl ratio is 10^{-2}–10^{-6}. Similar ratios are found for analogous reactions of acyl halides. The lower reactivity of the fluorides does not necessarily require a higher heat of formation of the activated complex, however; in the reactions of the methyl halides with water or hydroxyl ion, for example, the lower value of ΔH°_{act} for the fluoride (tending to make the reaction faster) is outweighed by a less positive ΔS°_{act} (tending to make the reaction slower) (Moelwyn-Hughes and Fells 1959). This can be interpreted qualitatively by saying that, relative to the reactants, the transition state $Y-R-X$ orientates solvent molecules to a greater extent when $X = F$ than when $X = Cl$. This example may serve to show that, simple as the ideas of transition state theory are, the quantitative applica-tion of them may prove to be a matter calling for nice judgement and, if possible, extensive acquaintance with the kinetics of other reactions in solution.

Fluorobenzene and chlorobenzene are both unreactive towards OH^-, OEt^- and amines under conditions in which kinetic studies can conveniently be made, but their 2-nitro, 2:4-dinitro, and 4-trifluoromethyl derivatives react readily at 50–100 °C, and reactions with OEt^- and amines have been examined quanti-tatively. For these second-order reactions in ethanol, the F/Cl ratios are 10^1–10^3, from which it is inferred that there has been little $C-X$ bond stretching in the transition state, which is probably of the type:

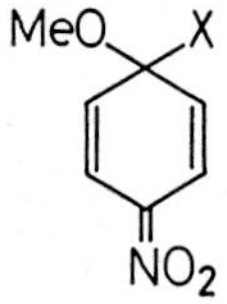

MeO   X

NO$_2$

The part played by the solvent in such reactions is, however, appreciable. If, for example, ethanol is replaced by ethyl acetate (a poorer hydrogen-bonding and solvating medium) as solvent for reactions with amines, some of the fluoro compounds now exhibit third-order kinetics (first order with respect to the aromatic fluoro compound, second order with respect to aniline). Much therefore remains to be discovered about these reactions and the fascinating cleavage of an aromatic C−F bond by hydrogen peroxide/peroxidase mentioned by Dr Saunders (1972).

SUMMARY

   According to modern chemical thermodynamics and kinetics, the extent (as measured by the equilibrium constant) and the speed (as measured by the rate constant) of a reaction depend on the free energy differences between reactants and products, and between reactants and the activated complex (or transition state), respectively. Physical properties which can be used to calculate, or at least to lead to a qualitative understanding of, such free energy differences (or to the corresponding heat content and entropy differences) therefore deserve special consideration. Unfortunately, what can be measured is often related only indirectly to what needs to be known. Physicochemical data for the C−F bond are reviewed critically with respect to both reliability and applicability to problems in the chemistry of the C−F bond under mild reaction conditions.
   Topics covered include bond energies, bond lengths, vibrational spectra, assessments of bond polarities, polarizability, the inductive and mesomeric effects, and the interpretation of kinetic data for reactions of the C−F bond in solution. The paucity of reliable physical data, the difficulty of obtaining them, and the subtlety of the interplay of heat and entropy factors suggest that for many years predictions based on the shortest possible extrapolation from existing information will contribute more to progress than a theoretical physicochemical approach.

References

BENSON, S. W., F. R. CRUICKSHANK, D. M. GOLDEN, G. R. HAUGEN, H. E. O'NEAL, A. S. RODGERS, R. SHAW and R. WALSH (1969) *Chem. Rev.* **69**, 279.

BROWN, J. K. and K. J. MORGAN (1965) *Adv. Fluorine Chem.* **4**, 253.

CONLEY, R. T. (1966) *Infrared Spectroscopy*. Boston: Allyn and Bacon.

CRUICKSHANK, D. W. J., D. W. JONES and G. WALKER (1964) *J. Chem. Soc.* 1303.

DAVIES, M. (1965) *Electrical and Optical Aspects of Molecular Behaviour*. Oxford: Pergamon Press.

FERGUSON, G. and K. M. S. ISLAM (1966) *Acta Crystallogr.* **21**, 1000.

GILLIOM, R. D. (1969) *Physical Organic Chemistry*. London: Addison-Wesley.

GOULD, E. S. (1959) *Mechanism and Structure in Organic Chemistry*. New York: Holt, Rinehart & Winston.

GUCKER, F. T. and R. L. SEIFERT (1967) *Physical Chemistry*. London: English Universities Press.

HENCHER, J. L. and S. H. BAUER (1967) *J. Am. Chem. Soc.* **89**, 5527.

HIROTA, E. (1962) *J. Chem. Phys.* **37**, 283.

HUGHES, D. O. and R. W. H. SMALL (1962) *Acta Crystallogr.* **15**, 933.

INGOLD, C. K. (1969) *Structure and Mechanism in Organic Chemistry*, 2nd ed. Ithaca: Cornell University Press.

KRAUSE, J. and A. C. DUNKEN (1966) *Acta Crystallogr.* **20**, 67.

LACHER, J. R. and H. A. SKINNER (1968) *J. Chem. Soc. A* 1034.

LINNETT, J. W. (1947) *Q. Rev. Chem. Soc.* **1**, 73.

MOELWYN-HUGHES, E. A. and I. FELLS (1959) *J. Chem. Soc.* 398.

MOORE, W. J. (1963) *Physical Chemistry*. London: Longmans.

PARKER, R. E. (1963) *Adv. Fluorine Chem.* **3**, 63.

PATRICK, C. R. (1961) *Adv. Fluorine Chem.* **2**, 1.

PAULING, L. C. (1940) *The Nature of the Chemical Bond*. Ithaca: Cornell University Press.

REED, T. M. (1964) In *Fluorine Chemistry*, vol. V, p. 133, ed. J. H. Simons. New York: Academic Press.

ROELOFSON, G., J. A. KANTERS, J. KROON and J. A. VLIEGENTHART (1971) *Acta Crystallogr.* **B27**, 702.

SAUNDERS, B. C. (1972) *This volume*, pp. 9–27.

SHARPE, A. G. (1967) In *Halogen Chemistry*, vol. I, p. 1, ed. V. Gutmann. London: Academic Press.

SHEPPARD, W. A. and C. M. SHARTS (1969) *Organic Fluorine Chemistry*. New York: Benjamin.

Discussion

Saunders: We found that ethyl 2-fluoroethoxyacetate could not be prepared by the action of ethyl diazoacetate on pure redistilled 2-fluoroethanol. However, fluoroethanol which had not been specially dried reacted immediately with ethyl diazoacetate with a vigorous evolution of nitrogen. This has always been a puzzle to us, but what you have said about moisture, Dr Sharpe, may go some way towards explaining it.

Armstrong: Dr Sharpe, can you give a comparison of the approximate bond strengths of the fluoro-carbon and hydroxy-carbon bonds?

Sharpe: The C—F bonds I have mentioned have energies of 450–485 kJ (108–116 kcal); the general value for the C—O bond in alcohols is 360 kJ (86 kcal).

Heidelberger: In my view, in biological terms the two most important properties of fluorine are the small size of the fluorine atom, which is so similar to the hydrogen atom that it fools enzymes, and the inductive effect that changes the acidity and hence the affinity of molecules for enzymes. For example, as shown in Fig. 1 (p. 126), at carbon-5 in the pyrimidine series we find that the dissociation constant of the proton on N-3 is lowered. In uracil, for example, the pK_a is about 9.6, and it is about the same for thymine. If there is a fluorine atom on carbon-5, the dissociation constant is lowered to about 8.1. We believe that it is this difference in the acid dissociation constants that gives rise to the fact that the fluoro compound has a thousand times more affinity for the target enzyme than the hydrogen compound. We also think that this indicates that nitrogen-3 is a binding site to the enzyme.

If we think in terms of the inductive effect of the trifluoromethyl group, we would expect that this proton on H-3 would be even more acidic. In fact, it turns out to have a pK_a of about 7.6 and is a biologically very active compound. But in 5-trifluoromethyl-6-azauracil the pK_a of this proton is 5.8, and this compound has no biological activity (Dipple and Heidelberger 1966). We conclude that this inactivity results from the fact that the H-3 proton is completely ionized at physiological pH, and that in the ionized form it does not biologically resemble the normal non-ionized form.

Sharpe: A difference of two units in pK_a corresponds to a difference of about 11 kJ (3 kcal) in free energy of ionization. Since both heat and entropy effects may contribute to this difference, it is extremely difficult to discuss the inductive effect rigidly unless pK_a measurements have been made over a range of temperatures.

Kun: The inductive effect of fluorine on the acid strength of carboxylic acid does not provide a good explanation for the inhibitory interactions of these carboxylic acids with enzymes. At the pH at which these enzymic reactions occur, these carboxyl ions are almost completely ionized. So the effect of fluorine on the enzyme-inhibitor complex must be of a different nature and one may assume it to be an entropy effect as Dr Sharpe proposed.

Kent: Dr Sharpe, would you care to speculate about the effect of the solvent in driving the reactions of fluorine compounds in which F^- is expelled? One wonders whether the effect of solvent is concerned with the solvation of the departing fluoride ion or with the stabilization of the whole transition complex.

To be able to distinguish clearly between these two would be a way to distinguish fluoride removal from other types of halogen decomposition.

Sharpe: You are really asking that one should first consider the equilibrium constants of the series of reactions and then the rate constants. Unfortunately there just are not enough data to do this. In simple inorganic systems the solvation of the departing fluoride ion is often sufficient to account for what happens. For example, when chloroplatinate is compared with the corresponding fluoro compound, the chloro compound is found to be more stable. This could be explained entirely in terms of the solvation of the fluorine ion. But you are asking for information that could be obtained only by studying the rate constants at different temperatures. As far as I know the hydrolysis of methyl fluoride is the only instance in which there has been a kinetic study at different temperatures of the reactions of a C—F compound (Moelwyn-Hughes and Fells 1959). Nearly all Chapman's work (see Parker 1963) has been on chlorine/fluorine ratios and has been carried out at one temperature only.

Kent: One can think of solvent as being a special environment in which the reaction occurs. The positioning of the fluorine in a particular environment may well influence its bond characteristics and hence its reactivity.

Hall: Dr Sharpe, there is a practical conformational outcome that can attend the introduction of the C—F bond into an organic compound. For example, the C—F bond of a glycosyl fluoride can dictate which conformation the sugar ring will adopt in solution (Hall and Manville 1967, 1969; Hall *et al.* 1971). Glycosyl fluoride derivations of the sugar, xylose, are currently the limiting example of this phenomenon. One might expect that the α-anomer (I) would adopt in solution that conformation that has all four of the bulky substituents equatorially oriented. However, the preference is almost exclusively for the conformation (II) that has the substituents axially oriented. For this class of molecule, a chlorine substituent exerts the same type of conformational effect (Holland, Horton and Jewell 1967).

In marked contrast, the oxygen heterocycle 1,3-dioxan, which again has the possibility of two chair conformations (III and IV), shows a difference between

(I) (II) (III) (IV)

the preference of a fluorine substituent at C-5 and a chlorine substituent (Eliel and Kaloustian 1970). The fluorine substituent shows a marked preference for an axial orientation (Eliel and Kaloustian 1970; L. D. Hall and R. N. Johnson 1969, unpublished results) whereas the chlorine substituent favours the equatorial orientation (Eliel and Kaloustian 1970). Clearly, for this organic system there must be a subtle interplay of forces, balancing the size of the substituent against some type of polar interaction. You have already said the amounts of energy involved here are quite small; could you suggest how we should start thinking about this type of conformational control?

Sharpe: I think it would be extremely difficult to interpret the relative stabilities of different conformations of sugars quantitatively. For a simple molecule like 1,2-difluoroethane in the vapour phase it does not seem surprising that interactions between bonds should be small, but for a sugar molecule in solution this would be unlikely to be true. However, the accumulation of data must precede their interpretation, and as a beginning it is highly desirable that the infrared or nuclear magnetic resonance spectra (from which deductions about conformations are made) should be recorded at different temperatures, so that heats of transition between different conformations can be evaluated.

Hall: I mention these examples because there is a tendency to equate bond lengths and atomic size with the effects these may have on the interaction between, for example, an enzyme and a particular inhibitor or substrate. The above findings indicate that it is possible for the conformation of a fluorinated compound to be fundamentally different from that of either a hydrogen analogue, a hydroxyl analogue or even a chlorine analogue. So before one starts relating difference in biological activity with such things as bond size, one should check that the gross features, such as molecular conformation, are identical through the series of compounds under investigation.

Kent: The C—F bond of the (C-1)-F axial group in glycosyl fluorides has been determined by X-ray crystal analysis to be 1.42 Å (Campbell *et al.* 1969).

Fowden: Dr Sharpe, you mentioned that the simple reaction between methanol and HF giving methyl fluoride might be possible with a very minimal energy supply. Would a similar reaction between fluoride and glycollic acid also have a low energy requirement and would this reaction occur more readily than one between fluoride and acetic acid? Both might yield fluoroacetic acid in biological systems.

Sharpe: If the C—F bond energy in fluoroacetic acid or fluoroacetate ion is the same as that in an alkyl fluoride, it should be possible to convert glycollic acid into fluoroacetic acid (or the glycollate ion into the fluoroacetate ion) by the absorption of only a small amount of energy, provided all bonds other than those involved in the reaction are unchanged. But because of the sensitivity of the value of K to a change in that of $\Delta G°$, extrapolation for this reaction from the reliable data in other cases I have cited in my paper is not necessarily valid. But where does fluoroacetic acid in a plant come from? Presumably from fluoride in the soil?

Goldman: We tried to demonstrate the formation of fluoroacetate from glycollate and fluoride by the bacterial enzyme that carries out the reverse reaction, namely:

$$FCH_2COO^- + H_2O \rightarrow HOCH_2COO^- + HF$$

But we never got any evidence that carbon-fluorine bond formation occurs by a reversal of this cleavage reaction (Goldman and Milne 1966).

Gál: I agree with Dr Goldman that one way an animal can degrade the $C-F$ bond is to form glycollic acid. After administration of radioactive fluoroacetate, we could demonstrate a compound chromatographically identical with radio-active glycollic acid in the urine of the animals.

Foster: Sir Rudolph, is anything known about the fluoride content of the soil in the areas where these plants that produce fluorinated acid are found?

Peters: Some of these plants are toxic and some that grow nearby are not. Mr L. Murray worked on the soil in Australia and could not find that it made any difference what soil they grew in (Murray and Woolley 1968). The roots of the Australian gidyea go down only about 10 feet, but *Dichapetalum cymosum* is supposed to have roots that go down 50 feet.

Saunders: The concentration of fluoride, not fluoroacetate, in the soil in South Africa is very high.

Peters: Hall (1970) reported that he had found fluoroacetate in the soil surrounding the toxic plants.

Armstrong: Most soils contain a few thousand parts per million of total fluoride. Most of the fluoride of soils is not in a soluble or ionic form since it is bound to apatite structures. It is a curious and interesting circumstance that only a few plants are able to absorb fluoride in appreciable amounts from soil. The tea plant is notable for its ability to absorb fluoride; most plants either reject or are unable to absorb fluoride.

Peters: The camellia can also absorb fluoride.

Saunders: Sir John Le Rougetell wrote that *Dichapetalum* grows best where the water has an abnormally high fluoride content, especially around warm baths, where there is so much fluoride in the water that it rots the teeth of the inhabitants and the whole town chews on its gums or dentures (see Saunders 1957).

Peters: That may be why the botanical gardens at Kew cannot grow it—not enough fluoride in the soil.

References

CAMPBELL, J. C., R. A. DWEK, P. W. KENT and C. K. PROUT (1969) *Carbohydr. Res.* **10**, 71.
DIPPLE, A. and C. HEIDELBERGER (1966) *J. Med. Chem.* **9**, 715.
ELIEL, E. L. and M. K. KALOUSTIAN (1970) *Chem. Commun.* 290.
GOLDMAN, P. and G. W. A. MILNE (1966) *J. Biol. Chem.* **241**, 5557–5559.
HALL, L. D., R. N. JOHNSON, A. B. FOSTER and J. H. WESTWOOD (1971) *Can. J. Chem.* **49**, 236–240.
HALL, L. D. and J. F. MANVILLE (1967) *Carbohydr. Res.* **4**, 512–513.
HALL, L. D. and J. F. MANVILLE (1969) *Can. J. Chem.* **47**, 19–30.
HALL, R. J. (1970) M.Sc. Thesis. University of Newcastle-upon-Tyne.
HOLLAND, C. V., D. HORTON and J. S. JEWELL (1967) *J. Org. Chem.* **32**, 1818.
MOELWYN-HUGHES, E. A. and I. FELLS (1959) *J. Chem. Soc.* 398.
MURRAY, L. R. and D. R. WOOLLEY (1968) *Aust. J. Soil Res.* **6**, 203–208.
PARKER, R. E. (1963) *Adv. Fluorine Chem.* **3**, 63.
SAUNDERS, B. C. (1957) *Some Aspects of the Chemistry and Toxic Action of Organic Compounds Containing Phosphorus and Fluorine*, p. 145. London: Cambridge University Press.

Some metabolic aspects of fluoroacetate especially related to fluorocitrate

Sir RUDOLPH PETERS

Department of Biochemistry, University of Cambridge

In the early 1940's it was known that fluoroacetate ($F \cdot CH_2 \cdot COO^-$) had no effect on any isolated enzyme and that it was the active principle of the toxic South African plant, *Dichapetalum cymosum* or gifblaar (Marais 1943, 1944: for reviews of this and related work see Chenoweth 1949; Peters 1957, 1963; Pattison and Peters 1966; Smith 1970). More recently, Lahiri and Quastel (1963) have detected an indirect effect on glutamine synthetase in brain tissue. This has been confirmed by Berl and colleagues (Clarke, Nicklas and Berl 1970) who think it arises from formation of fluorocitrate in special compartments of the brain. During the last 15 years, fluoroacetate, sometimes in large amounts, and in one case fluoro-oleic acid and other long chain fatty acids (Peters *et al.* 1960), have been detected in several other species of plant (see pp. 1–7).

LETHAL SYNTHESIS TO FLUOROCITRATE: THE BIOCHEMICAL LESION

In 1948 and 1949 Liébecq and I, using washed kidney particles from the guinea-pig, showed that fluoroacetate had no effect on acetate metabolism; but it induced accumulation of citrate, which later Buffa and I (1949) found also occurred *in vivo*. This has been extensively confirmed. The rise in citrate concentration in the kidney may reach 3000 µg/g instead of a normal 30–50 µg/g: there is less accumulation in the liver. We postulated the formation of an inhibitor of the enzymes of the citric acid cycle at a time when Martius (1949) independently drew similar conclusions from experiments with ox heart tissue. The reality of a 'lethal synthesis' to fluorocitrate (I) (Fig. 1) was finally proved by its isolation from kidney homogenates treated with fluoroacetate (Peters *et al.* 1953), which also indicated the extreme activity of fluorocitrate in inhibiting the enzyme aconitase (aconitate hydratase) in mitochondria (Peters and Wilson 1952). The synthesis occurs because the condensing enzyme, synthase, can accept fluoroacetyl-CoA in place of acetyl-CoA. This is not the place to enlarge upon the work on the inhibition of aconitase. Much work has

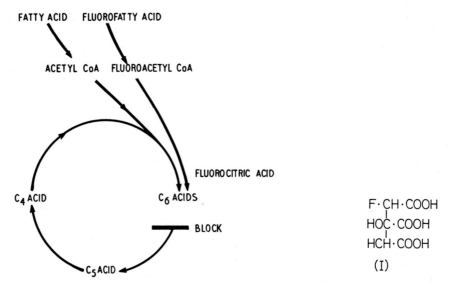

FIG. 1. Diagram showing the pathway of synthesis of fluoroacetate to fluorocitrate. The fluorocitrate blocks the metabolism of the C_6 acids at the aconitase stage of the citric acid cycle. The reactions of these are: citrate $\rightleftarrows$ *cis*-aconitate $\rightleftarrows$ isocitrate.

It is to be noted that the increasing citrate concentration also blocks the enzyme phosphofructokinase.

been done and more will be required to understand fully this enzyme complex and its inhibition. The enzyme is multiple (Fig. 2) (Peters and Shorthouse 1970; Guarriera-Bobyleva and Buffa 1969; Eanes and Kun 1971) and we think now that it consists of two active centres (Fig. 3) (Peters and Shorthouse 1970), which is a return to an older view. There is a reversible as well as an irreversible effect. To explain irreversibility it has been suggested (see Carrell *et al.* 1970) that F^- is split off on combination with fluorocitrate, but we were unable to confirm this experimentally (Peters and Shorthouse, unpublished). Two facts are quite clear: preparations of aconitase can be reached which are very sensitive to the inhibitor, and 50% inhibition has been observed with a 6.5×10^{-9} M solution of the active inhibitor (Peters 1961). There is always some aconitase present in aconitase preparations which is not inhibited. In one instance aconitase from single cell cultures of sycamore was 2000 times less sensitive to fluorocitrate than a partly purified animal aconitase (Treble, Lamport and

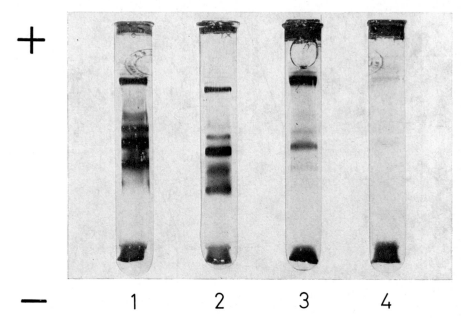

Fig. 2. Demonstration of the multiple nature of the aconitase enzymes and the fact that fluorocitrate has more inhibitory effect on some of the enzymes than others (from Peters and Shorthouse 1970).

Peters 1962), and Palmer (1964) found no iron in aconitase from mustard leaves, all of which may explain in part the apparent non-toxicity of fluoroacetate in the 'poison' plants.

Fluorocitrate has four isomers and has been synthesized by Rivett (1953) and Brown and Saunders (1962), the latter synthesis leading to the extensive work of Kun and colleagues (see Dummel and Kun 1969) who have isolated the active and other isomers and identified the active component as L-erythrofluorocitric acid. Early in my work I found that it was inadvisable to subject the biologically active isomer from kidney to warm acid solutions. The acid is now known to cause racemization (Peters, unpublished; Kun, personal communication). This adds to the difficulties, already great, of extracting the active fluorocitrate from tissues. Recently Mrs Shorthouse and I have confirmed that the pseudomonad of Kirk and Goldman (1970) cleaves the carbon-fluorine bond of the active isomer only.

TOXICITY: DOSES AND SIGNS IN ANIMALS

There is a wide variation in the dose of fluoroacetate which is toxic; for

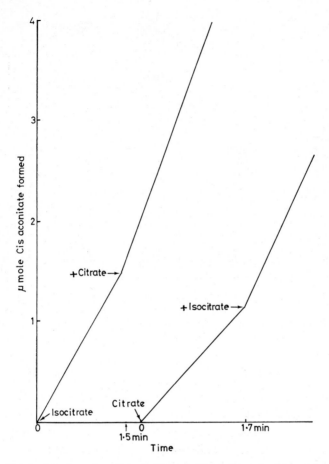

FIG. 3. Aconitase (aconitate hydratase). The effect of adding either citrate to the enzyme metabolizing isocitrate to *cis*-aconitate (isocitrate hydrolyase), or isocitrate to the enzyme metabolizing citrate (citrate hydrolyase). The initial substrates were added to give maximum velocity (from Peters and Shorthouse 1970).

instance the dog can be killed by 0.05 mg/kg, the sheep by 0.4 mg/kg and the rat by 5.0 mg/kg, whereas the toxic dose for the toad or frog is 300–500 mg/kg. The variation is best explained by the relative ease of conversion of fluoroacetate to fluorocitrate, because this appears to be the main cause of the toxicity. But it is known that fluoroacetate can be metabolized in other ways, as, for instance, in the liver (Phillips and Langdon 1955), and also to an unknown compound (Gál, Drewes and Taylor 1961); we cannot be certain that small amounts of other compounds are not formed in the animal. There are indications in the literature that some carbon-fluorine compounds such as fluoroaspartate (Kun,

Fanshier and Grassetti 1960) and fluorofumarate (Brodie and Nicholls 1970) may rapidly lose their F^- on formation. This makes the stability of the complex with aconitase even more astonishing.

The main signs of toxicity induced by giving fluoroacetate are seen in the heart or in the nervous system; the emphasis varies in different animals. In the rat the toxic signs start with tonic convulsions, resembling strychnine poisoning, 20–40 minutes after injection (Fig. 4). If the animal survives the convulsive period, a progressive muscular weakness sets in with a falling heart rate and lowered temperature; the collapse leads to death in 12–18 hours and lung

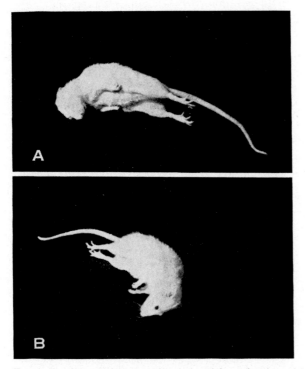

FIG. 4. Rat in a tonic convulsion induced by poisoning with fluoroacetate: (*a*) convulsion beginning; (*b*) convulsion ending (photos by P. Buffa).

oedema is often found on autopsy. In spite of much work these signs have not been explained. It might be thought that the convulsions are due to an initial overaction of acetylcholine, followed by a deficit of this due to failure of synthesis. Tests, however, showed no ameliorative effect with atropine, and acetylcholine injections did not improve the asthenia (Peters and Shorthouse, unpublished). Further, there is no evidence that fluoroacetylcholine has any abnormal effect, even if formed (see Peters 1957). Another possibility might be that the interruption of the citric acid cycle leads to a lowered concentration of γ-aminobutyrate; we are still investigating this, though it does not seem hopeful in view of the earlier work of Dawson (1953), who found little change, and of the extensive work of Koenig (1969) and Koenig and Patel (1967). These authors found remarkable phenomena after injection of very small amounts (9 μg) of fluorocitrate into the lumbo-sacral subarachnoid space of the cat. Clonic contractions of the hindlimb extensor muscles began after 1–2 hours and reached a rate of one per second. Alternations of extensor spasms with rhythmic myoclonus could continue for 24 hours. Among the many changes seen, including ballooning of the axons, they recorded the following changes: 25% reduction in ATP content; 21–41% reduction in glutamate, and also in glutamine and aspartate; an increase in alanine and ammonia content; but no change in γ-aminobutyrate or glycine. At first they thought that the amount of ammonia increased before the onset of convulsions, but now they have withdrawn this idea. Dr H. Koenig (personal communication) believes that convulsions may be due to a reduction of Mg^{2+} concentration due to citrate formation, with a consequent interference with the ATPase system. This is plausible, and certainly the idea that inhibition of the ATPase system is ultimately involved for the present appears to be the best working hypothesis. However, any hypothesis must take into account the fact that administration

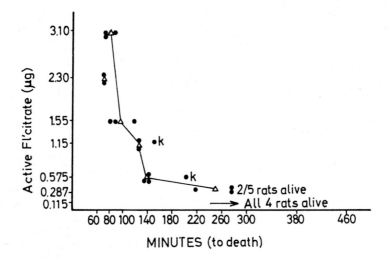

FIG. 5. Effects of intracerebral injection of fluorocitrate in rats (200 g) under ether. Ordinate: amount of sodium fluorocitrate injected, in terms of biologically active isomer; abscissa: minutes to death (from Morselli *et al.* 1968).

of compounds coming after citric acid in the citric acid cycle, even when given with divalent ions, does not stop the convulsive state (Hastings, Peters and Wakelin 1953, 1955).

Fluorocitrate is at least 200 times more toxic than fluoroacetate when injected into the subarachnoid space of a 200 g rat (Morselli *et al.* 1968), killing every rat with a dosage of less than 1.0 µg of the active isomer (Fig. 5), which is as toxic as diphtheria toxin. It is even more toxic in parts of the thalamus. This extreme toxicity in the brain of rats or in a kidney mitochondrial preparation from guinea-pigs is to be contrasted with its effects by the intraperitoneal route in rats. By this it is at least 200 times less toxic; by the mouth it is even less toxic (Peters and Shorthouse 1971*a*). On the other hand fluoroacetate is as toxic by the oral as by the intraperitoneal route. Presumably the smaller fluoroacetate molecule penetrates cell barriers and is synthesized in the mitochondria to fluorocitrate. This certainly happens in the heart; rat atria in a perfused preparation are not affected by fluorocitrate, but fluoroacetate penetrates and poisons them (Cerutti and Peters 1969). It is interesting that fluoroacetate injected into the subarachnoid space of the rat is barely toxic, possibly because it leaves the skull within one minute (Morselli *et al.* 1968). Evidently the blood-brain barrier can work in the reverse direction sometimes.

Attempts to find an antidote have not yet succeeded. Monoacetin has curative effects against fluoroacetate if introduced early enough (Chenoweth *et al.* 1951). Presumably it stops the conversion of fluoroacetate to fluorocitrate which we proved to happen in guinea-pig kidney *in vitro* (Peters and Wakelin 1957). In an extensive series of experiments (Peters and colleagues, mainly unpublished), it was found that malonates and substituted malonates could reverse the action of fluorocitrate on soluble aconitase (Fig. 6). Diethyl chloromalonate (II),

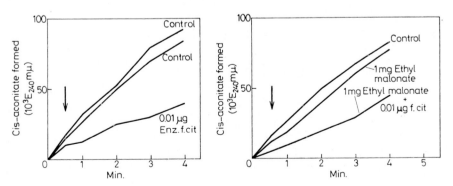

FIG. 6. Reversal of the inhibitory action of fluorocitrate on soluble aconitase by ethyl malonate.
Left: controls and effect of 0.01 µg fluorocitrate; right: reversal produced by ethyl malonate.

especially, could prevent the convulsions; but in spite of this the animals died in some hours with general weakness. This condition suggests the failure of some hormone, and we are still continuing to work on this possibility. It is

$$
\text{Cl·C·H}
\begin{array}{l}
\diagup\text{COOEt} \\
\diagdown\text{COOEt}
\end{array}
$$

(II)

interesting to note that Bacq, Liébecq-Hutter and Liébecq (1960) used fluoro-acetate to protect against X-rays, and that Dietrich and Shapiro (1956) found fluoroacetate and fluorocitrate to show carcinostatic activity against adeno-carcinoma; fluorocitrate showed low intraperitoneal toxicity.

FIG. 7. Seedling of *Acacia georginae*.

PLANT STUDIES

The failure to find a direct antidote *in vivo* led us to look into the metabolism of fluoride in the toxic plants in the hope of obtaining some hint for an antidote. We could not grow the South African *Dichapetalum cymosum* in Cambridge, but we were able to grow the Australian *Acacia georginae* from the seeds kindly supplied by Mr L. R. Murray from Alice Springs. Briefly, we found that homogenates of the leaves of seedlings of *A. georginae* (Fig. 7), suitably reinforced with ATP, pyruvate, Mn^{2+} and buffer phosphate (pH 7.0), can metabolize fluoride to an organic form, which is determined as the difference between F^- estimated in a straight diffusion from perchloric acid and of a similar diffusion after an alkaline combustion; the fluoride content is estimated by Hall's method which has a sensitivity of 0.2 µg F^- (Hall 1960). The acid diffusion does not distinguish between inorganic F^- and F^- split off by acid. The method combines ingeniously the diffusion introduced by Singer and Armstrong (1959) with the colorimetric method of Belcher, Leonard and West (1959). While studying the fluoride metabolism of *A. georginae* we found discrepancies between the controls and the samples of homogenates receiving additional inorganic fluoride; some of the added inorganic fluoride disappeared on combustion. It was our impression that watering with fluoride increased this phenomenon (Table 1), but we have not proved this. Similar disappearances of fluoride occurred with homogenates of ordinary plants like buttercups and daisies (Table 2) (Peters and Shorthouse 1967). We interpreted this as due to

TABLE 1

Disappearance of fluoride in plant homogenates

(*a*) *Acacia georginae* seedling watered with fluoride (approx. 200 µg/week) for 25 months. Reinforced homogenates from the leaves were divided into two parts; one of these was treated with extra fluoride. Both were shaken for one hour at 30 °C.

	F^- µg/g wet weight	
	Control	*+50 µg F^-/g*
F_i	15	32
Total F	35	11.3

(*b*) *Thea sinensis* (tea plant) watered with fluoride. Analysis of supernatant from homogenate of the leaves.

	F^- µg/g wet weight
F_i	200 µg
Total F	140 µg
Disappearing on combustion	60 µg

(Data of Peters and Shorthouse.)

TABLE 2

'Vanishing' fluoride from homogenates of various plants (selected data)*

	F^- (μg) added/g	% F^- disappearing
Acacia georginae	40	33.8
Pea	50.5	28.0
Grass	10.4	43.5
Daisy	15.9	41.5
Asclepias currasavisa	$\begin{cases} 55.3 \\ 26.2 \end{cases}$	31.0 / 52.5

*   From Peters and Shorthouse 1967.

the presence of a volatile compound lost on combustion, and have recently identified one such compound coming from homogenates of *A. georginae*, namely monofluoroacetone ($FCH_2 \cdot CO \cdot CH_3$), which was isolated by Miura and Hori (1961) from the livers of rats dosed with fluoroacetate. We know that there are others, as yet not identified (Peters and Shorthouse 1971*b*). [As it has been suggested (Hall 1969) that we may be dealing with compounds with silicate we added this in some experiments, but with no effect.]

Our discovery of a metabolism of fluoride in plants prepared our minds for the fascinating work of Gene Miller and colleagues in Utah (Lovelace, Miller and Welkie 1968), in which they found fluorocitrate and fluoroacetate in forage plants treated with or exposed to fluoride, and upon which Dr Miller will report later (see p. 117). Preuss, Lemmens and Weinstein (1968) have seen metabolism of fluoroacetate in *A. georginae* and other plants (see also Dr Ward p. 119). The presence of trace amounts of fluorocitrate now seems to us to be rather general. Previously we had not found organically combined fluoride in the leaves of a tea plant growing in Cambridge. Absence of combined fluoride in tea has also been reported by Kakabadse and co-workers (1971). Recently, using gas chromatography, we have seen a peak of fluorocitrate, not amounting to more than 30 $\mu g/g$, in an extract of a commercial tea. Anyone drinking tea made with this might get 0.24 mg in eight cups—a small amount. We have also found fluoride in organic form in single cell cultures of soya bean grown on fluoride, as well as in oatmeal, where it has been identified as fluorocitrate; and we have confirmed the formation of fluoroacetate in single cell cultures of *A. georginae* reported by Preuss, Birkhahn and Bergmann (1970) (Peters and Shorthouse, unpublished). Hence we are coming to the conclusion that small amounts of fluorocitrate are common constituents of our food. The determination of F^- in human urine by the specific fluorine electrode (Cernik, Cooke and Hall 1970) suggests the presence of some acid-labile fluoride. Presumably the fluorocitrate is formed via fluoroacetate. We still do not know how this latter is synthesized. We have tried to determine whether the carbon-fluorine bond is made via a chloro compound in *A. georginae*, but have found no evidence for this (Peters, Murray and Shorthouse 1965); or via some double bond compound like fumarate, or via fluorophosphate; but again our results were negative. A reasonable hypothesis would be that hydrofluoric acid adds on to ethylene ($CH_2:CH_2$), which is present; this would lead to either vinyl fluoride or fluoroethane as a first step. In fact we have seen (Peters and Shorthouse, unpublished) a small peak in the position of vinyl fluoride on a gas chromatogram but an attempt to find this by the workers in the Norwich Food Research Station failed.

Thus, from the above, it appears that plants can metabolize fluoride to an organic form; it is no longer correct to think that inorganic fluoride necessarily enters and leaves plants as such.

METABOLISM IN ANIMALS

The behaviour of fluoride in plants raises the question of whether the carbon-fluorine bond can come to pieces in animals, especially as it is known that there are compounds like fluoroaspartate which liberate inorganic fluoride, and that fluoropyruvate liberates F^- in contact with thiols (Avi-Dor and Mager 1956; Peters and Hall 1957). There is also the question of the presence of fluorocitrate in bones which will be discussed later in the meeting (see pp. 357–383).

FLUOROCITRATE AND ANIMALS

Recognition of the metabolism of fluoride in plants raised the question of the metabolism of fluorocitrate in animals. Some early work suggested that fluoride could be split off from fluoroacetate, for example in bones (Miller and Phillips 1955) and also from fluorovalerate in teeth (Ott, Piller and Schmidt 1956). These facts may have been wrongly interpreted; the carbon-fluorine compounds might have been turned into fluorocitrate. Because of our finding that the toxicity of fluorocitrate was relatively low when administered by mouth, and also because we had found that sub-lethal doses could be given to rats in their drinking water for several months without harm, we thought we should study the effect of the liver and kidney on fluorocitrate (Peters, Shorthouse and Ward, unpublished). We found no effect of liver microsomes *in vitro*. Using

[^{14}C] fluorocitrate labelled in the $-CH_2$ group of $-CH_2 \cdot COOH$ (kindly given to us by Professor Gál), we found, *in vivo* in a male Wistar rat, that within 48 hours of an oral or intraperitoneal dose, some 90% of the radioactivity was excreted in the urine. We are now trying to determine whether the whole molecule is excreted intact, because initial results do not exclude the possibility that there has been some degradation to a radioactively labelled C_2 or C_4 compound. In a recent experiment inorganic fluoride was excreted after an oral dose of fluorocitrate, indicating a removal of fluoride (Peters and Shorthouse, unpublished).

FLUOROACETATE AS A MARKER OF TOXICITY

The indirect effect of fluoroacetate leading to citrate accumulation by way of fluorocitrate formation has been much used as a marker of toxicity. We found some interesting cases of this with fluoroglycerols, which Dr Kent gave to us. l-Deoxy-l-fluoro-glycerol (1-FG) (III) was toxic to mice and induced citrate accumulation in the heart and kidney, as well as convulsions and

$$
\begin{array}{l}
\text{F} \cdot \text{CH}_2 \\
\quad | \\
\text{H} \cdot \text{C} \cdot \text{OH} \\
\quad | \\
\text{H}_2\text{C} \cdot \text{OH}
\end{array}
$$

(III)

bradycardia. But there was a puzzle: slices of kidney, brain, liver or muscle treated with this compound showed no citrate accumulation, nor was there any when 1-FG was perfused through the heart. Furthermore, there was no citrate accumulation *in vivo* after removal of the liver (O'Brien and Peters 1958*a*) (Fig. 8). But it was found that when the compound was perfused through the liver it was converted to another compound capable of inducing citrate accumulation both in the perfused heart and in kidney particles; this compound was shown to be fluoroacetate. A double lethal synthesis therefore occurred; and it is still not explained why the first stage required the intact liver, though it may be related to the problems raised by Clarke, Nicklas and Berl (1970) in brain tissue. The 2-deoxy-2-fluoroglyceraldehyde, however, was converted to fluoroacetate by a different route; after conversion to the 2-deoxy-2-fluoro-glycerate this compound appears to be converted to fluoroacetate by the L-serine hydroxymethyltransferase (O'Brien and Peters 1958*b*; Treble and Peters 1962).

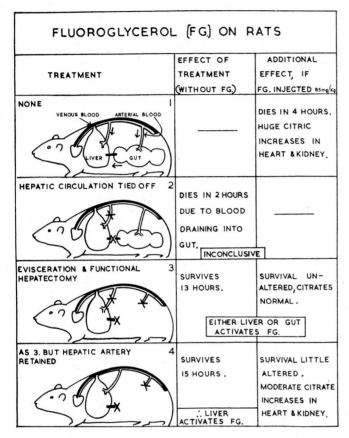

FIG. 8. Diagram illustrating the metabolism of fluoroglycerol (1-deoxy-1-fluoroglycerol) in rats (from O'Brien and Peters 1958*a*).

CONCLUDING REMARKS

Some of the vistas opened may be clarified by a few remarks upon the significance of our work as we see it. Apart from any possible applications to enzyme work and the nervous system, there is first the principle of 'lethal synthesis', the repercussions of which have spread far beyond the bounds of carbon-fluorine compounds. In regard to fluorine itself, there is the important discovery that fluoride does not enter living tissue merely to become an inorganic salt; it also takes part in the metabolism as a compound with carbon. Furthermore, it is evident that a consumption of small amounts of monofluoro-carbon compounds in food is quite normal. Are we to conclude that fluoride resembles selenium, being dangerous in large amounts but essential in small quantities? The puzzle raised by the large accumulations of fluoroacetate in plants and their apparent non-toxicity to the plant may be explained, at any rate in part, by the relative insensitivity of some plant aconitases to fluorocitrate.

SUMMARY

A short review is given of the salient features of our knowledge of the toxicity and behaviour of fluoroacetate in animals and plants. The main biochemical lesion causing toxicity in animals is due to a 'lethal synthesis' of fluoroacetate to fluorocitrate, which can block aconitase, an enzyme of the citric acid cycle, leading to large accumulations of citrate in various organs. The enzyme aconitase is a complex carrying out the reactions: citrate$\rightleftarrows$*cis*-aconitate$\rightleftarrows$iso-citrate. It exists in multiple forms, as well as having two centres. Extra-mitochondrial aconitase is much less inhibited by fluorocitrate than intramito-chondrial aconitase. By intracranial injection fluorocitrate is at least 200 times more toxic to rats than by intraperitoneal injection or by ingestion. The genesis of the tonic convulsion and subsequent asthenia due to these compounds is not yet certainly explained. No antidote to the poisoning has been found which acts *in vivo*. There is evidence that the animal can remove inorganic fluoride from fluorocitrate.

Fluoroacetate, fluorocitrate and fluoroacetone can be formed in plants, even in single cell cultures, from inorganic fluoride; and the insensitivity of the plant to fluoroacetate raises an interesting problem.

Other monofluorocarbon compounds can be converted to fluoroacetate and cause citrate accumulations, e.g. 1-deoxy-1-fluoroglycerol undergoes two stages of lethal synthesis. Acetate inhibits synthesis of fluorocitrate.

Fluorocitric acid, which exists in four isomers, has been synthesized, and the four isomers have been isolated in recent important work.

ACKNOWLEDGEMENTS

I am indebted to Mrs M. Shorthouse for reading the manuscript, to the Wellcome Trustees for grants in aid of the work, to Professor F. G. Young, FRS, for facilities provided, and to the Royal Society for a grant for gas chromatography.

References

Avi-Dor, Y. and J. Mager (1956) *Biochem. J.* **63**, 613-618.

Bacq, Z. M., S. Liébecq-Hutter and C. Liébecq (1960) *Radiat. Res.* **13**, 286-297.

Belcher, R., M. A. Leonard and T. S. West (1959) *J. Chem. Soc.* 3577-3579.

Brodie, J. B. and P. Nicholls (1970) *Biochim. & Biophys. Acta* **198**, 423-437.

Brown, P. J. and B. C. Saunders (1962) *Chem. & Ind.* 307-308.

Buffa, P. and R. A. Peters (1949) *J. Physiol. (Lond.)* **110**, 488-500.

Carrell, H. J., J. P. Glusker, J. J. Villafranca, A. S. Mildvan, R. J. Dummel and E. Kun (1970) *Science* **170**, 1412-1414.

Cernik, A. A., J. A. Cooke and R. J. Hall (1970) *Nature (Lond.)* **227**, 1260.

Cerutti, V. F. and R. A. Peters (1969) *Biochem. Pharmacol.* **18**, 2264-2267.

Chenoweth, M. B. (1949) *Pharmacol. Rev.* **1**, 383-424.

Chenoweth, M. B., A. Kandel, L. B. Johnson and D. R. Bennett (1951) *J. Pharmacol. & Exp. Ther.* **102**, 31-54.

Clarke, D. D., W. J. Nicklas and S. Berl (1970) *Biochem. J.* **120**, 345-351.

Dawson, R. M. C. (1953) *Biochim. & Biophys. Acta* **11**, 548-552.

Dietrich, L. S. and D. M. Shapiro (1956) *Cancer Res.* **16**, 585-588.

Dummel, R. J. and E. Kun (1969) *J. Biol. Chem.* **244**, 2966-2969.

Eanes, R. Z. and E. Kun (1971) *Biochim. & Biophys. Acta* **227**, 204-210.

Gál, E. M., P. A. Drewes and N. F. Taylor (1961) *Arch. Biochem. & Biophys.* **93**, 1-14.

Guarriera-Bobyleva, V. and P. Buffa (1969) *Biochem. J.* **113**, 853-860.

Hall, R. J. (1960) *Analyst (Lond.)* **85**, 560-563.

Hall, R. J. (1969) Thesis, University of Newcastle-upon-Tyne.

Hastings, A. B., R. A. Peters and R. W. Wakelin (1953) *J. Physiol. (Lond.)* **120**, 50 P.

Hastings, A. B., R. A. Peters and R. W. Wakelin (1955) *Q. J. Physiol.* **40**, 258-268.

Kakabadse, G. J., B. Manohin, J. M. Bather, E. C. Weller and P. Woodbridge (1971) *Nature (Lond.)* **229**, 626-627.

Kirk, K. and P. Goldman (1970) *Biochem. J.* **117**, 409-410.

Koenig, H. (1969) *Science* **164**, 310-312.

Koenig, H. and A. Patel (1967) *J. Cell Biol.* **35**, 72A.

Kun, E., D. W. Fanshier and D. R. Grassetti (1960) *J. Biol. Chem.* **235**, 416-419.

Lahiri, S. and J. H. Quastel (1963) *Biochem. J.* **89**, 157-163.

Liébecq, C. and R. A. Peters (1948) *J. Physiol. (Lond.)* **108**, 11P.

Liébecq, C. and R. A. Peters (1949) *Biochim. & Biophys. Acta* **3**, 215-230.

Lovelace, C. J., G. W. Miller and G. W. Welkie (1968) *Atmosph. Environ.* **2**, 187-190.

Marais, J. S. C. (1943) *Onderstepoort J. Vet. Sci. Anim. Ind.* **18**, 203-206.

Marais, J. S. C. (1944) *Onderstepoort J. Vet. Sci. Anim. Ind.* **20**, 67-73.

Martius, C. (1949) *Justus Liebigs Ann. Chem.* **561**, 227-232.

Miller, R. F. and P. H. Phillips (1955) *Proc. Soc. Exp. Biol. & Med.* **89**, 411-413.

Miura, K. and T. Hori (1961) *Agric. & Biol. Chem. (Jap.)* **25**, 94-99.

Morselli, P. L., S. Garattini, F. Marcucci, E. Mussini, W. Rewersky, L. Valzelli and R. A. Peters (1968) *Biochem. Pharmacol.* **17**, 195-202.

O'Brien, R. D. and R. A. Peters (1958a) *Biochem. J.* **70**, 188-195.

O'Brien, R. D. and R. A. Peters (1958b) *Biochem. Pharmacol.* **1**, 1-18.

Ott, E., G. Piller and H. J. Schmidt (1956) *Helv. Chim. Acta* **39**, 682-685.

Palmer, M. I. (1964) *Biochem. J.* **92**, 404-410.

Pattison, F. L. M. and R. A. Peters (1966) *Handb. Exp. Pharmakol.* **20**, pt 1, 387-458.

Peters, R. A. (1957) *Adv. Enzymol.* **18**, 113-159.

Peters, R. A. (1961) *Biochem. J.* **79**, 261-268.

Peters, R. A. (1963) In *Biochemical Lesions and Lethal Synthesis*, pp. 88-130, ed. P. Alexander and Z. M. Bacq. Oxford: Pergamon Press.

Peters, R. A., P. Buffa, R. W. Wakelin and L. C. Thomas (1953) *Proc. R. Soc. B* **140**, 497-507.

PETERS, R. A. and R. J. HALL (1957) *Biochim. & Biophys. Acta* **26**, 433–434.
PETERS, R. A., R. J. HALL, P. F. V. WARD and N. SHEPPARD (1960) *Biochem. J.* **77**, 17–23.
PETERS, R. A., L. R. MURRAY and M. SHORTHOUSE (1965) *Biochem. J.* **95**, 724–730.
PETERS, R. A. and M. SHORTHOUSE (1967) *Nature (Lond.)* **216**, 80–81.
PETERS, R. A. and M. SHORTHOUSE (1970) *Biochim. & Biophys. Acta* **220**, 569–579.
PETERS, R. A. and M. SHORTHOUSE (1971a) *J. Physiol. (Lond.)* **216**, 40–41P.
PETERS, R. A. and M. SHORTHOUSE (1971b) *Nature (Lond.)* **231**, 123–124.
PETERS, R. A. and R. W. WAKELIN (1957) *Biochem. J.* **67**, 280–286.
PETERS, R. A. and T. H. WILSON (1952) *Biochim. & Biophys. Acta* **9**, 310–315.
PHILLIPS, A. H. and R. G. LANGDON (1955) *Arch. Biochem. & Biophys.* **58**, 247–249.
PREUSS, P. W., R. BIRKHAHN and E. D. BERGMANN (1970) *Israel J. Bot.* **19**, 609–619.
PREUSS, P. W., A. G. LEMMENS and L. H. WEINSTEIN (1968) *Contrib. Boyce Thompson Inst.
    Plant Res.* **24**, 25–32.
RIVETT, D. E. A. (1953) *J. Chem. Soc.* 3710–3711.
SINGER, L. and W. D. ARMSTRONG (1959) *Anal. Chem.* **31**, 105–109.
SMITH, F. A. (1970) *Handb. Exp. Pharmakol.* **20**, pt. 2, 253–408.
TREBLE, D. H., D. T. A. LAMPORT and R. A. PETERS (1962) *Biochem. J.* **85**, 113–115.
TREBLE, D. H. and R. A. PETERS (1962) *Biochem. Pharmacol.* **11**, 891–899.

Discussion

Kun: I wish to talk about the molecular mechanism of fluorocitrate action
(Eanes 1971; Eanes, Skilleter and Kun 1972). The experiments I will discuss are

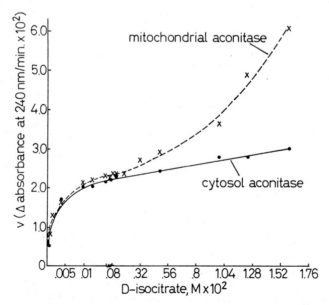

FIG. 1 (Kun). The effect of the concentration of D-isocitrate on the observed velocity of
formation of *cis*-aconitate catalysed by cytosol and mitochondrial aconitase.

partly enzymology and partly work with intact mitochondria. Normally one would not expect a reversible inhibitor, like fluorocitrate acting on aconitase, to be a lethal agent because it is clearly dissociable from the enzyme. This dilemma has driven us to more detailed experimentation. Fig. 1 shows a simple kinetic study of two aconitases. The cytosol enzyme is molecularly pure, has a molecular weight of 107 000 and shows essentially a monotonous enzyme–substrate relationship. The mitochondrial enzyme, however, shows a peculiar sigmoidal characteristic. The activation at higher concentrations of isocitrate is probably the same phenomenon as Sir Rudolph and collaborators also observed at higher concentrations of substrate. Table 1 shows various constants, although with the

TABLE 1 (Kun)

The apparent dissociation constants of the isoenzymes of aconitase for their substrates and for $(-)$-2-fluorocitrate

Aconitase Isoenzyme	K_m (M)		K_i (M)	
	Citrate	D-*Isocitrate*	Citrate	D-*Isocitrate*
Mitochondrial	4.3×10^{-4}	1.7×10^{-5}	6.6×10^{-5}	3.5×10^{-5}
Cytosol	4.3×10^{-4}	1.7×10^{-5}	3.4×10^{-5}	2.0×10^{-5}

mitochondrial enzyme K_m values are virtual (calculated), because the S/V relationship is sigmoidal and true constants cannot be directly evaluated. These inhibitory constants are much too large to explain the lethal action of fluorocitrate. Inhibitory constants should be at least 0.1 μM or less to correspond with *in vivo* toxicity.

A spectrophotometric test was devised which can be used to determine the rate of entry of tricarboxylic acid into mitochondria. A suspension of liver mitochondria is placed in a cuvette and isocitrate dehydrogenase (ICD) and NADP$^+$ are added to the extramitochondrial medium. If we inhibit the respiration of mitochondria (by rotenone, 1 μM) and add citrate, this will enter the mitochondria, isocitrate will leave mitochondria and reduce NADP$^+$ in the presence of added ICD. This reaction provides a kinetic test for citrate entry and isocitrate exit and permits accurate measurement of the rate of NADPH formation. Fig. 2 shows the rate of isocitrate exit when we add *cis*-aconitate to mitochondria. As shown in Fig. 2, if we add an agent that destroys the structure of mitochondria (Triton X-100), a much greater rate of reaction is obtained, indicating that the rate of entry of *cis*-aconitate (or citrate) is limited by the mitochondrial anion transfer system. If fluorocitrate is added, the rate of exit of isocitrate is greatly inhibited. However, if we destroy the mitochondrial structure by Triton X-100, the rate of NADPH formation is almost the same as in the

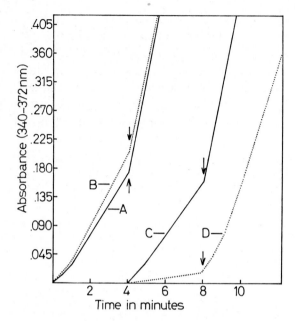

FIG. 2 (Kun). A study of the inhibition of the tricarboxylate carrier by 5×10^{-6} M-fluorocitrate. Curve (*A*): 1 mM-*cis*-aconitate was present from the onset; curve (*B*): 1 mM-*cis*-aconitate + 5×10^{-6} M-fluorocitrate was present from the onset. Curve (*C*): mitochondria incubated for 4 min without substrate, then 1 mM-*cis*-aconitate was added. Curve (*D*): mitochondria incubated for 4 min with 5×10^{-6} M-fluorocitrate, then 1 mM-*cis*-aconitate was added. At the arrows the mitochondria of each system were ruptured by addition of 30 µl of 10% Triton.

control (containing no fluorocitrate). The concentration of fluorocitrate sufficient for this inhibition can be as low as 10^{-8}M, which is 500 to 1000 times less than required for aconitase inhibition. This suggests that a mode of action of fluorocitrate is through the tricarboxylic acid carrier discovered originally by Chappell (1968), and that this inhibitor reacts in the inner membrane of mitochondria and fluorocitrate binds to it in an irreversible manner. This explains the irreversible toxic effect of fluorocitrate at the molecular level.

Peters: Our results were obtained with a soluble aconitase and we did not have the mitochondrial membrane present. We added enough isocitrate to ensure maximum enzyme velocity for this substrate, then added the other substrate, citrate. But you are claiming that you have a complicated system of the cell and the mitochondria inside it: we do not have that.

Kun: We can observe anomalous kinetics with soluble mitochondrial aconitase, but inhibition of cytosol aconitase by fluorocitrate is reversible. Addition of citrate would reverse the fluorocitric inhibition. With intact mitochondria, addition of citrate does not reverse the effect of fluorocitrate.

Peters: Mrs Shorthouse and I were not studying fluorocitrate in this experiment (Fig. 3, p. 58). We first added citrate and then, after some minutes, isocitrate.

Kun: The V_{max} of that reaction (i.e. with isocitrate) is slightly greater than for the citrate reaction. Therefore you did get an apparent activation.

Peters: The activation is not due to further addition of the one substrate, which was present already at an amount giving maximum velocity. The point is that the phenomenon occurred with initial addition of either citrate or isocitrate. We have other supporting information: (i) as mentioned, aconitase in our pig heart preparation is multiple; (ii) if our enzymes stand at $0\,°C$ rather than $-19\,°C$ for a certain length of time, the citrate reaction is reduced in activity much more than the isocitrate reaction. This is why we have returned to the original idea that aconitase is an enzyme with two active centres: (*a*) citrate → *cis*-aconitate (*b*) isocitrate → *cis*-aconitate (Peters and Shorthouse 1969).

Kun: I disagree with this, because one has to prepare a pure enzyme before performing kinetic studies. If the pure enzyme produces anomalous kinetics, then it is possible to postulate modification of catalysis at specific sites.

Peters: Were your experiments done with the pure enzyme?

Kun: Yes.

Gál: Some of our results (see pp. 77–90) support what Professor Kun has observed in intact mitochondria. I can now find an explanation for the toxic effect and the very small amount of fluorocitrate that is required in the mitochondria even in the absence of measurable citrate accumulation.

Saunders: Sir Rudolph, what are your views on Chenoweth's (1949) glyceryl monoacetate as an antidote to fluoroacetate poisoning? The compound exhibits both water solubility and fat solubility and acetate comes off slowly and thus competes with fluoroacetate.

Peters: I agree that monoacetin is an antidote if given quickly enough. Certain reactions in which inorganic fluoride can come off are known. Professor Kun and his colleagues (Kun, Fanshier and Grassetti 1960) have found that when they make fluoroaspartate, the fluorine splits off. This has been claimed recently by Brodie and Nicholls (1970) for fluorofumarate, and it occurs in the reaction I mentioned with fluoropyruvate. When the fluoropyruvate combines with SH, off comes fluorine.

Heidelberger: Professor Kun, could you elaborate on this tricarboxylic acid carrier?

Kun: The tricarboxylic acid carrier is a protein which is in the inner membrane and presumably acts by binding tricarboxylic acids; it only reacts with the substrate, which it somehow transfers to the inside of the inner membrane, so that it becomes available for metabolism in the matrix space. It is activated by inorganic phosphate and certain dicarboxylic acids. I would suggest that an antidote to fluorocitric acid poisoning should be an activator of the transfer protein. We have found several activators. Fluoromalic acid is a very good activator which partially reverses the apparent toxicity of fluorocitrate in mitochondria.

Gál: How does fluoromalate act *in vivo*?

Kun: I don't know. We have prepared a pure isomer of fluoromalate by enzymic synthesis, but we do not have enough to perform experiments on animals.

Kent: Sir Rudolph, is it possible that intestinal bacteria may be involved in these interesting differences between the toxic effects of intravenously administered fluoro compounds and those given orally?

Shorthouse: We are going to investigate that possibility.

Kent: Is fluoroacetate less effective when given orally?

Shorthouse: No, it is exactly the same whether given intraperitoneally or orally. When we gave fluoroacetate, 4 mg/kg, orally to rats whose liver enzymes had been induced by intraperitoneal injections of phenobarbitone, the response was very rapid and extremely violent: the animals had severe tonic convulsions. It is interesting to speculate whether changes in the brain brought about by the phenobarbitone had made the fluorocitrate synthesized from the fluoroacetate more accessible or whether acetylcholine was involved. Tentative experiments *in vivo* and *in vitro* on the effect of fluorocitrate on acetylcholinesterase gave no indication that this could be a reason for the violence of the convulsions after administration of phenobarbitone.

Kent: We injected *N*-fluoroacetyl glucosamine into mice and found that 40 to 50 mg/kg were toxic, which is similar to your results for fluoroacetate without phenobarbitone. However, it does not produce demonstrable toxic effects on growing mouse cells in tissue culture, at 1 mM or less.

Heidelberger: If mitochondria in the brain were the targets for the toxic action of fluorocitrate, then one could very readily understand why injecting it directly into the brain produces such great toxicity. This compound is inordinately beautifully designed not to get through the blood-brain barrier, which generally excludes negatively charged compounds. Is there any difference in sensitivity between brain mitochondria and mitochondria from other tissues?

Gál: The only difference I find for citrate accumulation is a quantitative one. The brain accumulates less than kidneys do.

Peters: Fluoroacetate is about 200 times less toxic than fluorocitrate when put inside the brain case (Morselli *et al.* 1968). We investigated this by gas chromatography and found that within one minute, 90% or more of the fluoroacetate that we had injected into the brain case came out into the circulation. This really did surprise us.

Taves: I am puzzled by the results in Table 1 (p. 63). How do you explain getting lower total amounts of fluoride than the original amounts of inorganic fluoride?

Peters: We think that during the process of taking homogenates to dryness, volatile forms of fluoride go off without taking the carbon-fluorine bond to pieces. This certainly must be happening with fluoroacetone, which is fairly volatile.

Taves: But you are seeing that process in the inorganic fluoride analysis. Are you saying that an organic fluoride is breaking down in the first acid diffusion?

Peters: Presumably it does not come out in that because it is combined in such a way that it would be acid-labile rather than inorganic. The first diffusion in acid will pick up inorganic fluoride and also any fluoride that is liberated by acid.

Taves: So really what you call inorganic fluoride ought to be called acid-labile fluoride.

Peters: No, because if a volatile compound goes off between 50°C and 100°C the chances are that the C—F bond will not be broken up before it comes off. We are not doing our combustion in a closed system. We gave up the bomb calorimeter in despair.

Taves: I agree that there are compounds like that. We worked with one, methoxy-difluoroacetic acid, where we found more fluoride by acid diffusion than by ashing. But to me it seems more appropriate to call what you get by the acid diffusion method an acid-labile fluoride, and use a fluoride electrode to measure the inorganic fluoride.

Armstrong: Sir Rudolph, for how long did you incubate the leaves of the different species of plants referred to in Tables 1 and 2 (pp. 63, 64)?

Peters: We have always used one hour at 30°C with shaking.

Armstrong: Was fluoroacetone found in all those species?

Peters: We have only found it in *Acacia*, but we have not looked for it in other leaves yet.

Kun: It was reported from S. Ochoa's laboratory that crystalline enolase from muscle can produce fluorophosphate, so possibly inorganic fluoride might enter into a fluorophosphate type of compound. Perhaps this is what you observed as an incorporation of fluoride.

Peters: We did find some fluorophosphate on paper chromatograms from reinforced *Acacia* homogenates, but in the fresh plant itself we could not detect any fluorophosphate.

References

BRODIE, J. B. and P. NICHOLLS (1970) *Biochim. & Biophys. Acta* **198**, 423–437.
CHAPPELL, J. B. (1968) *Br. Med. Bull.* **24**, 150.
CHENOWETH, M. B. (1949) *J. Pharmacol. Exp. Ther.* **97**, 383–424.
EANES, R. Z. (1971) Ph.D. Thesis. University of California.
EANES, R. Z., D. M. SKILLETER and E. KUN (1972) *Biochem. & Biophys. Res. Commun* in press.
KUN, E., D. W. FANSHIER and D. R. GRASSETTI (1960) *J. Biol. Chem.* **235**, 416–429.
MORSELLI, P. L., S. GARATTINI, F. MARCUCCI, E. MUSSINI, W. REWERSKI, L. VALZELLI and
    R. A. PETERS (1968) *Biochem. Pharmacol.* **17**, 195–202.
PETERS, R. A. and M. SHORTHOUSE (1969) *Nature (Lond.)* **221**, 774–775.

Effect of fluoro compounds on metabolic control in brain mitochondria

E. MARTIN GÁL

Laboratory of Neurobiochemistry, Department of Psychiatry, University of Iowa

The last twenty years witnessed the emergence of research related to the chemistry and biological action of fluoroinhibitors. Although preoccupation with these inhibitors may have been classed by some as esoteric, its net results have yielded a wealth of information profitable to all, as for example the final proof for the position of citrate in the tricarboxylic acid cycle. Fortunately, an excellent review (Pattison and Peters 1966) provides ample support for the above statements. One of the most significant contributions of the last five years, and long overdue, was the resolution of fluorocitric acid into its isomers (Dummel and Kun 1969).

This paper is biased towards brain mitochondria and the effect of fluoro compounds on them. This is because of our desire to understand the mechanism whereby convulsions are elicited by these compounds. Our research was limited to two monofluoro compounds—fluoroacetate and fluorocitrate. Among the various problems we attempted to explore with the aid of radioactive fluoroacetate (FAc) and fluorocitrate (FC) were the distribution, subcellular localization and mitochondrial penetration of these inhibitors, and their effect on mitochondrial control of Ca^{2+} uptake and retention. Earlier work with Ca^{2+} (Hastings, Peters and Wakelin 1953, 1955) had not revealed a correlation between Ca^{2+} movement and citrate accumulation.

Our working hypothesis for the production of convulsions via lethal synthesis is based on the fact that the mitochondria are an important part of synaptic morphology and function. Indeed, to quote Palay (1967), that mitochondria may be something special to the synapse is 'indicated by the fact that other cytoplasmic particulates like lysosomes and multivesicular bodies are rarely to be found in the terminal'. Our own studies (Millard, Kubose and Gál 1969) suggest that at least 10% of the mitochondrial population of the brain may be recovered from the synaptic area.

If the above is translated into simplistic terms, one is attracted by the idea that the inhibition of aconitase with a concomitant rise of citrate at the synaptic areas could critically alter Ca^{2+} binding to membrane subunits and also greatly affect the Ca^{2+} and Na^+ exchange required during excitation and recovery. I, for one, am sceptical of Chenoweth's view (1949) that citrate accumulation bears little relation to the appearance of the toxic syndrome. It seems that many of the problems raised could be answered more reliably through experimentation with the aid of isotopically labelled material. The findings reported here relate to such studies.

DISTRIBUTION OF FLUOROINHIBITORS *in vivo*

We begin our report by considering the distribution of radioactivity after intraperitoneal administration of sodium [^{14}C]fluoroacetate or trisodium [^{14}C]fluorocitrate to Sprague-Dawley male rats (150–250 g). Before the injection the animals were fasted overnight with water *ad libitum*. Urine, expiratory CO_2 and tissue radioactivity were determined as described previously (Gál, Drewes and Taylor 1961), except that aliquots of tissue homogenates were digested before being counted by liquid scintillation. Samples of $BaCO_3$ were mixed with Cabosil before counting. In some instances, expiratory CO_2, collected as $BaCO_3$, was released by addition of sulphuric acid and trapped in ethanolamine for counting.

A significant difference in the distribution of the radioactive FAc and FC is apparent in Table 1. It is reasonable to expect considerably less tricarboxylic acid to penetrate into the organs than FAc. The rapid elimination of FC *in vivo* is therefore almost predictable; but the appearance of radioactivity in the

TABLE 1

Radioactivity recovered 4 hours after intraperitoneal administration of inhibitor

Organ	[2-^{14}C]Fluoroacetate	[^{14}C]Fluorocitrate
	% Recovery	
Brain	2.3	0.1
Liver	12.0	1.2
Kidneys	2.5	0.3
Respiratory CO_2	1.8	0.5
Urine	0.5	50.0
	(3)	(4)

1.53 mg/kg sodium fluoroacetate (2 µCi/kg) or 15 mg/kg sodium fluorocitrate (0.44 µCi/kg) were injected. Number of animals is given in parentheses.

expiratory CO_2 after administration of FC was unexpected. There are two possible explanations. One is the presence of a small amount of metabolizable organic impurity. The second, more tenuous, assumes that some of the FC will be acted on by citrate synthase to yield fluoroacetyl-CoA and oxoacetic acid. This latter would contain the radioactive carbon which would ultimately appear in respiratory CO_2. Presumably this could come only from the [−]-erythro-fluorocitrate (Dummel and Kun 1969; Kirk and Goldman 1970).

In another series of experiments a non-convulsive dose of radioactive FC was administered intraperitoneally to rats. A biphasic rate of disappearance of radioactivity was found in the liver, with a half-life of 3 hours for the rapid phase and $28\frac{1}{2}$ hours for the slow phase. In the brain and the kidneys the half-lives were 72 and 10 hours respectively. Disappearance of FAc from the brain is biphasic with a half-life of $5\frac{1}{2}$ hours for the rapid phase and about 28 hours for the slow phase. This corresponds to the rates found for fluorocitrate.

No reliable data existed on the amount of FC in cerebral mitochondria after peripheral administration of convulsive and non-convulsive doses of FAc or FC. Acquisition of such information is the logical starting point for comparing effects *in vivo* and *in vitro*. One has little justification for using concentrations *in vitro* wholly out of proportion to those having physiological effects. (Perhaps exceptions can be made for kinetic studies with isolated enzymes.) The results in Table 2 demonstrate close agreement between the total amounts of FC found in cerebral mitochondria from animals treated with convulsive doses of both FAc and FC. Technically it was not feasible to determine the very low amounts of the inhibitor present in the mitochondrial fraction after non-convulsive doses of FC. The twice-washed mitochondrial fraction was isolated by the method of Schneider and Hogeboom (1950). Radioactive FAc and FC

TABLE 2

Comparison of mitochondrial concentration of fluoroinhibitors after convulsive and non-convulsive doses

Inhibitor	Injected		Time of sacri-fice min	Liver % [14C] recovered		Brain % [14C] recovered		Total FC	
	mg/kg	d.p.m. ×10⁻⁶		in organ	in mito-chondria	in organ	in mito-chondria	theory µg	found
Fluoroacetate (4)	4*	5.77	30	3.4	0.11	0.4	0.02	0.54	0.20
Fluorocitrate (6)	24	9.18	60	1.7	0.014	0.03	0.001	0.07	—
Fluorocitrate (6)	77*	19.9	60	2.0	0.029	0.11	0.0024	0.46	0.26

* Convulsive dose.   Numbers of animals in parentheses.

were separated and identified by steam distillation and paper chromatography as described elsewhere (Gál, Drewes and Taylor 1961).

At convulsive concentrations of either compound the total fluorocitrate found in cerebral mitochondria is on the average 0.25 µg (about 25 ng/mg mitochondrial protein). Since synthetic FC (containing all four isomers) was used, it is open to speculation whether the intramitochondrial FC represents all four isomers or whether a selective binding by the mitochondrial aconitase occurs for the natural isomer only. This speculation is not improbable if one considers that there is a difference between the rates of influx as well as the rates of efflux of the D- and L-amino acids (Lajtha 1962). It is precisely this type of question which might be answered by investigations using the resolved isomers of radioactively labelled FC.

To obtain the final distribution and localization of the fluoroinhibitors in the crude mitochondrial pellet, this was submitted to discontinuous sucrose density-gradient centrifugation (Rodríguez, Alberici and DeRobertis 1967). In these experiments the radioactive inhibitors were administered intracerebrally through holes drilled in the skull 24 hours earlier (Millard and Gál 1971). The animals (fasted overnight) were killed at the onset of convulsions. The rate of efflux of FAc from the brain after the intracerebral injection seems to be biphasic. First there is a rapid elimination of FAc *per se*, which supports similar observations (Morselli *et al.* 1968). The rate of efflux during the second, very slow, phase parallels that of FC. At the time of the first convulsion, at most 3% of the injected FAc and about 25% of FC were in the brain. Of this 3.8% was in the crude mitochondrial fraction from the FAc-injected animals against 12.4% in that from animals which had received intracerebral FC. The distribution of activity in the crude cerebral mitochondria after separation of the subfractions by density-gradient centrifugation indicates a recovery of FC calculated to be 5.7 ng per mg of mitochondrial protein after injection of FAc and 9 ng of FC per mg protein after FC. These figures, albeit somewhat smaller than those for the concentration of FC in the cerebral mitochondria after intraperitoneal administration of the inhibitors, are of the same order of magnitude. It is pertinent to note that all the animals receiving intracerebral injections of FC and FAc at the doses indicated (Table 3) convulsed at 30 and 60 minutes respectively. Analyses of the brains of these animals revealed 144 µg citrate/g in the FAc and 139 µg citrate/g in the FC injected rats. These concentrations of citrate agree remarkably well with those obtained after intraperitoneal administration of the inhibitors as well as with those in the brain after intracerebral administration of the enzymically synthesized FC (Gál, Peters and Wakelin 1956).

It is known that the so-called 'synaptosomal' fraction obtained after density-gradient centrifugation contains mitochondria. Therefore, the 14% of the radioactivity from FC found in this fraction (Table 3) could well be localized in these mitochondria. These synaptosomal mitochondria appear to show a

TABLE 3

Radioactivity recovered after density-gradient centrifugation of crude cerebral mitochondria

Fractions	*d.p.m./mg protein*		*% of d.p.m. applied**	
	FAc	*FC*	*FAc*	*FC*
Myelin	7	30	18.4	19.3
Synaptosomal membrane	3	11	4.7	4.4
Synaptosome	4	20	6.1	14.3
Mitochondria	29	173	16.1	41.2
Solution	—	—	54.7	20.9

Intracerebral injection of fluoroacetate (250 µg and 1.06×10^6 d.p.m.); fluorocitrate (13.9 µg and 1.91×10^5 d.p.m.), in 200 g animals. Rats were killed at the onset of convulsion (60 min for FAc and 30 min for FC). Protein by the method of Lowry and co-workers (1951).

* Corrected for losses.　　　　Four animals in each group.

different utilization of the intermediates of the tricarboxylic acid cycle from the brain mitochondria from other cellular areas. Also they appear to accumulate less calcium (Kerpel-Fronius and Hajos 1971). These facts support our thesis that synaptosomal mitochondria may be functionally important in the development of the convulsions brought about by FC.

STUDIES *in vitro* ON THE UPTAKE OF FLUOROINHIBITORS BY BRAIN MITOCHONDRIA

As noted earlier, the presence of the fluoroinhibitors in the mitochondria might be implicated in mechanisms other than those immediately affecting the functioning of the tricarboxylic acid cycle. However, for us to understand any of the structural or metabolic changes which occur in the mitochondria as a result of the action of these inhibitors, a quasi-quantitative accounting of their transport and presence in mitochondria seemed desirable.

In a series of experiments mitochondria were isolated from brains of un-treated adult male rats. These mitochondria, twice–washed with 0.25 M-sucrose, were suspended in the medium containing radioactive FAc or FC. The concentrations of the fluoroinhibitors in the assay system were 250 µg per 3 ml for sodium fluoroacetate and 13.9 µg per 3 ml for trisodium fluorocitrate. These amounts were chosen because they correspond to those concentrations in the whole brain which elicit a rise in cerebral citrate and the onset of convulsions *in vivo* when administered intracerebrally. However, in the *in vitro* experiments the erythro racemate of [^{14}C]fluorocitric acid was used (where indicated, Table 4). This was prepared by minor modification of the electrophoretic technique described by Fanshier, Gottwald and Kun (1964).

TABLE 4

Uptake of fluoroinhibitors by brain mitochondria *in vitro*

	Added		*Incu-*	*State of respiratory control*			
Inhibitor	*d.p.m.* $\times 10^{-3}$	*μM*	*bation* *min*	*3*	*4*	*3*	*4*
				d.p.m./mg protein		*Rate constant* *(efflux min^{-1})*	
Fluoroacetate	1.35	100	1	182 (4.2)*	312 (7.2)*	—	—
			30	103 (2.4)	110 (2.5)	0.02	0.036
Fluorocitrate**	1.47	0.2	1	359 (0.31)	435 (0.37)	—	—
			30	334 (0.28)	416 (0.35)	very small	

Assay system contained: 10 mM-tris hydrochloride, pH 7.6; 80 mM-KCl; 10 mM-MgCl$_2$; 10 mM-succinic acid; and mitochondrial suspension in a total volume of 3 ml. In state 3, 5 mM-inorganic phosphate and 0.5 mM-ADP were also added. Incubation at 37 °C in oxygen. Protein was determined according to Waddell (1956).

* Figures in parentheses denote μg inhibitor per mg of mitochondrial protein.
** Erythro racemate.

Ionic movements in mitochondria *in vitro* are for the most part dependent on the steady state of electron flow in the respiratory system and are controlled through the reactions of the energy transfer systems. We have therefore investigated the differences in uptake and efflux of FAc and FC in state 4 (controlled state, lacking ADP and P$_i$) and state 3 (active state, with exogenous ADP and P$_i$ added).

Under the conditions of our assay both fluoroinhibitors displayed a rapid penetration into the mitochondria within one minute. Of course, external fluoroacetate, like acetate, may equilibrate with the intramitochondrial space in a fraction of a second. The uptake of both inhibitors was greater in state 4 than in state 3. The difference was quite marked with FAc where 39 % of the total FAc in the medium was recovered intramitochondrially in state 4 and 23 % in state 3. It may safely be assumed that nearly all the intramitochondrial label corresponds to FAc since synthesis of FC by cerebral mitochondria *in vitro* is minimal under the conditions of our assay. It is particularly interesting, however, that 40–50 % of the FC entered the mitochondria within one minute. Although the FAc showed a measurable rate of efflux over 30 minutes (measured at 5-minute intervals by rapid high-speed centrifugation of the mitochondria at 2°C—the mitochondria were washed twice with 3 ml portions of 0.25 M-sucrose) the rate of efflux of FC, not unexpectedly, was exceedingly slow—to the point of being undetectable. The results of some of our experiments are presented in Table 4. Since labelled fluorocitrate as the permeant substance is not metabolized we are cautiously optimistic that the distribution of label has given us reliable quantitative information.

At present there is no direct experimental evidence for the active translocation of FC in the manner of citrate. However, two observations suggest such a manner of transport for FC. First, in our experiments summarized in Table 4 we have used succinic acid alone as a substrate in generating the citric acid cycle and this led to a marked accumulation of citric acid in the presence of FC. Thus the 1.7 nmol of malate necessary for the translocation of about 0.36 µg FC per mg of mitochondrial protein within the first minute was available. In addition, there was an efflux of inorganic phosphate in state 4, which is known to be a priming factor in the activation of the transport of di- and tricarboxylic acids into the mitochondria (Klingenberg 1970). Second, we have noted that pretreatment of rat liver mitochondria with 0.33 mM-fluoro-pyruvate inhibited the uptake of FC. Fluoropyruvate is an SH inhibitor (Avi-Dor and Mager 1956). SH-reagents are known to inhibit primarily the P_i shuttle in the mitochondria, thereby involving the transport of the other sub-

TABLE 5

Accumulation of citrate and transport of inhibitors in brain mitochondria in the presence of ATP

Additions	Concentration		State of respiratory control			
			3		4	
	d.p.m. $\times 10^{-3}$	µM	Inhibitor d.p.m./mg	Citrate** nmole/mg protein	Inhibitor d.p.m./mg	Citrate** nmole/mg protein
None	—	—	—	32	—	32
ATP	—	5	—	—	—	5
[14C]Fluoroacetate	9.95	830	505 (9.9)	35	633 (12.4)	56
[14C]Fluoroacetate and ATP	9.95	830	—	—	518 (10.2)	9
[14C]Fluorocitrate*	6.28	16	261 (0.42)	130	275 (0.44)	124
[14C]Fluorocitrate* and ATP	6.28	16	—	—	303 (0.48)	8

System as given in Table 4 except that 20 mM each of sodium pyruvate and sodium malate replaced succinate. Incubation was for 30 min at 37°C in oxygen.
Figures in parentheses denote µg inhibitor per mg mitochondrial protein.

*   Erythro racemate.
** Values given for 30 minutes.

strates (Fonyo and Bessman 1966; Tyler 1968, 1969; Johnson and Chappell 1970).

It is interesting to note from Table 5 that there is some activation of FAc in state 4 but not in state 3. Citrate accumulation in the presence of FC in brain mitochondria was not dependent on the system being controlled or active.

Addition of ATP in state 4 completely abolished citrate synthesis by cerebral mitochondria both in the control and in the presence of fluoroinhibitors (Table 6). This confirms earlier reports that ATP inhibits citrate synthase (Hathaway and Atkinson 1965; Shepherd and Garland 1966). Even though the FC (added simultaneously with ATP) will stimulate mitochondrial ATPase to hydrolyse 50% of the added nucleotide (Fairhurst, Smith and Gál 1958),

TABLE 6

The effect of fluorocitrate on intramitochondrial ATP accumulation and citrate synthesis

| Addition | Concentration | | Citric acid | Total AdN* | | ATP |
	d.p.m. $\times 10^{-3}$	mM	nmole/mg protein	d.p.m./mg	nmol/mg	nmol/mg
[^{14}C]ATP	19.2	5	27	21.5	16.9	8.8
[^{14}C]ATP and fluorocitrate, 0.016 mM	19.2	5	23	24.0	18.9	7.2

System as given in Table 4 for state 4 except that 20 mM each of sodium pyruvate and sodium malate replaced succinate.

* AdN = adenine nucleotides.

enough intramitochondrial ATP was recovered to account for the inhibition of citrate synthase. It is to be remembered that in the experiments given in Tables 5 and 6, pyruvate and malate replaced succinate as the substrate and that the formation of oxaloacetic acid proceeded uninhibited even in the presence of FC. Fluorocitrate does not inhibit the malic enzyme even at 3.3×10^{-4} M (Gál 1960). These results do not support the view, at least as far as cerebral mitochondria are concerned, that the citrate synthase activity is controlled primarily by the availability of oxaloacetic acid (Williamson *et al.* 1967) rather than by intramitochondrial ATP. As we pointed out above there is no difference in the synthesis of citrate between state 3 and state 4. The addition of FC simply affects accumulation. Addition of ATP to state 4, even if only 40–50% of it is recoverable as ATP after 30 minutes, will inhibit citrate synthesis. Addition of P_i (5 mM) in state 4 stimulates citrate synthesis by 20–30%. Addition of 0.5–5 mM-ADP in the presence of P_i stimulates citrate synthesis in our assay system and in the system described by Kun and Volfin (1966).

EFFECT OF FLUOROCITRATE ON SUCCINIC DEHYDROGENASE ACTIVITY OF RAT
BRAIN MITOCHONDRIA

Studies with kidney mitochondria of the rat indicate a competitive inhibition
of succinic dehydrogenase by enzymically synthesized FC (K_i of 6×10^{-4} M FC)
which gives 64% inhibition at 7.8×10^{-4} M (Fanshier, Gottwald and Kun 1964).
Therefore, not surprisingly, Mehlman (1968) found no effect with synthetic FC
at 1.33×10^{-4} M in intact kidney mitochondria. Earlier in this study we showed
that the concentration of intramitochondrial FC which is present at the time
of convulsion and which will produce marked elevation of citrate in the cells
and also effectively block the aconitase of cerebral mitochondria *in vitro*, is
considerably less than that required to inhibit succinic dehydrogenase. Therefore, using the system and methodology of Mehlman (1968), we examined the
effect of synthetic FC at 1.6×10^{-5} M on succinic dehydrogenase of rat brain
mitochondria. The results in Table 7 indicated no significant difference from
the control.

We feel that our results *in vivo* and *in vitro* very closely support the finding
that, at 10^{-7} M, FC brings about complete irreversible inhibition of aconitase
(Dummel and Kun 1969).

TABLE 7

Effect of fluorocitrate on succinic dehydrogenase activity of brain mitochondria

Additions	Malate		Citrate		P/O
	Formed	[^{14}C]incor-porated	Formed	[^{14}C]incor-porated	
		μmol/3 ml			
[^{14}C]Succinic acid, 15 mM	4.6	0.38	0.19	0.40	2.18
[^{14}C]Succinic acid, 15 mM and fluorocitrate, 0.016 mM	4.1	0.57	0.80	0.91	1.81
KH[^{14}C]O$_3$, 10 mM; succinate, 15 mM	4.7	0.90	0.08	0.15	—
KH[^{14}C]O$_3$, 10 mM; fluorocitrate, 0.016 mM and succinate, 15 mM	4.3	0.86	1.41	0.39	—

System contained 10 mM-phosphate buffer, pH 7.4; 150 mM-KCl; 2 mM-MgCl$_2$; 20 mM-Na-
pyruvate; and 2 mM-ADP; mitochondrial suspension (5.5 mg protein); in a total volume of
3 ml. Incubation at 37°C in oxygen for 30 minutes.

FLUOROINHIBITORS AND CALCIUM TRANSPORT IN CEREBRAL MITOCHONDRIA

The significance of the transport of calcium to mitochondrial metabolism is
well established and has led to a search for the postulated mitochondrial carrier

of this cation (Lehninger 1970). It was of interest to examine whether transport of calcium might be altered in the presence of the fluoroinhibitors in view of their competition with substrates such as acetyl-CoA or citrate at the appropriate enzymic site; that is of citrate synthase and aconitase. In other words, is the inhibition of specific intramitochondrial enzymes by fluoroinhibitors related to changes in mitochondrial transport of calcium?

The methodology employed in our experiments followed the experimental designs of Carafoli and Lehninger (1971), except that the mitochondria were collected by high-speed centrifugation at a low temperature instead of by rapid filtration and were washed twice with ice-cold 0.25 M-sucrose before analysis. Also, only trace amounts of $^{45}Ca^{2+}$ were used. This was done to avoid major perturbations of mitochondrial calcium and to enable us to look at changes in calcium transport and efflux under endogenous conditions.

Several types of experiments *in vivo* and *in vitro* were undertaken.

In vivo. $^{45}Ca^{2+}$ (150 µCi/kg) was given intraperitoneally to two groups of rats. One hour later one group received 5 mg/kg sodium fluoroacetate. At the time of convulsion, 60 minutes later, both groups of animals were killed and the mitochondria isolated from individual animals. The presence of intramitochondrial $^{45}Ca^{2+}$ was assayed by liquid scintillation counting and atomic absorption spectrometry. The results of these studies indicate that the calcium content of brains of convulsing animals was not different from that of normal animals, and represented about 0.02% of the total $^{45}Ca^{2+}$ injected. Atomic absorption spectroscopy showed that about 35 ng-atoms per mg of mitochondrial protein did not appreciably change with administration of FAc.

In vitro. Mitochondria from brains of control and convulsing (FAc) animals were assayed in controlled and active states after the addition of 5.4 ng-atoms Ca^{2+} (corresponding to 88 600 d.p.m. $^{45}Ca^{2+}$). We found that administration of convulsive doses of FAc did not alter the mitochondrial penetration of calcium *in vitro* in either respiratory state when the system was incubated in oxygen at 37°C for 30 minutes. Of course, the transport of calcium was 100% greater in state 3 than in state 4, and this difference was maintained in spite of prior administration of FAc *in vivo*.

In a second series of experiments the mitochondria isolated from the brains of control and FAc-treated rats were prelabelled by incubation for one minute in a medium containing $^{45}Ca^{2+}$. The mitochondria were harvested, washed and resuspended in media corresponding to respiratory states 3 and 4, with succinate as substrate. The results, compared to the effect of 0.1 mM-dinitrophenol, are given in Fig. 1. The discharge of calcium from mitochondria seems to be state-dependent as in both groups it was faster in state 4 than in state 3. The prior administration of FAc increased the rate of discharge of $^{45}Ca^{2+}$ from the mitochondria; presumably they contain intramitochondrial fluorocitrate. The rate does not approach that found in the presence of 2,4-dinitrophenol. In our experimental conditions we have not found that the calcium content remains constant over seven minutes in the control mitochondria, as observed by Carafoli and Lehninger (1971) for rabbit brain mitochondria. However, their medium contained 80 ng-atoms of Ca^{2+} per mg of mitochondrial protein. A semi-logarithmic plot of intramitochondrial $^{45}Ca^{2+}$ versus time gave us a rate constant of efflux of 0.052 min^{-1} for the control and 0.075 min^{-1} for FAc in state 4, and 0.037 min^{-1} for the control and 0.090 min^{-1} for FAc in state 3.

Cerebral mitochondria from untreated rats were prelabelled by incubation

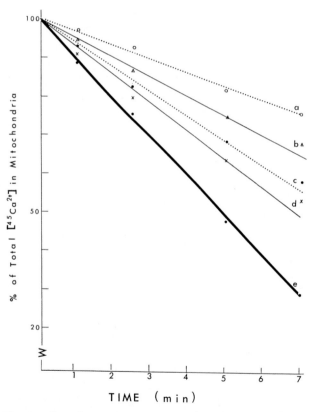

FIG. 1. Efflux of endogenous Ca^{2+} from cerebral mitochondria. Lines *a* and *b* are control mitochondria, state 3 and state 4, respectively; *c* and *d* are mitochondria from FAc-treated rats, state 3 and state 4, respectively; line *e* represents mitochondria in the presence of 2,4-dinitrophenol.

for one minute in a medium containing $^{45}Ca^{2+}$. After washing they were delivered into the flasks containing inhibitor-free medium or added sodium fluoroacetate (250 µg) or synthetic sodium fluorocitrate (13.9 µg). At the end of seven minutes no difference was found between the rate of efflux of calcium from control samples and those incubated in the presence of fluoroinhibitors in either respiratory state. It is likely that the fluoroinhibitors, particularly FC in spite of its rapid penetration as demonstrated above, may require longer than seven minutes to achieve complete irreversible inhibition of aconitase [15–20 minutes are reported by Dummel and Kun (1969)], or that intramitochondrial structural changes are relevant to an increased rate of efflux of calcium, such as found with mitochondria from brains of FAc-treated animals, obtained at the time of convulsion.

There is an apparent discrepancy between the results obtained from mitochondria labelled by $^{45}Ca^{2+}$ administration *in vivo* and our conditions *in vitro*, in which the intramitochondrial calcium is discharged practically against a concentration gradient of zero. Under these conditions any change in the affinity of mitochondrial binding sites for calcium as a result of the presence of FC would become more readily apparent. Obviously these points require further experimental work, in particular with mitochondria obtained from the synaptosomal regions.

These studies do not presume to do more than initiate the research needed to prove our working hypothesis outlined earlier (p. 77). Hopefully, they have done so. It is particularly gratifying to me to note how many of our observations complement results obtained in different contexts by other researchers engaged in unravelling the biochemistry of the fluoroinhibitors.

SUMMARY

Since the discovery of 'lethal synthesis' by Peters great effort has been expended on understanding the details of this mechanism and the total effect of the fluoroinhibitors in biological systems.

Here we present data from our *in vivo* and *in vitro* experiments with radioactive fluoroinhibitors. Major points discussed are: (1) comparison of the subcellular distribution of fluoroinhibitors in the liver and brain at convulsive and non-convulsive doses; (2) recovery of inhibitors from brain mitochondria by density-gradient sedimentation; (3) effect of DL-erythro-fluorocitrate on the ability of brain mitochondria to accumulate calcium *in vitro* and changes in the respiration-dependent mitochondrial calcium after administration of fluoroacetate; (4) penetration of fluoro compounds into cerebral mitochondria;

and (5) changes in the metabolism and membrane transport of cerebral mito-chondria as a function of enzyme inhibition by fluoro compounds.

ACKNOWLEDGEMENTS

The author gratefully acknowledges the technical assistance of J. Armstrong, M. Bouschlicher, and S. Horning. We thank Dr. Harold P. Schedl and D. Miller for the atomic absorption determinations supported by Grant USPHS-HM–02534. The author is particularly indebted to Dr. S. A. Millard for assistance and criticism in reviewing the data and the manuscript.

References

AVI-DOR, Y. and J. MAGER (1956) *Biochem. J.* **63**, 613–618.
CARAFOLI, E. and A. L. LEHNINGER (1971) *Biochem. J.* **122**, 681–690.
CHENOWETH, M. B. (1949) *Pharmacol. Rev.* **1**, 383–424.
DUMMEL, R. J. and E. KUN (1969) *J. Biol. Chem.* **244**, 2966–2969.
FAIRHURST, A. S., R. E. SMITH and E. M. GÁL (1958) *Biochem. Pharmacol.* **1**, 280–287.
FANSHIER, D. W., L. K. GOTTWALD and E. KUN (1964) *J. Biol. Chem.* **239**, 425–434.
FONYO, A. and S. P. BESSMAN (1966) *Biochem. & Biophys. Res. Commun.* **24**, 61–66.
GÁL, E. M. (1960) *Arch. Biochem. & Biophys.* **90**, 278–287.
GÁL, E. M., P. A. DREWES and N. F. TAYLOR (1961) *Arch. Biochem. & Biophys.* **93**, 1–14.
GÁL, E. M., R. A. PETERS and R. W. WAKELIN (1956) *Biochem. J.* **64**, 161–168.
HASTINGS, A. B., R. A. PETERS and R. W. WAKELIN (1953) *J. Physiol. (Lond.)* **120**, 50P.
HASTINGS, A. B., R. A. PETERS and R. W. WAKELIN (1955) *Q. J. Exp. Physiol.* **40**, 258–268.
HATHAWAY, J. A. and D. E. ATKINSON (1965) *Biochem. & Biophys. Res. Commun.* **20**, 661–665.
JOHNSON, R. N. and J. B. CHAPPELL (1970) *Biochem. J.* **116**, 37P.
KERPEL-FRONIUS, S. and F. HAJOS (1971) *III Int. Meet. Int. Soc. Neurochem.*, p. 411, ed. J. Domonkos, A. Fonyo, I. Huszak and J. Szentágothai. Budapest: Akadémiai Kiadó.
KIRK, K. and P. GOLDMAN (1970) *Biochem. J.* **117**, 409–410.
KLINGENBERG, M. (1970) In *Assays in Biochemistry*, pp. 119–159, ed. P. N. Campbell and F. Dickens, London: Academic Press.
KUN, E. and P. VOLFIN (1966) *Biochem. & Biophys. Res. Commun.* **23**, 696–701.
LAJTHA, A. (1962) In *Properties of Membranes and Diseases of the Nervous System*, pp. 43–54, ed. M. D. Yahr. New York: Springer-Verlag.
LEHNINGER, A. L. (1970) *Biochem. J.* **119**, 129–138.
LOWRY, O. H., N. J. ROSEBROUGH, A. L. FARR and J. RANDALL (1951) *J. Biol. Chem.* **193**, 265–275.
MEHLMAN, M. A. (1968) *J. Biol. Chem.* **243**, 1919–1925.
MILLARD, S. A. and E. M. GÁL (1971) *Int. J. Neurosci.* **1**, 211–218.
MILLARD, S. A., A. KUBOSE and E. M. GÁL (1969) *J. Biol. Chem.* **244**, 2511–2515.
MORSELLI, P. L., S. GARATTINI, F. MARCUCCI, E. MUSSINI, W. REWERSKY, L. VALZELLI and R. A. PETERS (1968) *Biochem. Pharmacol.* **17**, 195–202.
PALAY, S. L. (1967) In *Neuroscience*, pp. 24–31, ed. G. C. Quarton, P. H. Melnechuk and F. O. Schmitt. New York: Rockefeller University Press.

PATTISON, F. L. M. and R. A. PETERS (1966) *Handb. Exp. Pharmakol.* **20**, pt. 1, 387–458.
RODRÍGUEZ, G., M. ALBERICI and E. DEROBERTIS (1967) *J. Neurochem.* **14**, 215–225.
SCHNEIDER, W. C. and G. H. HOGEBOOM (1950) *J. Biol. Chem.* **183**, 123–128.
SHEPHERD, D. and P. B. GARLAND (1966) *Biochem. & Biophys. Res. Commun.* **22**, 89–93.
TYLER, D. D. (1968) *Biochem. J.* **107**, 121–123.
TYLER, D. D. (1969) *Biochem. J.* **111**, 665–678.
WADDELL, W. J. (1956) *J. Lab. & Clin. Med.* **48**, 311–314.
WILLIAMSON, J. R., M. S. OLSON, B. E. HERCZEG and H. S. COLES (1967) *Biochem. & Biophys. Res. Commun.* **27**, 595–600.

Discussion

Peters: I agree that one must give up the idea that the fluorocitrate is synthesized outside the brain. Berl and colleagues (Clarke, Nicklas and Berl 1970) clearly showed that there are compartments in the brain that do synthesize fluorocitrate from fluoroacetate. But I am glad that you confirmed that the fluoroacetate does not accumulate, which was a great surprise to us, and that the fluorocitrate is there in minute quantities inhibiting the system.

We had great difficulty in isolating fluorocitrate from rat urine, since in the presence of urea in acid solution we could not extract the fluorocitrate with ether. In the end we had to evaporate and extract the urea with methanol. Did you have this difficulty, Professor Gál?

Gál: No. We filter and acidify the urine, and then concentrate it at low temperature under vacuum.

Peters: You estimated the fluorocitrate, but would you not have got the same result if the fluoride had just come off and you had estimated the citrate, because citrate would travel at the same rate on your paper chromatograph?

Gál: No. Citrate and fluorocitrate do not have exactly the same R_f. Besides, citric acid does not present any problem since in our experiments all the radioactivity is in fluorocitrate. Loss of fluorine from fluorocitrate, if any, is minimal. It is perhaps 0.05% if one can extrapolate from the labelled CO_2 recovered after [^{14}C]fluorocitrate administration.

Peters: Would you explain the states you mentioned?

Gál: The steady states of electron transport are established by respiratory control via energy transfer (Chance and Williams 1956). State 4 or 'controlled state' is that in which there is substrate but no exogenous coupling energy added, such as ADP and inorganic phosphate. State 3 is the 'active state' in which an excess of exogenous ADP and inorganic phosphate are added to the mitochondria to drive the coupled phosphorylation. There are, of course, other states, but I will not mention them here.

Buffa: In Table 1 (p. 78) where you gave radioactive fluoroacetate to a starved rat, was the 12% activity found in the liver due to fluoroacetate or to fluorocitrate?

Gál: Most of this activity in the liver is due to fluoroacetate.

Buffa: Is this because there is no activation?

Gál: I have not discerned any activation. It appears to be a sex-dependent reaction, in the sense that only the liver of female rats will synthesize fluorocitrate.

Buffa: How do you explain the inhibition (shown in Table 5, p. 83) of citrate oxidation by ATP in brain mitochondria with either fluoroacetate or fluorocitrate?

Gál: ATP seems to be the most specific inhibitor of citrate synthase.

Kun: We prepared fluoropyruvate in an unorthodox and highly expensive fashion by decarboxylating crystalline fluoro-oxaloacetate. This gave us the most pure and most precisely defined fluoropyruvate. But to our amazement, this fluoropyruvate was *not* a vigorous alkylating agent (see Dummel, Berry and Kun 1971).

Gál: A. S. Fairhurst and I found (unpublished observations) that when we vacuum distilled fluoropyruvate we got two forms, each with a different melting point. One alkylated SH and the other did not. Perhaps the melting points of your product and mine would correspond.

Of the two forms obtained by Bergman's synthesis after the last stage of purification, I thought one was due to polymerization and the other was actually the pure fluoropyruvate; now I don't know which one is which.

Saunders: Are the melting points very far apart?

Kun: The melting point of our product has yet to be determined.

Peters: Both you and Dr Taylor found CO_2 coming off, but where does it come from (Gál, Drewes and Taylor 1961)?

Gál: I have two ideas about this. First, if one injects radioactive fluoroacetate, CO_2 may be obtained from the metabolism of one of the possible fluoroacetoacetates, $(CH_3COCHFCOO)^-$, in which there is a labilization of fluorine. Some of this compound, after losing its fluorine, could be the source of radioactive CO_2. However, a more likely event might be the carboxylation of fluoroacetyl-CoA to fluoromalonyl-CoA in which the stability of the C—F bond will become greatly reduced. This would be another explanation for the label in fluoroacetate appearing in CO_2. As to the origin of the labelled CO_2 from fluorocitrate, at present I can't go beyond the speculations already advanced in my presentation.

Kent: Professor Gál, in the recovery experiments referred to in Table 1 (p. 78), could either fluoroacetate or fluorocitrate be implicated in other biochemical transformations?

Gál: I did not present the total breakdown, but most of the fluorocitrate appears in the urine very fast. It can be found in other organs besides those given in Table 1.

Kent: Is there any further evidence of redistribution of fluoroacetate to other organic forms? Is it transferred into other lipids?

Gál: Yes, we also showed the appearance of a label in what might be fluorocholesterol (Gál, Drewes and Taylor 1961). I think it would be interesting to re-examine this problem with modern techniques, such as n.m.r. or neutron activation.

Armstrong: It is well known that the bone mineral accumulates ordinary citric acid. Did you examine the skeletons of the animals for ^{14}C-labelled citrate?

Gál: No, I did not.

Peters: I wonder whether fluorocitrate, if it is in the circulation, goes into the bone or whether it gets there because it starts as fluoroacetate and then gets converted in the bone.

Taylor: Professor Gál, have you considered using 2,2'-difluorocitrate, which inhibits aconitase, to see whether there are transport effects in mitochondria similar to those you get with fluorocitrate? It would also be interesting to know whether the transport effects were similar for difluorocitrate and monofluorocitrate.

Gál: No, my first concern was to separate the isomers and test whether they displayed any difference in transport and also whether the mitochondrial retention of fluorocitrate was stereospecific for the toxic isomer.

Taylor: Is there a mitochondrial swelling associated with this transport effect?

Gál: I didn't look at that.

Kun: Dr Taylor, what do you mean by difluorocitrate? Are there two fluorines on one carbon atom?

Taylor: Yes, two fluorines replacing the two hydrogens at the one α-carbon atom.

Kun: We found this totally ineffective as a specific enzyme inhibitor. At what concentration does it inhibit?

Taylor: Experiments in our laboratories indicate that 2,2'-difluorocitrate inhibits aconitase extracted from rat liver mitochondria at 3 mM (T. C. Stephens and R. Eisenthal 1971, unpublished results). Monofluorocitrate inhibits the same preparation of aconitase at 50 μM.

Kun: Isolated aconitase is sensitive to fluorocitrate at concentrations of 10 to 50 μM. At relatively high concentrations, various tricarboxylic acids inhibit aconitase (around 0.1 mM), but fluorocitrate is effective at much lower concentrations.

Taylor: Our results with isolated aconitase (mammalian) and 2,2'-difluorocitrate appear to be similar to those you have obtained with monofluorocitrate. With whole yeast mitochondria we find that much lower concentrations of 2,2'-difluorocitrate will inhibit citrate oxidation than those required to inhibit aconitase (Brunt, Eisenthal and Symonds 1971). This is considered to be due to the inhibition of the transport system in yeast mitochondria.

Gál: You have to remember that in yeast mitochondria there is no active Ca^{2+} transport.

Peters: I feel that calcium must be quite important, although Professor Hastings and I never found that calcium, magnesium or any divalent ion put into the brain reversed the poisoning in pigeons (Hastings, Peters and Wakelin 1953). Do you think that the lysosomes which Koenig (1969) talks about could be concerned in this?

Gál: No. I think it is Professor Kun's transport factor that may be concerned with this. Fluorocitrate obviously has changed the affinity for calcium at the high affinity sites of the mitochondria.

References

BRUNT, R. V., R. EISENTHAL and S. A. SYMONS (1971) *FEBS Lett.* **13**, 89–91.
CHANCE, B. and G. R. WILLIAMS (1956) *Adv. Enzymol.* **17**, 65–134.
CLARKE, D. D., W. J. NICKLAS and S. BERL (1970) *Biochem. J.* **120**, 345–351.
DUMMEL, R. J., M. N. BERRY and E. KUN (1971) *Mol. Pharmacol.* **7**, 367.
GÁL, E. M., P. A. DREWES and N. F. TAYLOR (1961) *Arch. Biochem. Biophys.* **93**, 1–14.
HASTINGS, A. B., R. A. PETERS and R. W. WAKELIN (1953) *J. Physiol. (Lond.)* **120**, 50P.
KOENIG, H. (1969) *Science* **164**, 310–312.

Fluorine as a substituent for oxygen in biological systems: examples in mammalian membrane transport and glycosidase action

J. E. G. BARNETT

Department of Physiology and Biochemistry, University of Southampton

Previous speakers have discussed the physicochemical properties of the carbon-fluorine bond. Because of its small size and the stability of its bond with carbon, fluorine has been a very successful substituent for hydrogen in biological systems. The similar size and electronegativity of fluorine and oxygen (Table 1) suggest that substitution of fluorine into an organic molecule in place of oxygen might also give retention of biological activity. The fluorine atom might replace an oxygen atom involved in the binding of the molecule to an enzyme, membrane or receptor site, or it might replace an oxygen atom subject to carbon-oxygen bond cleavage. W. T. S. Jarvis, A. Ralph, G. O. Holman and I, in association with K. A. Munday, have been able to demonstrate successfully such a retention of biological activity on substitution of oxygen by fluorine in both these contexts, and in this paper I shall discuss the use of fluorine analogues of sugars to investigate the binding of sugars to membrane carriers and their use as substrates for glycosidases.

TABLE 1

Comparison of the size and electronegativity of some elements

Element	Bond length* (CH_3-X) $(\mathring{A})$	Van der Waals** radius $(\mathring{A})$	Total $(\mathring{A})$	Electro-negativity***
H	1.09	1.20	2.29	2.1
F	1.39	1.35	2.74	4.0
O (in OH)	1.43	1.40	2.83	3.5
Cl	1.77	1.80	3.57	3.0
S (in SH)	1.82	1.85	3.67	2.5

*   Chemical Society (1958); Chemical Society (1965).
**   Pauling (1940a).
***   Pauling (1940b).

Elements with an electronegativity of about 3.0, or greater, form the hydrogen bond.

BINDING OF FLUORINATED SUGARS TO MEMBRANES

Introduction

We were interested in investigating the structural and binding requirements
for the transport of sugars across mammalian membranes, particularly by the
processes of facilitated diffusion or active transport. In both cases the sugar
appears to bind to a protein 'carrier' which in combination with the sugar can
traverse the membrane. In the case of active transport this is achieved against
an electrochemical gradient.

A substance such as glucose may be expected to bind to the carrier by
hydrogen bonds. A hydrogen bond between a hydroxyl group and a protein
may use either the hydrogen of the hydroxyl group or that of the protein as the
bridge hydrogen (Fig. 1). If the hydroxyl group is replaced by a fluorine atom,
or a similar electronegative group lacking a hydrogen atom, this may still form
a hydrogen bond if the bridge hydrogen is provided by the protein. If the
hydroxyl group is replaced by hydrogen or a similar group which does not
form hydrogen bonds, there will be no sugar–protein binding at this position.

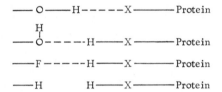

FIG. 1. Possible hydrogen bonds between carrier and sugar.

Where a sensitive and reliable quantitative test of this binding can be found,
the existence of a hydrogen bond at each position of the sugar can be investigated by comparing the binding of sugars in which the hydroxyl group is
substituted by fluorine or hydrogen. Since the binding may be expected to be
fairly stereospecific, the latter sugars include those in which the hydroxyl group
has been inverted.

The idea of a multipoint contact between sugar and protein through hydrogen
bonding is of long standing (see Gottschalk 1950), and the use of halogenated
sugars to explore the bonds between sugar and enzyme was pioneered by
Helferich, Grünler and Gnüchtel (1937). Crane (1962) compared the binding
of deoxy analogues and D-glucose as substrates or inhibitors of hexokinase, but
systematic substitution of fluorine for the hydroxyl group at each sugar position
to explore the binding of sugars to protein does not appear to have previously

been used, although So and Goldstein (1967) have independently suggested that fluorine substitution might be used to explore the bonding.

Isolated reports of the use of halogenated sugars as inhibitors of, or as substrates for, membrane transport have appeared. 6-Deoxy-6-fluoro-D-glucose has been shown to be an inhibitor of glucose transport into yeast cells (Blakley and Boyer 1955) and rat diaphragm (Wick *et al.* 1959), and is a substrate for active transport in the hamster intestine (Wilson and Landau 1960). In this paper we investigate the binding of fluorinated or deoxy-sugars to the carrier for active sugar transport in the hamster intestine, and the red blood cell, in an attempt to rationalize the structural requirements for transport, and to investigate the carrier–sugar binding.

Binding requirements for active sugar transport in hamster intestine

Crane (1960) proposed a minimum structure for active transport in the hamster intestine (Fig. 2). Most sugars with this minimum structure which had then been tested were actively transported. Soon, however, anomalies began to appear. Thus although 6-deoxy-D-glucose and 1-deoxy-D-glucose were actively transported, 1,6-dideoxy-D-glucose was not (Bihler, Hawkins and Crane 1962).

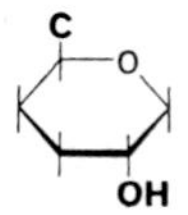

FIG. 2. Minimum structural requirements for active sugar transport in the hamster intestine (Crane 1960).

Similarly, D-allose (hydroxyl group inverted at C-3) and D-galactose (inverted at C-4) were transported, but D-gulose (inverted at C-3 and C-4) was not (Wilson and Landau 1960). We therefore considered that D-glucose might be the ideal substrate for binding to the carrier, which would be hydrogen bonded at all its hydroxyl groups. Generally inversion of one hydroxyl group, which leads to the loss of one hydrogen bond, would still allow transport, but inversion of two hydroxyl groups would result in poor transport with insufficient binding to the carrier.

Since D-galactose already has one departure from the D-glucose structure, we decided to test this hypothesis using D-galactose derivatives substituted at C-6. If bonding at C-6 was maintained the sugar should be transported well, but if not transport should be poor.

Experimental procedure

The transport of the sugar was tested in two ways. Initially accumulation of the sugar into everted hamster intestinal rings was measured. The entire small intestines of two hamsters were washed *in situ*, dissected out and everted. Slices about 2–3 mm wide were cut under ice-cold, oxygenated buffer and distributed between 10 flasks, each flask taking every tenth segment. The slices were incubated with the sugar (1 mM) at 37 °C for 30 minutes and the ratio of tissue fluid concentration to medium concentration at the end of the experiment measured. This tissue/medium ratio is a test for overall transport and does not distinguish the rate of transport from the efficiency of binding to the carrier. The latter was measured in the following way.

All sugars which are actively transported into hamster intestine share the same carrier. It is therefore possible for one to test the relative affinities for the carrier by using the sugars to be tested as competitive inhibitors of a sugar known to be transported by the active transport system. Such a sugar displays Michaelis-Menten kinetics for entry into the tissue rings, and an inhibition constant, K_i, can easily be found for the competing sugar. The experimental procedure was similar to that described above, except that the accumulation of the transported sugar (1–5 mM) was measured after 10 minutes in the presence of the inhibitor at a constant concentration (usually 20 mM). The experimental techniques are described in detail elsewhere (Barnett, Jarvis and Munday 1968; Barnett, Ralph and Munday 1969; Barnett, Ralph and Munday 1970).

Results and discussion

Substitution at C-6. The tissue/medium ratios for D-galactose derivatives substituted at C-6 are shown in Table 2. D-Galactose was well transported and 6-deoxy-D-galactose (D-fucose) was poorly transported. 6-Deoxy-6-fluoro-D-

TABLE 2

The effect of substitution at C-6 of the sugar

Sugar	Tissue/medium ratio (initial concentration 1 mM)	K_i (mM)*
D-Galactose	8.3 ± 0.25	3
6-Deoxy-D-galactose	1.5 ± 0.15	16
6-Deoxy-6-fluoro-D-galactose	4.3 ± 0.8	7

* Using methyl α-D-glucopyranoside as substrate.

galactose was transported to an intermediate extent. This situation was reflected in the relative binding capacities to the carrier, which we determined by using
the sugars as competitive inhibitors of methyl α-D-glucopyranoside transport
(Fig. 3). The low K_i for D-galactose indicates strong binding of the sugar to the
carrier, whereas that of the deoxy-sugar is much weaker and that of the fluoro

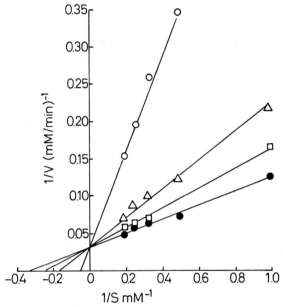

FIG. 3. Lineweaver-Burk plot showing competitive inhibition of methyl α-D-glucopyranoside
accumulation by 6-substituted D-galactoses. ●: No addition; □: 20 mM-6-deoxy-D-galactose;
△: 20 mM-6-deoxy-6-fluoro-D-galactose; ○: 20 mM-D-galactose.

compound is intermediate. It can be concluded that there may possibly be a
hydrogen bond from the protein to the hydroxyl group at C-6 and, since the
binding of the fluoro analogue is greater than that of the deoxy-sugar, that
this may involve the protein hydrogen.

Substitution at C-3. Next we tested the effect of substitution at C-3. We used
analogues of D-glucose since an easy synthesis of 3-deoxy-3-fluoro-D-glucose
was reported in the literature (Foster, Hems and Webber 1967). The experimental procedure was as before, except that fluoride ion (2 mM) was added to the
medium to reduce the metabolism of D-glucose. This concentration of F⁻ did
not affect the transport of D-galactose. The competitive inhibition constants, K_i,
were measured with D-galactose as the substrate. The results were similar to
those for C-6, except that in this position substitution by fluorine did not reduce

TABLE 3

The effect of substitution at C-3 and C-5 of the sugar

Sugar	Tissue/medium ratio (initial concentration 1 mM)	K_i (mM)*
D-Glucose	16.4 ± 0.4**	2.3
3-Deoxy-D-glucose	2.1 ± 0.2	24
D-Allose	1.7 ± 0.1	40
3-Deoxy-3-fluoro-D-glucose	26.6 ± 3.6	2.5
3,6-Dideoxy-D-glucose	0.5 ± 0.05	—
3-Deoxy-D-galactose	1.0 ± 0.04	—
D-Gulose	0.3 ± 0.01	—
5-Thio-D-glucose	12.5 ± 0.7**	2.9

*   Using D-galactose as substrate.
**   In the presence of 2 mM-fluoride.

the accumulation and the 3-fluoro analogue bound just as strongly to the carrier as the parent D-glucose (Table 3). In contrast 3-deoxy-D-glucose and D-allose, in which the hydroxyl group in the D-*gluco* configuration at C-3 is replaced by hydrogen, were more weakly transported and were weaker competitive inhibitors (Fig. 4). Two deviations from the D-glucose structure as in D-gulose, 3-deoxy-D-galactose and 3,6-dideoxy-D-glucose (Table 3), gave almost complete

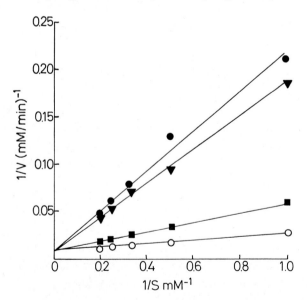

FIG. 4. Lineweaver-Burk plot showing competitive inhibition of D-galactose accumulation by 3-substituted D-glucoses. ○: No addition; ■: 20 mM-3-deoxy-D-glucose; ▼: 17.5 mM-3-deoxy-3-fluoro-D-glucose; ●: 20 mM-D-glucose.

loss of transport. It seems that a hydrogen bond is also formed between the carrier and the C-3 hydroxyl group of D-glucose, and that the protein again donates the hydrogen.

Substitution at C-1 and C-5. In similar experiments at C-1 we compared the transport of methyl β-D-galactopyranoside, which can hydrogen bond at C-1 by using the oxygen of the methoxyl group, with methyl β-D-thiogalactopyranoside, which would not be expected to do so since sulphur does not usually form a strong hydrogen bond. The results are shown in Table 4. A D-galactose derivative with oxygen in either the α- or β-configuration gave good accumulation, but those galactose derivatives lacking oxygen at C-1 were poorly accumulated. Substitution of a fluorine atom into the α-configuration only marginally improved the transport. Again there seems to be a hydrogen bond from the carrier to the oxygen at C-1 using the protein hydrogen, which would be required for

TABLE 4

The effect of substitution at C-1 of the sugar

Sugar	Tissue/medium ratio (initial concentration 1 mM)
D-galactose	8.3 ± 0.25
methyl α-D-galactopyranoside	10.0 ± 1.5
methyl β-D-galactopyranoside	5.6 ± 0.2
methyl β-D-thiogalactopyranoside	0.5 ± 0.03
1-deoxy-D-galactose	0.4 ± 0.1
α-D-galactopyranosyl fluoride	1.2 ± 0.3

binding the methyl galactosides. The relatively poor binding of α-D-galactosyl fluoride is therefore of some interest, since it shows that here the fluorine atom does not form a good hydrogen bond, even when oxygen does. Factors other than electronegativity, such as the steric arrangement, must also be important.

Substitution of the ring oxygen by sulphur to give 5-thio-D-glucose did not affect either the accumulation or the binding capacity (Table 3), showing that no hydrogen bond exists between the carrier and the ring oxygen.

We concluded that a hydrogen bond probably exists between the carrier and the C-1, C-3, and C-6 hydroxyl groups of D-glucose, and also at C-4, since sugars with the D-*galacto*-configuration were always transported less well than those with the D-*gluco*-configuration.

Substitution at C-2. When any one of these bonds was missing, the sugar was still accumulated quite well, but loss of two of these hydroxyl groups gave a poor accumulation of sugar. When we investigated changes at C-2, we found that the loss of this hydroxyl group resulted in absolute loss of transport (Table 5). The tissue/medium ratios for derivatives lacking this hydroxyl group were identical to that of D-mannitol, which does not bind to the active transport carrier. The

TABLE 5

The effect of substitution at C-2 of the sugar

Sugar	Tissue/medium ratio (initial concentration 1 mM)	K_i (mM)*	K_i (mM)**
D-Glucose	16.4 ± 0.4***	2.3	—
2-Deoxy-D-glucose	0.2 ± 0.02	—	65
D-Mannose	0.2 ± 0.01	∞	65
2-Deoxy-2-fluoro-D-glucose	0.4 ± 0.2	∞	65
2-Chloro-2-deoxy-D-glucose	0.2 ± 0.03	∞	65
D-Mannitol	0.2 ± 0.02	—	—

*   Using D-galactose as substrate.
**   Using L-glucose as substrate.
***   In the presence of 2 mM-fluoride.

higher apparent value for the fluoro analogue and the higher standard error are due to the difficulty of assaying this compound and probably do not denote any increased transport. None of the sugars inhibited D-galactose transport, although they showed very weak inhibition of the accumulation of the extremely poor substrate, L-glucose. The loss of transport was much more severe when C-2 was modified than when any other position was altered. When a fluorine atom was introduced at C-2 there was still no active accumulation of the sugar in contrast to the effect of fluorine substitution at other positions. We concluded that the bond to the membrane at C-2 was of a different nature from the others and was probably covalent.

A mechanism for active transport in hamster intestine. Crane (1965) has shown that the binding of sugars to the membrane is strong in the presence of sodium ions and weak in their absence. He has suggested that the sugar binds to the membrane strongly in the high sodium ion concentration outside the cell, and that the membrane rearranges so that the sugar and sodium ion are on the inside of the cell where the sodium ion diffuses off in the low sodium ion concentration. This in turn leads to release of the sugar even when the intracellular concentration of sugar is higher than that in the extracellular fluid. The sodium concentration is restored by the sodium pump.

Since this hypothesis for active transport demands that the sugar should

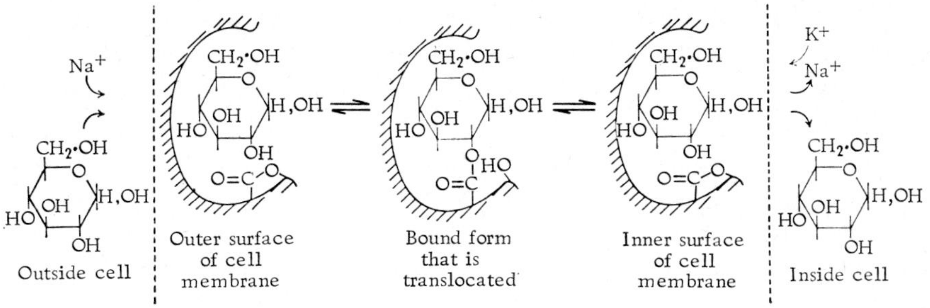

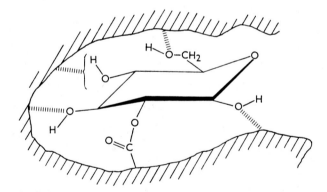

FIG. 5. Possible mechanism for the active transport of D-glucose into the hamster intestine.

bind reversibly to the membrane and without net utilization of energy, we envisage that the proposed covalent bond at C-2, possibly an ester, may be formed by cleavage of an intramolecular ester on the membrane after the initial absorption of the sugar, as shown in Fig. 5. On release of the sugar, the intramolecular bond in the membrane would be regenerated.

A hypothetical model for the binding of D-glucose to the carrier is shown in Fig. 6.

FIG. 6. Hypothetical model for the binding of D-glucose to the carrier for active sugar transport.

The mammalian red blood cell

Recently, we have turned our attention to the human red blood cell. The specificity of this facilitated diffusion system is different from that of the mammalian intestine (LeFevre 1961), in particular 2-deoxy-D-glucose is well transported. We again tested sugar analogues specifically modified by the introduction of hydrogen or fluorine in place of the hydroxyl group for their relative binding capacities to the carrier. In our preliminary work we used the method of Levine, Levine and Jones (1971) to determine the inhibition constants, K_i, of D-glucose and some analogues modified at C-3, for the entry of L-sorbose into the red blood cell (Table 6). Substitution by hydrogen led to a marked loss of binding capacity, whereas substitution by fluorine or methoxyl, both of which can hydrogen bond, gave a binding capacity close to that of D-glucose. It seems that the sugar–carrier complex for sugar transport in the human red cell may also have a hydrogen bond at C-3.

TABLE 6

The effect of substitution at C-3 of the sugar (red blood cell)

Sugar	K_i *(mM)**
D-Glucose	6.9
3-Deoxy-D-glucose	50
3-Deoxy-3-fluoro-D-glucose	7.1
3-*O*-Methyl-D-glucose	13.1
3-Chloro-3-deoxy-D-glucose	30
D-Allose	200 (approx.)

* Using L-sorbose as substrate.

CARBON-FLUORINE BOND CLEAVAGE

Hydrolysis of glycosyl fluorides by glycosidases

Hitherto fluorine has been discussed as a replacement for an oxygen in a position not involved in bond cleavage during the reaction. During some of the experiments using fluorinated analogues of D-glucose in the hamster intestine, we tested α-D-glucopyranosyl fluoride as a substrate for transport and found that it was rapidly hydrolysed by the intestinal α-glucosidase (maltase) to give D-glucose and hydrogen fluoride (Fig. 7). Here the fluorine atom was replacing the alkoxy radical as the leaving group. Careful review of the literature showed that α- and β-D-galactopyranosyl fluorides had been reported to be hydrolysed by bacteria (Hofsten 1961), and we soon showed that the reaction was a general one for glycosidases and for glucoamylase (Barnett,

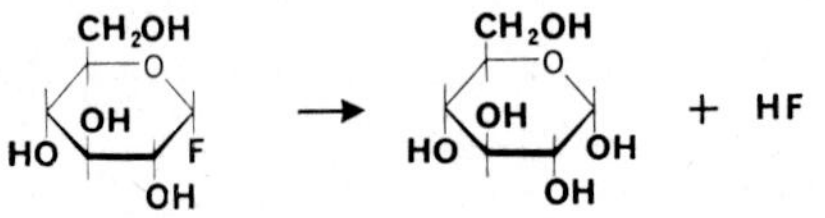

FIG. 7. Hydrolysis of α-D-glucopyranosyl fluoride by α-glucosidase.

Jarvis and Munday 1967). Furthermore the fluorides are invariably good substrates, the rate of hydrolysis is very fast, and the small size of the fluorine atom ensures that there is no steric hindrance to enzyme–substrate binding. This contrasts with the aryl glycosides which are often used as substrates for glycosidases and are sometimes only poorly hydrolysed. Examples are rat intestinal maltase (α-glucosidase) and *Aspergillus* glucoamylase where, at a concentration of 1 mM, the fluoride is hydrolysed by the respective enzymes 40 and 250 times more rapidly than *p*-nitrophenyl-α-D-glucopyranoside and 0.8 and 2.0 times as fast as maltose.

The mechanism of fluoride displacement

The mechanism for glycosidase action usually accepted is shown in Fig. 8. Protonation of the glycosidic oxygen is followed by nucleophilic attack, either of water leading to inversion of configuration, or, in the more usual situation shown in Fig. 8, by a nucleophilic group on the enzyme which is then displaced

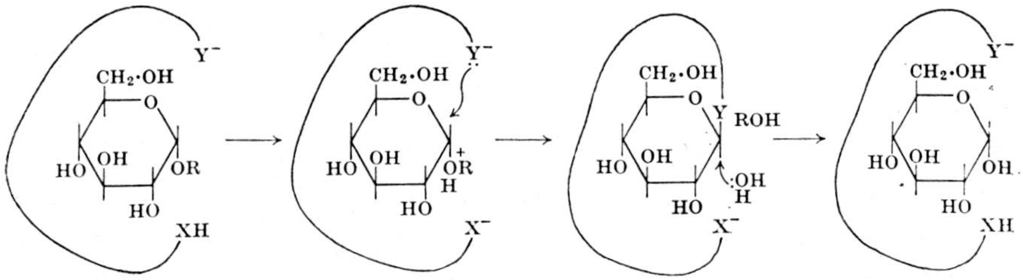

FIG. 8. Mechanism for α-D-glucosidase action involving retention of configuration.

by water to give retention of configuration. Glycosyl fluorides are more susceptible to chemical attack by hydroxide ion than the corresponding methyl glycosides (Barnett 1969). It is therefore possible that when the reaction is catalysed by glycosidase the fluorine also acts as a better leaving group than the alkoxyl group, and does not require protonation for displacement. However, the superimposition of the pH optimum curves for rat intestinal α-glucosidase hydrolysis with either α-D-glucopyranosyl fluoride or maltose as substrates (Fig. 9), suggests that the same mechanism is involved. The enzyme appears to protonate the fluorine atom as readily as it does an oxygen atom.

Uses of the glycosyl fluorides as substrates for glycosidases

We have been able to use the hydrolysis of glycosyl fluorides as the basis of

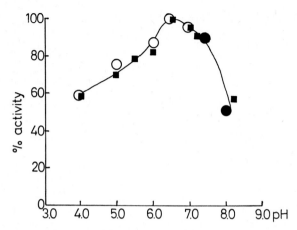

FIG. 9. Variation of rat intestinal α-glucosidase with pH. ○: α-ᴅ-glucopyranosyl fluoride as substrate, sodium maleate buffer; ●: α-ᴅ-glucopyranosyl fluoride as substrate, sodium carbonate buffer; ■: maltose as substrate, sodium phosphate buffer.

a titrimetric assay in which the hydrogen fluoride produced during hydrolysis is neutralized by sodium hydroxide in a pH-stat. This assay can be used with very crude, optically impure enzyme preparations (Barnett 1971).

The rapid rate of hydrolysis and the fact that the fluoride ion has no optical rotation make the glycosyl fluorides excellent substrates for the study of the anomeric nature of the product of glycosidase action. On hydrolysis the sugar released immediately begins to mutarotate, but if the hydrolysis is fast enough there will be a detectably greater concentration in the mixture of anomers of the one first produced when compared with the equilibrium mixture. The mixture can be analysed either polarimetrically, or by gas-liquid chromatography after freeze-drying and trimethylsilylation (Barnett 1971). A typical gas-liquid chromatograph is shown for the incubation of α-ᴅ-glucopyranosyl fluoride with yeast α-ᴅ-glucosidase for 10 minutes at room temperature. In this case, all the glycosyl fluoride was hydrolysed and the blank (no enzyme) is shown for comparison (Fig. 10). Some β/α ratios are shown in Table 7.

TABLE 7

Ratios of β- to α-ᴅ-glucose in hydrolysates of α- and β-ᴅ-glucopyranosyl fluorides by glycosidases, determined to investigate whether the reactions are catalysed with retention or inversion of configuration

Hydrolysate	β/α ratio	Reaction pathway
Equilibrium ᴅ-glucose	1.45	—
Yeast α-glucosidase	0.29	retention
Aspergillus amyloglucosidase	3.22	inversion
Almond emulsin β-glucosidase	3.18	retention

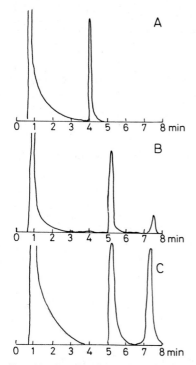

F<small>IG</small>. 10. Gas-liquid chromatographic separation of the trimethylsilyl ethers of the anomeric products of the action of yeast α-D-glucosidase on α-D-glucopyranosyl fluoride for 10 min at pH 6.0 and 25 °C. The samples were rapidly frozen, frozen-dried and converted to the trimethylsilyl ethers. The mixture was separated on a Pye 104 gas chromatograph in a 60 inch glass column of 3 % E.30 silicone oil as stationary phase on a 100–120 mesh Diatomite CQ support with N_2 as carrier (flow rate 50 ml/min) and temperature 180 °C.

A: Blank (no enzyme); B: experimental; C: equilibrium mixture of α- and β-D-glucose. Retention times: α-D-glucopyranosyl fluoride, 4 min; α-D-glucose, 5.1 min; β-D-glucose, 7.8 min.

Glycosyl fluoride hydrolysis by other enzymes

Recently it has been shown that the enzymic hydrolysis of glycosyl fluorides is not confined to glycosidases. Gold and Osber (1971) have shown that sucrose phosphorylase will utilize α-D-glucopyranosyl fluoride as a substrate. Again the fluoride hydrolysed more rapidly than the natural substrate.

$$\text{sucrose} + P_i \rightleftharpoons \text{glucose 1-phosphate} + \text{fructose}$$

$$\text{α-D-glucopyranosyl fluoride} + P_i \rightleftharpoons \text{glucose 1-phosphate} + \text{HF}$$

It seems probable that this type of reaction will also be given by the analogous maltose phosphorylase.

In some experiments on amylomaltase, which is an enzyme which had been reported to equilibrate glucose, maltose and amylose without hydrolysis (Wiesmeyer and Cohn 1960),

$$[\text{maltose} + (\text{glucose})_n \rightleftharpoons \text{glucose} + (\text{glucose})_{n+1}]$$

we tested α-D-glucopyranosyl fluoride as a substrate in the hope of generating amylose:

$$n \text{ α-D-glucopyranosyl fluoride} \rightarrow (\text{glucosyl})_n \text{ fluoride} + (n-1)\text{F}^-$$

Instead we found that the glucosyl fluoride was hydrolysed, so polymeric products did not accumulate. Although the enzyme had been purified by the method of the previous workers and was from the same source, we thought that perhaps the preparation was contaminated with a hydrolytic enzyme. However, recently it has been shown that highly purified amylomaltase does have some hydrolytic activity (Haeselbarth, Schulz and Schwinn 1971). Presumably the ratio of hydrolysis to transfer is disturbed by the use of the glucosyl fluoride as a substrate, possibly because of the absence of maltose as an acceptor.

In addition to the glucoamylase and amylomaltase described above, α-amylases have been shown to use α-D-glucopyranosyl fluoride as a substrate (Hehre, Genghof and Okada 1971) and these authors report that it is also a substrate for dextran sucrase.

Glycosyl fluorides must now be regarded as potential substrates for all enzymes involved with glycosyl transfer. The small size of the fluorine atom and its very rapid hydrolysis seem to confer an advantage on the fluorides over other unnatural substrates and even over the natural ones. The use of these compounds as biochemical tools should therefore continue to increase.

SUMMARY

The similar size and electronegativity of fluorine and oxygen suggest that substitution of fluorine for oxygen in molecules may sometimes allow retention of biological activity. The role of the substituted oxygen may be in physical binding between the substrate and protein, or the carbon-oxygen bond may be broken during an enzyme reaction. Examples of the first situation may be found in the active transport of sugars across the hamster intestine, and of the second in the hydrolysis of glycosyl fluorides by glycosidases.

The hydrogen bonding at each position of the glucose molecule has been explored by comparing the relative binding to the membrane carrier of sugars in which hydrogen and fluorine have been specifically substituted at each position for the hydroxyl group. A model for the structural and bonding requirements of the active carrier for sugars is presented. Preliminary results for the sugar carrier of the human red blood cell are discussed.

Glycosyl fluoride hydrolysis by glycosidases is rapid and general in contrast to that of aryl glycosides which are not substrates for some glycosidases. The assay of glycosidases by pH-stat, and the determination of the anomeric nature of the initial product of glycosidase action, either polarimetrically or by gas-liquid chromatography, with fluorides as substrates, is described. The enzymic hydrolysis of glycosyl fluorides is not restricted to glycosidases.

ACKNOWLEDGEMENTS

I thank my co-workers, Dr W. T. S. Jarvis, Dr A. Ralph and Mr G. O. Holman for their valuable contributions to this work; Professor K. A. Munday for his encouragement and advice; and Dr P. W. Kent for introducing me to the fascination of fluorine in biological molecules. I also thank Professor A. B. Foster, Professor R. L. Whistler and Dr N. F. Taylor for supplying some of the sugars used.

References

BARNETT, J. E. G. (1969) *Carbohydr. Res.* **9**, 21–31.
BARNETT, J. E. G. (1971) *Biochem. J.* **123**, 607–611.
BARNETT, J. E. G., W. T. S. JARVIS and K. A. MUNDAY (1967) *Biochem. J.* **105**, 669–672.
BARNETT, J. E. G., W. T. S. JARVIS and K. A. MUNDAY (1968) *Biochem. J.* **109**, 61–67.
BARNETT, J. E. G., A. RALPH and K. A. MUNDAY (1969) *Biochem. J.* **114**, 569–573.
BARNETT, J. E. G., A. RALPH and K. A. MUNDAY (1970) *Biochem. J.* **118**, 843–850.
BIHLER, I., K. A. HAWKINS and R. K. CRANE (1962) *Biochim. & Biophys. Acta* **59**, 94–102.
BLAKLEY, E. R. and P. D. BOYER (1955) *Biochim. & Biophys. Acta* **16**, 576–582.
CHEMICAL SOCIETY (1958) *Tables of Interatomic Distances and Configuration in Molecules and Ions*, Special Publication No. 11, p. S12, M19. London: Chemical Society.
CHEMICAL SOCIETY (1965) *Tables of Interatomic Distances and Configuration in Molecules and Ions*, Suppl. 1956–1959, Special Publication No. 18, p. S14s. London: Chemical Society.
CRANE, R. K. (1960) *Physiol. Rev.* **40**, 789–825.
CRANE, R. K. (1962) In *The Enzymes*, vol. 6, 2nd edn, pp. 47–64, ed. P. D. Boyer, H. Lardy and K. Myerbäck. New York: Academic Press.
CRANE, R. K. (1965) *Fed. Proc. Fed. Am. Soc. Exp. Biol.* **24**, 1000–1006.
FOSTER, A. B., R. HEMS and J. M. WEBBER (1967) *Carbohydr. Res.* **5**, 292–301.
GOLD, A. M. and M. P. OSBER (1971) *Biochem. & Biophys. Res. Commun.* **42**, 469–474.
GOTTSCHALK, A. (1950) *Adv. Carbohydr. Chem.* **5**, 49–76.
HAESELBARTH, V., G. V. SCHULZ and H. SCHWINN (1971) *Biochim. & Biophys. Acta* **227**, 296–312.
HEHRE, E. J., D. S. GFNGHOF and G. OKADA (1971) *Arch. Biochem. & Biophys.* **142**, 382–393.

HELFERICH, B., S. GRÜNLER and A. GNÜCHTEL (1937) *Hoppe-Seyler's Z. Physiol. Chem.* **248**, 85–95.
HOFSTEN, B. v. (1961) *Biochim. & Biophys. Acta* **48**, 159–163.
LeFEVRE, P. G. (1961) *Pharmacol. Rev.* **13**, 39–70.
LEVINE, M., S. LEVINE and M. N. JONES (1971) *Biochim. & Biophys. Acta* **225**, 291–300.
PAULING, L. (1940a) *The Nature of the Chemical Bond*, 2nd edn, p. 189. London: Oxford University Press.
PAULING, L. (1940b) *The Nature of the Chemical Bond*, 2nd edn, p. 64. London: Oxford University Press.
So, L. L. and I. J. GOLDSTEIN (1967) *J. Immunol.* **99**, 158–163.
WICK, A. N., G. S. SERIF, C. J. STEWART, H. I. NAKADA, E. R. LARSON and D. R. DRURY (1959) *Diabetes* **8**, 112–115.
WIESMEYER, H. and M. COHN (1960) *Biochim. & Biophys. Acta* **39**, 427–439.
WILSON, T. H. and B. R. LANDAU (1960) *Am. J. Physiol.* **198**, 99–102.

Discussion

Foster: There is increasing evidence that in the malignant transformation of certain cells mediated by viruses there is a significant change in transport of sugars, and one wonders whether this parameter might be exploited in designing new chemotherapeutic agents. This might be investigated by studying transport differences in the normal cell line and the virus-transformed cell line. So far we have looked at the effect of various fluorinated sugars on a lymphoma cell line in culture. In our lymphoma cells grown in the presence of glucose, the only sugars that seem to have any significant growth-inhibitory effect are those in which a fluorine substituent at position 2 replaces the hydroxyl group, for example 2-fluoroglucose, 2-fluoromannose and 2-fluorogalactose. However, if the hydroxyl group is replaced by hydrogen (2-deoxy sugar), methoxyl or chlorine, none of these compounds has any significant growth-inhibitory effect. It is possible that the 2-fluorosugars are active because of interference with sugar transport.

Dwek: Dr Barnett, in your hydrolysis of α-D-glucopyranosyl fluoride you mentioned that you obtained the same pH profile with the substrate as with maltose, but surely you cannot deduce the mechanism from that because pH is an equilibrium and not a kinetic parameter of the system?

Barnett: I agree that no conclusions as to the enzyme mechanism can be drawn from consideration of the pH profile alone, since the groups governing the profile might be involved in functions of the protein other than direct catalysis, such as the involvement of the amino group of chymotrypsin, pK 9, which maintains its structural integrity (Stryer 1968).

However, if we make the assumption that in intestinal maltase the pH profile is determined by the active site groups, then if α-D-glucopyranosyl fluoride were to hydrolyse without protonation, only the acid side of the profile would appear, corresponding to ionization of the alkaline nucleophilic group on the enzyme.

Dwek: That does not appear to be conclusive evidence.

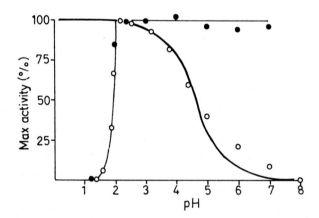

FIG. 1 (Barnett). The pH-activity (○) and pH-stability (●) curves for the ß-galactosidase of the albumen gland of *Helix aspersa*. [Theoretical ionization curve for $RCOOH=RCOO^- + H^+$ (pK=4.6).]

Barnett: I agree that it is by no means conclusive evidence, but perhaps I might elaborate. *Helix aspersa* β-galactosidase has a pH profile as shown in Fig. 1. The acid aspect corresponds to denaturation, and the alkaline aspect to the conversion of an acid, pK 4.6, into its conjugate base (Barnett 1971). One might expect the mirror image of this profile if the pH profile reflected the requirement for a single nucleophilic group at the active site.

Kun: If there is an oriented transfer of sugar through the cell surface without energy coupling (see Fig. 5, p. 103) one would normally expect an oriented process to be coupled to something. What exactly are you measuring?

Barnett: We measure the sugar accumulated into slices of everted intestine of hamster. When we measure the tissue/medium ratios (T/M ratios), we allow the sugar (1 mM) to accumulate at 37°C for 30 minutes and then measure the final tissue and medium concentrations. The ratio of these is the tissue/medium ratio.

Kun: Are you measuring the initial rate or the zero order part of the reaction?

Barnett: The initial rate. In the kinetic experiment to find the inhibition constants, K_i, we measured the accumulation for only 10 minutes, and we have shown that the uptake is linear in this system for about 60 minutes, so we are measuring initial rates. To find the K_i, we use substrate concentrations of 1–5 mM and plot the reciprocal of the rate of accumulation against the reciprocal of the substrate concentration to find 'K_m'. We then find the apparent 'K_m', using each of the modified sugars (20 mM) as a competitive inhibitor in turn, and from this we find an apparent K_i for each of the modified sugars.

Kun: What is your mechanism of energy coupling?

Barnett: This is discussed in my paper (p. 102). In the Crane hypothesis (Crane 1965), the energy coupling is due to the imbalance of sodium ions maintained by the sodium pump, which requires ATP.

Kun: But how can you distinguish between the binding effects and the ion exchange coupling, sodium coupling, transfer coupling, or the transfer process itself?

Barnett: The assumption we made was that if one always studies accumulation of the same sugar, for instance galactose, then if a series of modified sugars such as D-glucose, 3-deoxy-D-glucose and 3-deoxy-3-fluoro-D-glucose, are added to the medium, they would compete with it for the carrier binding site. Although one would not expect an accurate K_i, because they are themselves being transported, we would get the relative competition of these three sugars for galactose transport as measured by the K_i values. Qualitatively these should represent the relative binding of the 'inhibiting sugars' at the carrier binding site.

Kun: But ultimately the ATP drives this and the sodium pump.

Barnett: Yes. Indeed, Crane (1964) has incubated intestinal rings under anaerobic conditions to inhibit ATP formation, and has shown that under these conditions, if the sodium gradient across the cell membrane is reversed, then the sugar is accumulated in the opposite direction, that is from the inside to the outside of the cell.

Heidelberger: In your earlier series of experiments, Dr Barnett, you showed the effect of the 1-fluoro substituent (Barnett, Jarvis and Munday 1968). Do you have any information on the half-life of the α-D-galactosyl fluoride in that system?

Barnett: It is probably very long. All the sugars were isolated by chromatography from the tissue fluids after they had been transported inwards.

Heidelberger: Is this still in intestinal tissue, as in the second part of your paper where you show the hydrolysis of the same compound?

Barnett: Yes, but it is a different fluoride. The hydrolysed sugar is α-D-glucosyl fluoride. That is why we used α-D-galactosyl fluoride in the accumulation experiments.

Heidelberger: Is that split?

Barnett: No. There is no α-galactosidase in the intestine.

Sharpe: I have a slight reservation about the interpretation of your results as indicating two different types of binding, hydrogen bonding and covalent interaction, because of the differences in K_i. Normally the hydrogen bond energy would be taken as about 5 kcal, and the covalent bond energy considerably higher.

Barnett: Where we suggest a covalent bond, that is at C-2, modification of this position causes a complete absence of inhibition. K_i has the value infinity using D-galactose as the substrate and either 2-deoxy-D-glucose or 2-deoxy-2-fluoro-D-glucose as inhibitors. With D-galactose as substrate there appears to be no interaction of the C-2 modified sugars with the carrier, and they are not substrates for active intestinal transport.

Sharpe: Should one use the logarithm of K_i for comparisons?

Barnett: This would give a change in $\log_{10} K_i$ of about 3 for a hydrogen bond of 5 kcal at 37 °C, and we find that $\log_{10} K_i$ changes only by about 1. I have thought about the discrepancy. The species of sugar in solution is the hydrate, for example, glucose·nH$_2$O, but we do not know the value of n. On the membrane there are probably no water molecules bound to the sugar, which instead binds by the hydroxyl groups to the carrier:

$$\text{carrier} + \text{glucose·}n(\text{H}_2\text{O}) \rightleftharpoons \text{carrier·glucose} + n\text{H}_2\text{O}$$

So what is really being measured is some function determined by this equilibrium, that is the relative strengths of the hydrate/hydrogen bonds in the solution and, presumably, more favourable hydrogen bonds to carrier in the membrane. Therefore one would expect a change in $\log_{10} K_i$ of less than 3.

Sharpe: So you might be measuring a change in n.

Barnett: I doubt it, but there are probably water changes in the equilibrium between the lipid membrane and the free solution.

Wettstein: Dr Barnett, in your explanation of the experimental results did you consider that the fluorine atom decisively changes the electronic character of the hydroxyl at the nearby carbon atom, leading to a much stronger binding with a proton acceptor?

Barnett: You are suggesting that the presence of a fluorine α to a hydroxyl group could increase the binding of the hydroxyl group and so explain the observed increases in the binding of fluorine-containing sugars relative to the corresponding deoxy sugars. I do not think that this interpretation would fit our results.

Heidelberger: Are conformational changes produced in these fluorine compounds?

Barnett: We have not studied that.

Foster: We have been studying the effect of fluorine on various parameters in fluorinated carbohydrates. For the anomeric equilibrium, we sought to answer the questions: is the α/β ratio changed, is the furanose/pyranose ratio changed, and is the conformation changed? The evidence for fluoro derivatives of glucose is that none of these features is significantly changed. This is for solutions of acetylated fluorinated sugars in deuterochloroform and for the free sugars in aqueous media. The conformation was examined by detailed n.m.r. analysis (Barford *et al.* 1971).

Dwek: Is the furanose/pyranose equilibrium also examined by n.m.r. analysis?

Foster: Yes. Four ^{19}F resonances have been observed for 3-deoxy-3-fluoro-L-idose in aqueous solution and these correspond to the α and β, furanose and pyranose forms. The signals are clear and the percentage of each anomer, which one deduces from the ^{19}F resonance intensities, parallels very closely the percentages deduced on the basis of ^{1}H resonances for L-idose.

Dwek: What happens when you take these fluorosugars and put them into a complex with a macromolecule? Are there conformational changes?

Foster: This has not been done yet. The ground state of fluoroglucoses, as reflected by the conformation in aqueous solution, is not changed by the presence of the fluorine substituent. We do not know if there is a difference between the ground state conformation and the conformation of a fluorosugar in a complex with a macromolecule. A particular attraction of the fluorine nucleus would be if one could observe the splitting of the fluorine signal when the fluorosugar was associated with an enzyme. One might then be able to deduce the shape of sugar associated with an enzyme.

Dwek: In practice, I do not think this experiment is generally feasible. One may assume that conditions of rapid exchange will obtain and that the measured coupling constant will be an average of the two environments (bound and free). The value of *gem*-fluorine coupling constants rarely exceeds 30 Hz, and I think it would be very difficult to look for small deviations from the initial value. On the other hand, fluorine chemical shifts in different environments are much larger and the chances of observing changes in these are greater.

Kent: Dr Barnett, do those fluorosugars which enter the cell persist as the free sugar or are they phosphorylated? This is clearly pertinent to the sort of ratios that are being discussed, if the sugar is transported and then is further metabolized.

Barnett: I do not think that with the huge effects we see using hamster intestine, metabolism of the transported sugar is relevant. We have obtained recoveries of more than 90% for all the sugars except D-glucose. Addition of 2 mM-fluoride ion to the buffer also stopped the metabolism of glucose.

The rate of transport of the actively transported sugars is very high and the proportion phosphorylated tiny. We have never detected any sugar phosphates with the silver nitrate stain when we have chromatographed the tissue extracts.

Kent: Is there any evidence that 6-fluorogalactose, for instance, appears inside the cell as a phosphate?

Barnett: We did not find any chromatographic spots due to it.

Taylor: In whole cells where the uptake of glucose is not hexokinase limited, but dependent upon the mechanism of cell entry, Blakley and Boyer (1955) showed a competitive inhibition of uptake of glucose by 6-deoxy-6-fluoro-glucose in yeast cells similar to that described by you for intestine, Dr Barnett. Serif and Wick (1958) also obtained similar results with rat kidney slices. Did you ever find evidence for the release of free fluoride in your enzyme or transport studies?

Barnett: We have only tested for free fluoride in the case of α-D-glucosyl fluoride, which was extensively hydrolysed, as I have mentioned.

Gál: Have you thought of using the stroma of the red cell instead of the whole red cell to unravel this riddle of transport? Whittam and Ager (1964) used stroma and manipulated the ions to see whether conformational effects are related to transfer.

Barnett: Thank you for the suggestion.

Gál: Is anything known about fluorinated ribose?

Foster: We have no results on the fluorinated ribose derivatives.

Barnett: And I have done no work on it either.

References

BARFORD, A. D., A. B. FOSTER, J. H. WESTWOOD, L. D. HALL and R. N. JOHNSON (1971) *Carbohydr. Res.* **19**, 49.

BARNETT, J. E. G. (1971) *Comp. Biochem. Physiol.* **40**B, 585–592.

BARNETT, J. E. G., W. T. S. JARVIS and K. A. MUNDAY (1968) *Biochem. J.* **109**, 61–67.

BLAKLEY, E. R. and P. D. BOYER (1955) *Biochim. & Biophys. Acta* **16**, 576–582.

CRANE, R. K. (1964) *Biochem. & Biophys. Res. Commun.* **17**, 481–485.

CRANE, R. K. (1965) *Fed. Proc. Fed. Am. Soc. Exp. Biol.* **24**, 1000–1006.

SERIF, G. S. and A. N. WICK (1958) *J. Biol. Chem.* **233**, 559–561.

STRYER, L. (1968) *Annu. Rev. Biochem.* **37**, 25-50.

WHITTAM, R. and M. E. AGER (1964) *Biochem. J.* **93**, 337–348.

General discussion I

Miller: Higher plant species accumulate fluoride to different extents, so one cannot correlate the amount of fluoride accumulated with the amount of sensitivity to injury. We often find changes in organic acid metabolism in plants that accumulate fluoride and show injury. There are generally increases in organic acids, such as citric and malonic, and one wonders whether this could be related to changes in certain enzymes, such as aconitase. Aconitase has been known to be affected by fluorocitrate; it is, however, insensitive to inorganic fluoride.

Sir Rudolph indicated earlier that various plants are able to synthesize fluoroacetate. We reported previously (Cheng *et al.* 1968; Lovelace, Miller and Welkie 1968) that cultivated plants such as crested wheat grass and soya bean may have certain amounts of fluoroacetate and fluorocitrate. The experiments consisted of short strip chromatography, infrared spectroscopy, and the inhibition of aconitase extracted from the plants. Recently we analysed grass samples from various areas and were unable to detect the presence of fluoro-organic acids in any of them. The discrepancy between these results and those reported earlier may reflect differences in plant samples, such as location and other environmental factors or contamination of earlier samples. At the moment we are studying, by gas chromatography, the changes in organic acids that take place in plants. In the crested wheat grass we found increases in such acids as *cis*-aconitate and *trans*-aconitate. The concentration for *trans*-aconitate increased about threefold. The increase in this acid may account for *in vitro* inhibition of aconitase by extracts from fluoride-treated plants. We also found that glycollate was increased about twofold in the crested wheat grass.

Cain: Sir Rudolph, have you observed any differences between the *Acacia* grown in the wild around the Alice Springs area and the plants you grow in pots? The material that Murray sends us direct from the Alice Springs area is covered with a white wax. Mr R. J. Hall at Newcastle has detected some organic fluorine material in this wax cover (Hall 1970), but we never find this wax on pot-grown plants.

Also, as you mentioned earlier, Sir Rudolph, Murray and Woolley (1968), have found that they cannot correlate about 7 or 8 environmental parameters with the toxicity of the plant. We have yet to find a toxic *Acacia* plant amongst those we have raised from Murray's seeds. But fluoroacetate added to the soil in which these plants are growing, or to the same plants grown hydroponically in quartz, is very rapidly cleaved to inorganic fluoride. Consequently, we have had great difficulty in repeating your experiment of measuring an increase in organic fluoride in this plant, because when we add organic fluoride, it is cleaved before we can do anything with it.

Peters: We have had these plants that do not behave properly. We think, though we haven't proved it, that if they are watered with fluoride for some time, it improves the formation of the so-called organic fluoride.

Cain: Do you know whether the maturity of the plant affects this? We now have plants over an age range of a few months to several years. It may be that our seed batch came from a non-toxic tree.

Peters: This may be part of it.

Shorthouse: The formation of fluoroacetate is thought to be associated with the period of rapid growth. The farmers in particular feel this to be so, the trees being most toxic with the growth of new leaves. Murray's work (Murray and Woolley 1968) put some interesting theories to the test, but none seem to have survived. I personally would like to see the 'water stress' theory for the *Acacia* more thoroughly investigated.

Cain: So one ought to be able to detect it without difficulty if it is present in young seedlings?

Shorthouse: Yes, if you are fortunate to have seeds capable of producing plants that will synthesize fluoroacetate and they are in soil containing 5 p.p.m. fluoride.

Peters: A botanist at Alice Springs told me that toxic and non-toxic plants grow side by side and he could not tell the difference between them. I think this difference is probably a genetic one. We have had plants like Mr Hall's (Hall 1970), and when they get older they do not seem so good.

Preuss, Birkhahn and Bergmann (1970) found that fluoroacetate was formed in single cell cultures of *Acacia georginae*, and Mrs Shorthouse has confirmed this.

Cain: R. J. Hall and I were beginning to get the same impression but we and, I believe, some of the South African workers had started to think that perhaps the fluoroacetate was coming from the action of the microorganisms in the rhizosphere, and the plant was taking up preformed fluoroacetate. However, if you are picking it up in sterile tissue cultures, this is clearly not the case.

Gál: In tissue cultures, could this be connected with the light/dark phases of the photosynthetic mechanism?

Peters: No, these are just grown as single cell cultures by Dr D. H. Northcoate's technique.

Shorthouse: The tissue cultures were not green. Generally they are in the dark, but no attempt is made to keep out light when we handle them. So it is unlikely that it is a photosynthetic response.

METABOLISM OF FLUOROACETATE BY THE LETTUCE

Ward: I devised an *in vivo* type of experiment to see what would happen when a plant like the lettuce was given a sub-lethal dose of fluoroacetate. Lettuce plants injected with [1-^{14}C]- or [2-^{14}C]fluoroacetate were put in an 8 inch desiccator with about 300 ml water and air was passed through the desiccator (Fig. 1). We soon found that the radioactivity was translocated throughout the plant. This system was incubated for about 43 hours, after which the plant seemed perfectly normal and intact. The air that is passed through takes away any volatile metabolic products. This air is dried and passed through ethanolamine to absorb CO_2 and check its radioactivity. Only about 0.1 % of the radioactivity turned up in the CO_2 and this may be due to a small impurity in our

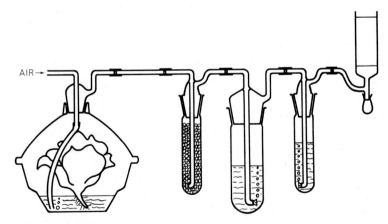

AIR →

FIG. 1 (Ward). Apparatus for the incubation of lettuces with acetate or fluoroacetate. Air was bubbled through 300 ml of water, in which the roots of the plant were suspended. The effluent gas from the desiccator was first dried with $CaCl_2$, then CO_2 was absorbed on two tubes of 2-methoxy-ethanol-ethanolamine, 2:1 v/v. The air flow was adjusted to 40–50 ml/min using the soap bubble flow meter.

fluoroacetate. When we injected radioactive acetate into the stem at the base of the leaves, we found about 20% of the radioactivity in the CO_2 after 43 hours, and when we injected non-radioactive fluoroacetate followed 24 hours later by radioactive acetate we again found about 20% of the radioactivity in the CO_2 after a further 43 hours. We injected 50 mg [^{14}C]fluoroacetate/kg wet weight of plant and judged from the above measurement of radioactivity in the CO_2 that the tricarboxylic acid cycle was not being blocked by this amount of fluoro-acetate. After 43 hours we found that about 50% of the fluoroacetate was unchanged, about 2% was fluorocitrate and the rest seemed to have suffered fission of the C—F bond, releasing a large amount of inorganic fluoride, but I do not know whether this exactly represents the amount that was split off from the fluoroacetate.

I had expected to find the carbon part of the fluoroacetate in many metabo-lites, but it was only in one of a pair of very closely related substances, S-carboxy-methylcysteine and S-carboxymethylglutathione. However, iodoacetate will react with either glutathione or cysteine very readily at normal pH and room temperature, producing the same derivatives, so perhaps the reaction with fluoroacetate is non-enzymic also. I could not get any reaction between fluoro-acetate and glutathione or cysteine outside the plant. So it is clear that enzyme systems take the C—F bond apart, though whether they take it all apart to F$^-$ or whether some of the fluorine is attached to another carbon atom, so forming a new non-radioactive metabolite, I am not sure.

Peters: So lettuce behaves in a similar way to what Boyland (1962; Boyland and Chasseaud 1969) suggested for the liver. Is it the S—H that is reactive?

Ward: Yes.

Buffa: I was very interested to hear that lettuce can convert fluoroacetate into fluorocitrate. Years ago I tried to obtain fluorocitrate from lemon trees, which make large amounts of citrate, by watering the plants with a solution of fluoro-acetate. However, the lemon trees were defoliated and killed by fluoroacetate within a few days.

Ward: Is this a general reaction or is it peculiar to lettuces?

Fowden: The situation in lettuce you described Dr Ward, differs markedly from that found with seedlings of *Acacia georginae*. After [^{14}C]fluoroacetate was supplied to *Acacia georginae*, ^{14}C was present in many compounds, including lipids and components of the cationic and anionic fractions (Preuss, Lemmens and Weinstein 1968). In my laboratory, we have shown that at least 50 labelled compounds arose from ^{14}C-labelled fluoroacetate supplied to 5-day-old seed-lings of *Acacia georginae*. These results suggest that *Acacia* seedlings may have developed a facile mechanism for degrading fluoroacetate. Presumably lettuce lacks this degradative mechanism.

In one of your experiments, did you first give unlabelled fluoroacetate followed by labelled fluoroacetate?

Ward: No, we gave unlabelled fluoroacetate followed by labelled acetate.

Fowden: A degradative enzyme system might be induced by giving unlabelled fluoroacetate. If [^{14}C]fluoroacetate were then supplied 24 hours later, breakdown of the labelled compound might be observed.

Ward: This is unlikely, because your suggestion is essentially the same as my 43-hour experiments with labelled fluoroacetate. I just wanted to show that the lettuce plant was capable of metabolizing acetate 24 hours after being given fluoroacetate, to make sure my system was viable.

Harper: Could you explain the synthesis of fluoroacetate in plants by reversal of this process?

Ward: I do not know. I was hoping that I would get some idea of how to synthesize the C—F bond.

Peters: In this particular case fluorine seems to be behaving as a halogen, for a change.

Gál: Did you notice whether there was an increase in NADPH?

Ward: No, I did not investigate that. What I did find was that the citrate level in the plant was just about double that in a normal plant.

Saunders: In the synthesis of fluoro sesqui-mustard that I mentioned earlier (p. 13), we were getting a reaction with SH.

Peters: It is very disappointing that fluoroacetate will not react with SH.

Goldman: This nice demonstration of the enzymic carboxymethylation of the thiol group by fluoroacetate helps to justify a mechanism we postulated (Goldman 1965) for the enzymic elimination of fluoride from fluoroacetate. In that proposal the halide was eliminated by the formation of a thiol ether in which the thiol group was donated by the enzyme, rather than by cysteine, as in Dr Ward's experiments with lettuce.

Although not applicable to these experiments, it might be worth mentioning an experience we had with a commercial batch of [^{14}C]fluoroacetate. That material gave a single peak on chromatography but unfortunately it was separated quite cleanly from authentic fluoroacetate.

Ward: We did not use a commercial product, but I agree that one would have to check that any commercial product one used really was fluoroacetate.

Kent: In *Escherichia coli*, *N*-iodoacetylglucosamine is a powerful inhibitor of active transport and growth. Bacteria, unlike man, contain enzymes which will cleave the *N*-haloacylglucosamine leaving the same compounds as Dr Ward finds in lettuce. Since then, *N*-fluoroacetylglucosamine has also been found to inhibit growth and active transport. Even though it is taken apart much more slowly in *E. coli*, it does seem to fit the carboxymethylation scheme (Kent, Ackers and White 1970).

Gál: Is it an amide?

Kent: Yes. The deaminase activity is very low in mammalian cells. In certain mouse cancer-transformed cells grown through one cell generation, *N*-fluoro-acetylglucosamine has the effect of changing the biosynthetic pattern of the membrane components, and this may also bear on the transport apparatus.

Bergmann (added in proof): Some data recently obtained in our laboratories may be of interest here. (*a*) The fluorine atom in the 2-fluoroallyl halide system (I. Shahak, Y. Itach and J. Azran, unpublished work), $R \cdot CH = CF \cdot CH_2Cl$ (and analogues containing bromine or the tosyloxy group, instead of the chlorine atom), prevents rearrangements during nucleophilic substitution (R = phenyl, pentyl, heptyl, octyl, nonyl). The corresponding allyl alcohols were obtained from the α-fluoroacrylates, $R \cdot CH = CF \cdot COOR$, by reduction with aluminium hydride; the alcohols gave the primary chlorides with concentrated hydro-chloric acid (for $R = C_6H_5$) or with thionyl chloride in ether (for R = alkyl). The following reagents have been studied: mercaptans, sodium benzene-sul-phinate, diethyl malonate, diethyl methylmalonate and diethyl acetamido-malonate, the enolate of cyclohexanone (giving I), potassio-phthalimide, sodium diethyl phosphite (leading to II), triethyl phosphonoacetate (leading to III), sodium iodide, phenylethynylmagnesium bromide (giving IV). The only appar-ent exception was the reaction with potassium thiocyanate which gave, by a Claisen rearrangement (Bergmann 1935), compound VI instead of the expected thiocyanate (V).

(*b*) We have also studied the spectral effect of fluorine substitution in con-jugated systems. The reaction of diethyl 2-fluoro-3-phenyl-allylphosphonate

$$R \cdot CH = CF \cdot CH_2 \text{—cyclohexanone}$$

(I)

$$R \cdot CH = CF \cdot CH_2 \cdot \underset{O}{\overset{}{P}}\overset{OEt}{\underset{OEt}{}}$$

(II)

$$R \cdot CH = CF \cdot CH_2 \cdot CH(COOEt) \cdot \underset{O}{\overset{}{P}}\overset{OEt}{\underset{OEt}{}}$$

(III)

$$R \cdot CH = CF \cdot CH_2C \equiv C \cdot C_6H_5$$

(IV)

$$R \cdot CH = CF \cdot CH_2 \cdot SCN$$

(V)

$$R \cdot CH \cdot CF = CH_2$$
$$\underset{N = C = S}{|}$$

(VI)

with benzaldehyde and cinnamaldehyde has led to 2-fluoro-1,4-diphenylbuta-1,3-diene (I) and 2-fluoro-1,6-diphenylhexa-1,3,5-triene (II) (I. Shahak, J. Azran and E. D. Bergmann, unpublished results). A comparison of their longest absorption bands with those of 1,4-diphenylbuta-1,3-diene (III) (Hirshberg, Bergmann and Bergmann 1950) and 1,6-diphenylhexa-1,3,5-triene (IV) (Koch 1948) shows—assuming that I and II are also all-*trans* compounds—that in I but not in II the fluorine atom causes a hypochromic shift.

		λ_{max}	$log\ \varepsilon$
(I)	$C_6H_5{\cdot}CH{=}CF{\cdot}CH{=}CH{\cdot}C_6H_5$	325 nm	4.70
(II)	$C_6H_5{\cdot}CH{=}CF{\cdot}CH{=}CH{\cdot}CH{=}CH{\cdot}C_6H_5$	349	4.74
(III)	$C_6H_5{\cdot}CH{=}CH{\cdot}CH{=}CH{\cdot}C_6H_5$	345	4.40
(IV)	$C_6H_5{\cdot}CH{=}CH{\cdot}CH{=}CH{\cdot}CH{=}CH{\cdot}C_6H_5$	349	4.83

If one compares equally the electronic spectra of cinnamic acid, cinnamyl-idene-acetic acid (V, VI) with those of their α-fluoro-analogues (VII, VIII) (Bergmann and Schwarcz 1956), one also finds a hypochromic effect.

		λ_{max}	$log\ \varepsilon$
(V)	$C_6H_5{\cdot}CH{=}CH{\cdot}COOH$	273 nm	4.32
(VI)	$C_6H_5{\cdot}CH{=}CH{\cdot}CH{=}CH{\cdot}COOH$	307	4.56
(VII)	$C_6H_5{\cdot}CH{=}CF{\cdot}COOH$	262	4.36
(VIII)	$C_6H_5CH{=}CH{\cdot}CH{=}CF{\cdot}COOH$	302	4.63

References

BERGMANN, E. (1935) *J. Chem. Soc.* 1361.
BERGMANN, E. D. and J. SCHWARCZ (1956) *J. Chem. Soc.* 1524.
BOYLAND, E. (1962) In *Proc. I. Int. Pharmacol. Meet.* vol. 6, p. 65, ed. B. B. Brodie and E. G. Erdos. Oxford: Pergamon.
BOYLAND, E. and L. F. CHASSEAUD (1969) *Adv. Enzymol.* **32**, 173–219.
CHENG, J. Y. O., M. H. YU, G. W. MILLER and G. W. WELKIE (1968) *Environ. Sci. & Technol.* **2**, 367–370.
GOLDMAN, P. (1965) *J. Biol. Chem.* **240**, 3434–3438.
HALL, R. J. (1970) M.Sc. Thesis. University of Newcastle-upon-Tyne.
HIRSHBERG, Y., E. D. BERGMANN and F. BERGMANN (1950) *J. Am. Chem. Soc.* **72**, 5120.
KENT, P. W., J. P. ACKERS and R. J. WHITE (1970) *Biochem. J.* **118**, 73–79.
KOCH, H. P. (1948) *J. Chem. Soc.* 1111.
LOVELACE, J., G. W. MILLER and G. W. WELKIE (1968) *Atmos. Environ.* **2**, 187–190.
MURRAY, L. R. and D. R. WOOLLEY (1968) *Aust. J. Soil Res.* **6**, 203–211.
PREUSS, P., R. BIRKHAHN and E. D. BERGMANN (1970) *Israel J. Bot.* **19**, 609–619.
PREUSS, P. W., A. G. LEMMENS and L. H. WEINSTEIN (1968) *Contrib. Boyce Thomson Inst. Plant. Res.* **24**, 25–31.

The nucleotides of fluorinated pyrimidines and their biological activities

CHARLES HEIDELBERGER

McArdle Laboratory for Cancer Research, University of Wisconsin, Madison, Wisconsin

It is a great honour for me to have been invited to participate in this symposium, and it is a particularly appropriate and pleasant opportunity for me to pay my deepest and warmest respects to Sir Rudolph Peters who initiated the field that we are meeting to discuss, and whose concept of lethal synthesis (Peters 1970) has been particularly fruitful in my own research.

I have been concerned with biologically active fluorine compounds since 1956. The discovery by Rutman, Cantarow and Paschkis (1954) of the enhanced utilization of uracil for DNA biosynthesis by tumours, led me to prepare an antimetabolite of uracil with a substitution of a fluorine atom for a hydrogen atom. The specific rationale behind the synthesis of 5-fluorouracil (FU) has been described many times (Heidelberger 1965, 1970) and is beyond the scope of this brief presentation. Suffice it to say that we predicted correctly its major biochemical mechanisms of action before the compound was synthesized. In our preliminary report (Heidelberger *et al.* 1957) we described the tumour-inhibitory properties, the biochemical mechanisms of action, and the initial clinical trial of 5-fluorouracil in patients with advanced cancer. This publication was closely followed by a description of the method of synthesis (Duschinsky, Pleven and Heidelberger 1957). 5-Fluorouracil has been widely used throughout the world for the palliative treatment of patients suffering from disseminated cancers, particularly those of the breast and colon (Ansfield *et al.* 1969). But it is not the purpose of this paper to discuss clinical results.

The structures of the naturally occurring pyrimidines, uracil and thymine, together with the analogues 5-fluorouracil and 5-trifluoromethyluracil, which we synthesized much later (Heidelberger, Parsons and Remy 1964), are shown in Fig. 1. The close similarity in size between the hydrogen and fluorine atoms is shown, as is the greater acid dissociation constant of the proton on nitrogen-3 of the fluoro analogues. The structures of the most important nucleoside

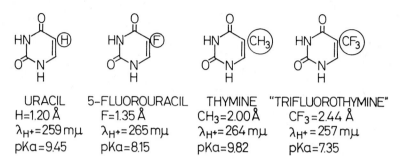

URACIL	5-FLUOROURACIL	THYMINE	"TRIFLUOROTHYMINE"
H=1.20 Å	F=1.35 Å	CH$_3$=2.00Å	CF$_3$=2.44 Å
λ_{H^+}=259 mμ	λ_{H^+}=265 mμ	λ_{H^+}=264 mμ	λ_{H^+}=257 mμ
pKa=9.45	pKa=8.15	pKa=9.82	pKa=7.35

FIG. 1. The structures of two fluorinated pyrimidines and their natural counterparts.

derivatives of FU are shown in Fig. 2. Fluorouridine (FUR) is generally more toxic and less effective than FU as a tumour inhibitor; however, 5-fluoro-2'-deoxyuridine (FUDR) is rather less toxic and often chemotherapeutically more effective than FU.

Once FU had been shown to have interesting biological properties, we labelled it with ^{14}C and studied its metabolism and incorporation into nucleic acids in suspensions of Ehrlich ascites tumour cells. As predicted from its close

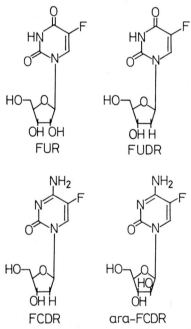

FUR

FUDR

FCDR

ara-FCDR

FIG. 2. The structures of pyrimidine nucleoside analogues. FUR: 5-fluorouridine; FUDR: 5-fluoro-2'-deoxyuridine; FCDR: 5-fluoro-2'-deoxycytidine; ara-FCDR: 1-β-D-arabinofuranosyl-5-fluorocytosine.

structural relationship to uracil, fluorouracil was found to be incorporated into RNA in non-terminal positions. This was demonstrated by the isolation of a fluorinated nucleotide, as shown in Fig. 3. There was no incorporation into DNA (Chaudhuri, Montag and Heidelberger 1958). Since then incorporation of FU into RNA has been confirmed in many organisms studied in other laboratories (see review by Mandel 1969). A number of interesting biological effects are a consequence of such incorporation, which occurs in ribosomal messenger and transfer RNA's (Lowrie and Bergquist 1968; Kaiser 1971).

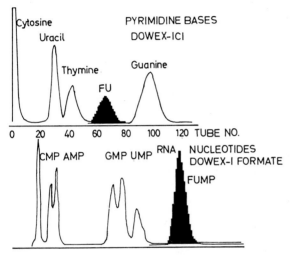

FIG. 3. The incorporation of labelled FU into RNA in suspensions of Ehrlich ascites carcinoma cells. The RNA was hydrolysed with alkali (lower curve) and the nucleotides were separated on Dowex-1-formate. The solid curves represent radioactivity, the open curves represent optical density at 260 nm. FUMP: 5-fluorouridine-5′-monophosphate (from Chaudhuri, Montag and Heidelberger 1958, with permission).

These effects include: mutagenesis in RNA viruses, changes in acceptor activities of transfer RNA's (Wagner and Heidelberger 1962), miscoding in protein synthesis, inhibition of the maturation of ribosomal RNA in bacterial and mammalian cells, and several other less well documented effects (Mandel 1969). One of the most striking effects is the inhibition of enzyme induction in bacteria (Nakada and Magasanik 1964) and rat liver (Cihak, Wilkinson and Pitot 1971). In bacteria this has been attributed to the incorporation of FU into messenger RNA (Nakada and Magasanik 1964), or, contrariwise, to catabolite repression (Horowitz and Kohlmeier 1967). We had postulated that these effects might be due to errors in transcription resulting from FU base-pairing as if it were cytosine instead of uracil. However, attempts to demonstrate such transcriptional errors failed (Bujard and Heidelberger 1966). The experiments of Champe and Benzer (1962) on the production by FU of phenotypic reversions in amber mutants of bacteriophage suggested that the error occurred during translation, and this was proved by the critical experiments of Rosen, Rothman and Weigert (1969). It is not yet known to what extent the incorporation of FU into RNA plays a role in the inhibition of tumour growth.

Because it was known that thymine, an essential building block of DNA, is made by the attachment of a methyl group to the 5-carbon of uracil, we predicted that the presence of a fluorine atom in the same location would inhibit that methylation. The confirmation of this prediction is shown in the experiment depicted in Fig. 4. Formate is a specific precursor of the methyl group of thymine, and when Ehrlich ascites cells were incubated with [^{14}C]-formate and various concentrations of fluorinated pyrimidines and their nucleosides, its incorporation into DNA was completely inhibited (Bosch, Harbers and Heidelberger 1958). FUDR was the most active in this respect, and 5-fluoro-orotic acid (FO) was the least effective inhibitor.

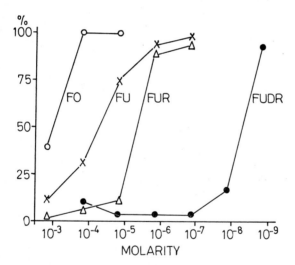

Fig. 4. The effect of fluorinated pyrimidines on the incorporation of [^{14}C]formate into the DNA thymine of suspensions of Ehrlich ascites carcinoma cells (from Bosch, Harbers and Heidelberger 1958, with permission).

The methylation reaction is considerably more complicated than I implied above. The substrate is deoxyuridylate, the coenzyme methylenetetrahydro-folate, the product thymidylate, and the enzyme that catalyses the conversion is thymidylate synthetase. Since all these compounds are phosphorylated, it was found (as expected) that FUDR does not inhibit this enzyme, but its

5′-monophosphate (FdUMP) does (Cohen *et al.* 1958). There is considerable evidence, based on the study of resistant tumours, that inhibition of thymidylate synthetase is responsible for the major tumour-inhibitory effect of the fluorinated pyrimidines (Heidelberger *et al.* 1960; Kessel and Hall 1969), although the contribution of incorporation of FU into RNA cannot be excluded in some cases.

Thus thymidylate synthetase is the main target of the action of these compounds, and the mode of inhibition of this enzyme by the fluorinated nucleotides becomes of interest. The kinetics of inhibition by FdUMP are shown in Fig. 5 and are clearly competitive (Reyes and Heidelberger 1965). To jump ahead, we found that trifluorothymine (F_3T) was biologically inactive in mammalian systems, but the nucleoside (F_3TDR, trifluorothymidine) was highly active against tumours and DNA viruses. We found that the nucleotide,

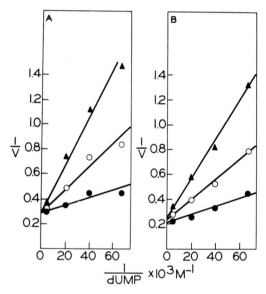

Fig. 5. Lineweaver-Burk plot of the inhibition of thymidylate synthetase by FdUMP, with and without preincubation. A: no preincubation; B: with 10 min preincubation; ●: control; ○: FdUMP 8×10^{-8} M; ▲: FdUMP 20×10^{-8} M. The numbers along the abscissa should be multiplied by 10^3 to obtain correct numerical values (from Reyes and Heidelberger 1965, with permission).

5-trifluoromethyl-2′-deoxyuridine-5′-monophosphate (F_3dTMP), inhibits thymidylate synthetase competitively (Fig. 6) if added simultaneously, but noncompetitively if preincubated with the enzyme. We have postulated the formation of a covalent bond between the carbon that carried the trifluoromethyl group and an amino group near to the active site (Reyes and Heidelberger 1965).

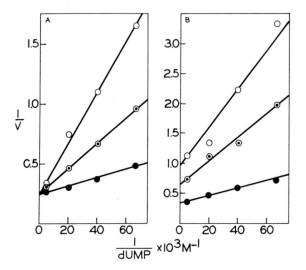

FIG. 6. Lineweaver-Burk plot of the inhibition of thymidylate synthetase by F_3dTMP, with and without preincubation. A: no preincubation; B: with 10 min preincubation; ●: control; ⊙: F_3dTMP 5×10^{-7} M; ○: F_3dTMP 10×10^{-7} M. The numbers along the abscissa should be multiplied by 10^3 (from Reyes and Heidelberger 1965, with permission).

If this is so, then we have accomplished the preparation of a built-in active-site-directed irreversible inhibitor, a desideratum of Baker (1967). To prove this point it became necessary to purify the enzyme, and we have recently effected the first complete purification to homogeneity of a mammalian thymidylate synthetase (Fridland, Langenbach and Heidelberger 1971).

Since the nucleotide, FdUMP, is the active form of the drug FU, we have thus established another case of Sir Rudolph Peters' principle of lethal synthesis (Peters 1970), and it was necessary to elucidate the pathways thereto. The scheme of metabolism, mostly worked out in our laboratory, is shown in Fig. 7. We have demonstrated that FU is catabolized to α-fluoro-β-alanine, carbon dioxide and ammonia (step 11); this catabolic degradation is greatly diminished in most tumours, including human colon carcinomas, and may be responsible for the selective action of these compounds against tumours (Mukherjee and Heidelberger 1960). By contrast, the ring of F_3T is not cleaved at all; the sole degradation product is 5-carboxyuracil. There are several routes by which FU can be converted into the lethal FdUMP: (*a*) FU is converted to FUR by a nucleoside phosphorylase (10) and FUR can be phosphorylated by a kinase (9) to FUMP; (*b*) FU can also react with phosphoribosylpyrophosphate (PRPP) to give FUMP directly (Reyes 1969); (*c*) FUMP can be phosphorylated to the di- and triphosphate and incorporated into RNA by RNA polymerase (8);

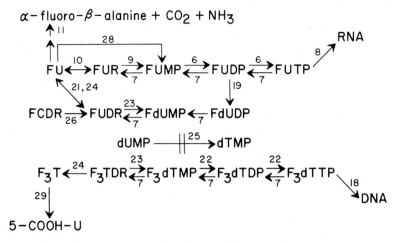

FIG. 7. Overall pathways of fluorinated pyrimidine metabolism.

(*d*) FU can be converted by a nucleoside phosphorylase (21, 24) into FUDR, which is also obtained by the deamination of 5-fluoro-2′-deoxycytidine (FCDR); (*e*) FUDR is phosphorylated to FdUMP by thymidine kinase (23); (*f*) it was necessary to postulate that in some cells FUDP was reduced to the deoxy series by ribonucleotide reductase (19). This conversion has recently been demonstrated for the first time (Kent and Heidelberger 1972). In the F_3TDR series, the nucleoside is irreversibly cleaved to the inactive F_3T, which is converted into 5-carboxyuracil with the release of inorganic fluoride. F_3TDR is phosphorylated to F_3dTMP by thymidine kinase (23), and further to the deoxyribonucleoside di- and triphosphates (22), the latter of which is incorporated by DNA polymerase into DNA (18) (Gottschling and Heidelberger 1963).

From this rather complicated scheme (Fig. 7), it is evident that the response of tumours is determined by the relative activities of the various enzymes required to accomplish the lethal synthesis by the different alternative metabolic pathways. It is clear, nevertheless, that the fluorinated pyrimidines are acting by inhibiting the *de novo* pathway of DNA thymine synthesis. There is also a 'salvage' pathway of DNA thymine synthesis, which involves the phosphorylation by thymidine kinase of any thymidine produced from the breakdown of tissues or exogenously added; thymidine prevents the inhibition produced both by FU and FUDR. Kessel and Wodinsky (1970) found in a series of mouse leukaemias that the chemotherapeutic response to FUDR varied inversely as the activity of thymidine kinase, even though this is the enzyme that activates FUDR. The explanation of this apparent paradox is that the K_m/K_i of FdUMP is about 1000, so whatever FUDR is phosphorylated is very

active at inhibiting the enzyme. Wolberg (1969) is studying some of these
pathways of metabolism in slices of human tumours in an attempt to develop
a prognostic test for the response of an individual patient to FU and FUDR.

A summary of the key enzymes involved in the mechanism of action of
FUDR is shown in Fig. 8. FUDR is much less effective in animals and patients
than expected; this is because it is rapidly cleaved to FU by the enemy nucleoside phosphorylases. Attempts to improve the efficacy of FUDR by inhibiting

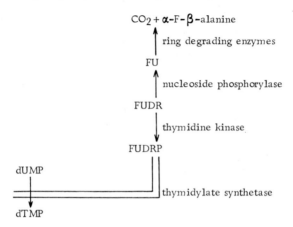

FIG. 8. The key enzymes involved in the mechanism of action of fluorinated pyrimidines and
their nucleosides.

this cleavage enzyme have not yet been successful. Thymidine kinase is the
activating enzyme for FUDR, and thymidylate synthetase is the target enzyme
for chemotherapeutic action. We have determined some structure-activity
relationships for FUDR and these three enzymes. We synthesized all possible
O- and *N*-methyl derivatives of FUDR and studied their affinity for those
enzymes (Kent, Khawaja and Heidelberger 1970). The conclusions are given
in Fig. 9. No derivative of FUDR so far prepared is as active as FUDR. However, 3'-*O*-methyl-FUDR is not cleaved; it is phosphorylated, and the nucleotide inhibits thymidylate synthetase, but not nearly as effectively as does
FdUMP. We are now in the process of synthesizing isosteres of the pyrimidine
ring to gain more information about structure-activity relationships, and
hopefully to produce a more effective drug.

I have already mentioned briefly that F_3TDR is incorporated into DNA and
is mutagenic to bacteriophage T4 (Gottschling and Heidelberger 1963). However, it is incorporated to only a very slight extent into the DNA of mammalian
cells, and exerts its tumour-inhibitory activity as a consequence of being converted by thymidine kinase into the nucleotide, F_3dTMP, which is an irreversible

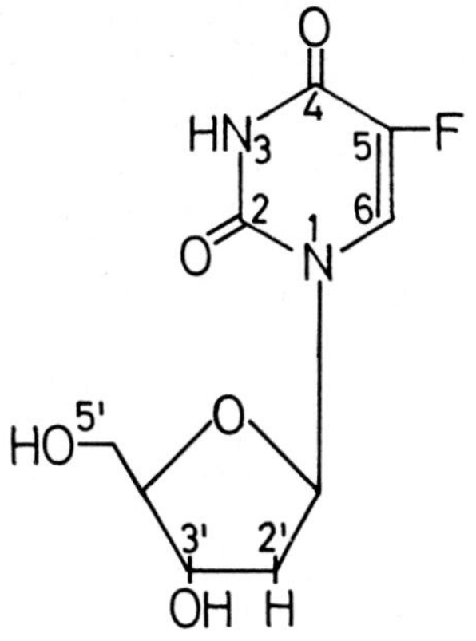

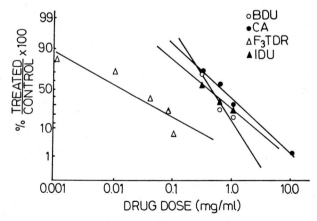

N-1 Necessity unknown: to be tested.

O-2 Necessary for activity.

N-3 Must be unsubstituted and with proton of correct pK_a.

O-4 Probably necessary. Cannot be methylated.

C-5 Size of substituent determines incorporation into DNA or RNA and electronic properties.

C-6 Must be unsubstituted. 6-AzaFU could not be synthesized. 6-AzaF$_3$TDR inactive: too acidic.

H-2′ H needed for DNA action, OH for RNA. 2′-Ara-hydroxyl gives 4-amino compounds desirable properties.

H-3′ Can be blocked with methoxyl group and retains slight TdR-kinase and TMP-synthetase activity, but prevents nucleoside phosphorylase cleavage.

H-5′ Must be unsubstituted. However, some 5′-halo nucleosides inhibit TMP-kinase.

Furanose-O. Probably necessary, but not adequately tested.

FIG. 9. Structure-activity relationships of the interaction of FUDR and its derivatives with the key enzymes involved in its mode of action (from Kent, Khawaja and Heidelberger 1970, with permission).

inhibitor of thymidylate synthetase (Reyes and Heidelberger 1965). However, based on the observation that 5-iodo-2′-deoxyuridine (IUDR) is incorporated into DNA and is used clinically for local treatment of herpes simplex keratitis of the eye, we tested F$_3$TDR for therapeutic antiviral activity. It was found to be very active against herpes simplex keratitis of the rabbit's eye (Kaufman and Heidelberger 1964). In fact, F$_3$TDR is considerably more active than the other pyrimidine nucleoside analogues tested, as shown in Fig. 10 (Kaufman 1965).

FIG. 10. Response to antiviral drugs of herpes keratitis of the rabbit's eye. BDU and IDU are Kaufman's abbreviations of BUDR, IUDR (from Kaufman 1965, with permission).

Since my philosophy has always been to try to follow the problem wherever it leads, we have studied the mechanism of action of F_3TDR against vaccinia virus replicating in HeLa cells. We chose vaccinia rather than herpes virus because it is safer to work with, and because it multiplies in the cytoplasm of cells; hence it is easy to isolate vaccinia virions in pure form, quite separate from the host-cell DNA in the nuclei. We compared the effects of four pyrimidine nucleoside analogues against vaccinia virus replicating in HeLa cells, and found (Fig. 11) that F_3TDR was tenfold more active than IUDR, FUDR and cytosine arabinoside (CA) (Umeda and Heidelberger 1969). In that study we also showed that although the inhibition produced by FUDR (an inhibitor of DNA synthesis) was reversed after delayed administration of thymidine, the

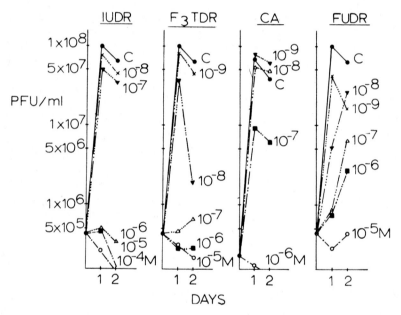

FIG. 11. The inhibition by pyrimidine nucleoside analogues of the replication of vaccinia virus in HeLa cells. PFU: plaque-forming units (from Umeda and Heidelberger 1969, with permission).

production of virus could not be so rescued from F_3TDR. This suggested that the irreversible inhibition might be a consequence of the incorporation of F_3TDR into the DNA of vaccinia virions. Using labelled F_3TDR we then studied this. As shown in Fig. 12, the analogue is incorporated into virion DNA, which DNA has a smaller molecular weight than normal, as shown by sucrose gradient sedimentation (Fujiwara and Heidelberger 1970). It was also found that virions containing as little as 5% replacement of thymidine by

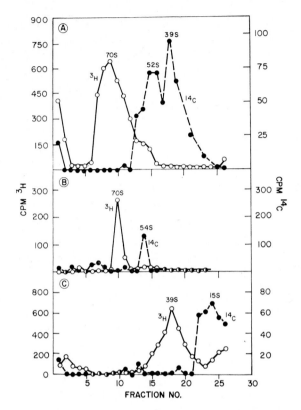

FIG. 12. Sucrose gradient sedimentation of the DNA isolated from purified vaccinia virions replicated in the presence of [³H]thymidine (○) or [¹⁴C]F₃TDR (●). A: neutral sucrose; B: resedimentation of the two ¹⁴C peaks from A; C: alkaline sucrose (from Fujiwara and Heidelberger 1970, with permission)

F₃TDR were non-infective and that they had a highly abnormal morphology (Fujiwara and Heidelberger 1970).

We then looked at viral messenger RNA production in the presence of the analogue. As determined by Oda and Joklik (1967), the early mRNA is coded for by the input virus and the late mRNA is coded for by the progeny viral DNA. In accord with expectation, we have found that early mRNA production occurs normally in the presence of F₃TDR, but that late mRNA production is quite abnormal (Oki and Heidelberger 1971). We are currently studying further details of this abnormality by the methods of DNA-RNA hybridization, and hope eventually to elucidate the complete details of the molecular mechanism of antiviral activity of F₃TDR.

In conclusion, I would like once again to pay tribute to Sir Rudolph Peters, to his concept of lethal synthesis, and to his pioneering studies on biologically active fluoro compounds. I believe that the work already accomplished has fully justified this approach, and because of the subtleties of antimetabolite design and the continual increase in our fundamental knowledge of cancer and viruses, it should be possible to continue to design and synthesize ever more effective and specific drugs for the treatment of these two human diseases (Heidelberger 1969).

SUMMARY

Although the pyrimidine analogue, 5-fluorouracil (FU), is widely used for the palliative treatment of patients with a variety of solid cancers, it must be converted by cellular enzymes into nucleotides, which are the active forms of the drug. This is another illustration of the principle of 'lethal synthesis', first discovered and enunciated by Sir Rudolph Peters. The various pathways of anabolic activation and catabolic degradation of FU are described. The deoxyribonucleotide of FU, FdUMP, is a powerful inhibitor of thymidylate synthetase and recent results on the purification of this enzyme and its inhibition are described. Structure-activity relationships relating to the catabolic enzyme nucleoside phosphorylase, the anabolic enzyme thymidine kinase, and the target enzyme thymidylate synthetase are considered. The synthesis and mechanism of antiviral action of trifluorothymidine (F_3TDR), the most powerful known drug against DNA viruses, are discussed.

References

ANSFIELD, F. J., G. RAMIREZ, S. MACKMAN, G. T. BRYAN and A. R. CURRERI (1969) *Cancer Res.* **29**, 1062–1066.

BAKER, B. R. (1967) *Design of Active Site-directed Irreversible Inhibitors.* New York: Wiley.

BOSCH, L., E. HARBERS and C. HEIDELBERGER (1958) *Cancer Res.* **18**, 335–343.

BUJARD, H. and C. HEIDELBERGER (1966) *Biochemistry* **5**, 3339–3345.

CHAMPE, S. P. and S. BENZER (1962) *Proc. Natl. Acad. Sci. (U.S.A.)* **48**, 532–546.

CHAUDHURI, N. K., B. J. MONTAG and C. HEIDELBERGER (1958) *Cancer Res.* **18**, 318–328.

CIHAK, A., D. WILKINSON and H. C. PITOT (1971) *Adv. Enzyme Regul.* **9**, 267–289.

COHEN, S. S., J. G. FLAKS, H. D. BARNER, M. R. LOEB and J. LICHTENSTEIN (1958) *Proc. Natl. Acad. Sci. (U.S.A.)* **44**, 1004–1012.

DUSCHINSKY, R., E. PLEVEN and C. HEIDELBERGER (1957) *J. Am. Chem. Soc.* **79**, 4559–4560.

FRIDLAND, A., R. L. LANGENBACH and C. HEIDELBERGER (1971) *J. Biol. Chem.* **246**, 7110–7114.

FUJIWARA, Y. and C. HEIDELBERGER (1970) *Mol. Pharmacol.* **6**, 281–291.

GOTTSCHLING, H. and C. HEIDELBERGER (1963) *J. Mol. Biol.* **7**, 541–560.

HEIDELBERGER, C. (1965) *Prog. Nucleic Acid Res. & Mol. Biol.* **4**, 1–50.
HEIDELBERGER, C. (1969) *Cancer Res.* **29**, 2435–2442.
HEIDELBERGER C. (1970) *Cancer Res.* **30**, 1549–1569.
HEIDELBERGER, C., N. K. CHAUDHURI, P. DANNEBERG, D. MOOREN, L. GRIESBACH, R. DUSCHINSKY, R. J. SCHNITZER, E. PLEVEN and J. SCHEINER (1957) *Nature (Lond.)* **179**, 663–666.
HEIDELBERGER, C., G. KALDOR, K. L. MUKHERJEE and P. B. DANNEBERG (1960) *Cancer Res.* **20**, 903–909.
HEIDELBERGER, C., D. G. PARSONS and D. C. REMY (1964) *J. Med. Chem.* **7**, 1–5.
HOROWITZ, J. and V. KOHLMEIER (1967) *Biochim. & Biophys. Acta* **142**, 208–218.
KAISER, I. I. (1971) *Biochemistry* **10**, 1540–1545.
KAUFMAN, H. E. (1965) *Ann. N. Y. Acad. Sci.* **130**, 168–180.
KAUFMAN, H. E. and C. HEIDELBERGER (1964) *Science* **154**, 585–586.
KENT, R. J. and C. HEIDELBERGER (1972) *Mol. Pharmacol.* in press.
KENT, R. J., T. A. KHAWAJA and C. HEIDELBERGER (1970) *J. Med. Chem.* **13**, 70–73.
KESSEL, D. and T. C. HALL (1969) *Cancer Res.* **29**, 1749–1754.
KESSEL, D. and I. WODINSKY (1970) *Mol. Pharmacol.* **6**, 251–254.
LOWRIE, R. J. and P. L. BERGQUIST (1968) *Biochemistry* **7**, 1761.
MANDEL, H. G. (1969) *Prog. Mol. Subcell. Biol.* **1**, 82–135.
MUKHERJEE, K. L. and C. HEIDELBERGER (1960) *J. Biol. Chem.* **235**, 433–437.
NAKADA, D. and B. MAGASANIK (1964) *J. Mol. Biol.* **8**, 105–117.
ODA, K. and W. K. JOKLIK (1967) *J. Mol. Biol.* **27**, 395–414.
OKI, T. and C. HEIDELBERGER (1971) *Mol. Pharmacol.* **7**, 653–662.
PETERS, R. A. (1970) *Bibliotheca Nutritio et Dieta No. 15: Biological Interrelations and Nutrition*, pp. 19–28, ed. J. C. Somogyi. Basel: Karger.
REYES, P. (1969) *Biochemistry* **8**, 2057–2062.
REYES, P. and C. HEIDELBERGER (1965) *Mol. Pharmacol.* **1**, 14–30.
ROSEN, B., F. ROTHMAN and M. G. WEIGERT (1969) *J. Mol. Biol.* **44**, 363–375.
RUTMAN, R. J., A. CANTAROW amd K. E. PASCHKIS (1954) *Cancer Res.* **14**, 119–134.
UMEDA, M. and C. HEIDELBERGER (1969) *Proc. Soc. Exp. Biol. & Med.* **130**, 24–29.
WAGNER, N. J. and C. HEIDELBERGER (1962) *Biochim. & Biophys. Acta* **61**, 373–379.
WOLBERG, W. H. (1969) *Cancer Res.* **29**, 2137–2144.

Discussion

Saunders: Professor Heidelberger, do your statements about F_3TDR and herpes of the eye apply equally to herpes simplex of the lips?

Heidelberger: At the moment this treatment has to be given locally. A solution or an ointment of the drug is put in the eye. F_3TDR has been quite successful in the treatment of herpes zoster (shingles), which a lot of patients with advanced cancer have. It has not been used for cold sores—it is probably too expensive for that.

Saunders: What about toxicity to the corneal epithelium?

Heidelberger: As I am not an ophthalmologist, I cannot say. Apparently there is very slow cell division in the corneal epithelium and an inhibition of DNA synthesis would produce damage. Or the damage may be due to incorporation. It is interesting that about 30–50% of the thymidine in the vaccinia DNA is replaced by IUDR, but only 5–10% is replaced with F_3TDR.

Kent: Have the triphosphates of FUDR any effect on RNA polymerase? Orgel and colleagues have suggested that 2-fluoro-2-deoxy-cytidine triphosphate is a powerful inhibitor of DNA-dependent RNA polymerase (Hendler *et al.* 1971).

Heidelberger: We have never studied the ribonucleoside triphosphates. I would think that they are not very good inhibitors because inhibition of RNA synthesis in cells in suspension requires concentrations of FU about 10 000-fold higher than for DNA synthesis. We are in the process of making the triphosphate of F_3TDR, which we are going to look at in a mammalian DNA polymerase system.

Gál: You have two different types of inhibitors; FdUMP and F_3dTMP. Did you use circular dichroism or optical rotatory dispersion to see whether there were any differences in the two compounds that might be reflected in some kinds of changes that you could pick up?

Heidelberger: No, not yet. At the moment we are doing the binding studies.

Gál: I was very interested in the ribonucleotide reductase. Can your compound FUDP be an effector or an inhibitor of other nucleotides?

Heidelberger: We have not studied that. But I doubt whether it would be an effector because we see no particular effect of these compounds on the reduction of other nucleotides.

Bergmann: There is a nucleoside antibiotic, nucleocidin, that contains a fluorine atom not in the purine moiety, but in the sugar (Morton *et al.* 1969, also see Kent, p. 169). Do you think this has a similar mechanism of action because it is also a 'fraudulent' nucleoside?

Heidelberger: Very little work has been done on the mechanism of nucleocidin. But the relationships between all these antimetabolites and the normal metabolites are very subtle.

Taylor: You showed arabinofuranosyl analogues on one of your slides. Could you say a few words about those, because they are interesting in connection with the arabinofuranosyl purine-6-mercapto analogues and antitumour agents?

Heidelberger: I have not worked with these compounds. FCDR (fluorocytosine deoxyribonucleoside) is equivalent in activity to FUDR because it is simply deaminated to FUDR. Jack Fox has made ara-FCDR, which is 5-fluoro-ara-C. It appears from Burchenal's studies in mouse leukaemias that ara-FC has about the same activity as ara-C (Burchenal *et al.* 1966). There has been a big polemic in the literature as to whether or not ara-C is incorporated into DNA. So, in connection with our work on the mutagenicity of nucleotide analogues to mammalian cells in culture, we looked at ara-C. It turned out to be very highly mutagenic to mammalian cells, hence it has to be incorporated into DNA (Huberman and Heidelberger 1972).

Barnett: I understand that the C-4 oxygen was a necessary structural requirement for binding of the uracil to the enzyme because the 4-methoxy analogue was inactive. Change of a phenolic hydroxyl group to a methoxyl group is a significant change in the molecule. Have you tried the sulphur analogue ($-SH$) which would be able to exist in a form equivalent to the keto form?

Heidelberger: The 4-thio compound has been prepared and does not seem to be very active.

Peters: Do compounds with the CF_3 group go into the bones?

Heidelberger: In the process of the hydrolysis of CF_3 groups, inorganic fluoride is produced. We measured the total amount of inorganic fluoride in the urine of humans and it turned out to be about 50% of the inorganic fluoride one would expect on the basis of that amount of carboxyuracil formed. We inferred from this that there is incorporation into the bone. We also gave a known dose of inorganic fluoride, equivalent to the amount produced in the metabolism of F_3TDR. Again, we recovered only 50% of the fluoride in the urine. From the literature, one would expect about 50% of it to go into the bones.

Peters: Was there perhaps also some organically combined fluoride in the urine, and did you compare this with the acid-labile fluoride?

Heidelberger: We did not study acid-labile fluoride. Analysis, including chromatography of the urine, showed that F_3TDR itself, the free base, F_3T, and 5-carboxyuracil are present. These are the only compounds derived from the drug which are detectable, except for the inorganic fluoride, for which we used the fluoride electrode (Dexter *et al.* 1972).

Peters: Dr Taves performed similar analyses with blood.

Heidelberger: We too looked for the other metabolites in the blood and have determined the half-times of all these processes in patients (Dexter *et al.* 1972).

Peters: Do you think that this is the start of being able to tackle the viruses?

Heidelberger: Most of the viruses that cause human disease are RNA viruses; the only two important DNA viruses, clinically, are vaccinia and herpes. Kaufman has shown that F_3TDR is clearly superior to IUDR for the treatment of DNA viruses (H. E. Kaufman, private communication). However, F_3TDR was originally synthesized for cancer chemotherapy and, as the U.S. Government owns the patent, the pharmaceutical companies are reluctant to make and market F_3TDR as an antiviral agent.

Barnett: Agents like F_3TDR, which alkylate an enzyme to inactivate it, would seem to be of less use in chemotherapy than the ones that just compete, because they do not have a built-in 'feedback system'. Once the enzyme is alkylated, only *de novo* synthesis can restore activity. Competitive inhibition will be reversed by a large build- up of substrate. I would predict that drugs that alkylate would be too toxic. Is this why F_3TDR is not being used in cancer chemotherapy?

Heidelberger: First, what is 'too toxic'? All of these drugs are toxic. F_3TDR does not exert toxicity on the gastrointestinal tract, whereas FU and FUDR do. Also, we think that it has some activity in cancer in children which the others do not. However, it is very expensive to make and it does not give spectacular results in cancer chemotherapy, and this is why it is not being used.

Bergmann: Some of the cancer-inducing viruses require arginine before they can multiply. Some 'anti-arginine compounds' have been found in nature or have been made synthetically. If one could make fluoroarginine, for example, this might be an anti-cancer compound, containing fluorine but not a nucleoside. We are trying to prepare this compound.

Taves: How much fluorine were you picking up in the urine after giving F_3TDR?

Heidelberger: We were giving 0.3 mg/day of F_3TDR, which is roughly 1 g/day. About 200 mg of that is converted to carboxyuracil, so there was roughly 10 mg/day of inorganic fluoride in the urine.

References

BURCHENAL, J. H., H. H. ADAMS, N. S. NEWELL and J. J. FOX (1966) *Cancer Res.* **26**, 370.
DEXTER, D. L., W. H. WOLBERG, F. J. ANSFIELD, L. H. HELSON and C. HEIDELBERGER (1972) *Cancer Res.* **32**, 247-253.
HENDLER, S. S., D. H. SHANNAHOFF, R. A. SANCHEZ and L. E. ORGEL (1971) *Abstr. Am. Chem. Soc. Conf.* Washington.
HUBERMAN, E. and C. HEIDELBERGER (1972) *Mutation Res.* **14**, 130–132.
MORTON, G. O., J. E. LANCASTER, G. E. VAN LEAR, W. FULMOR and W. E. MEYER (1969) *J. Am. Chem. Soc.* **91**, 1535.

Fluoroamino acids and protein synthesis

L. FOWDEN

Department of Botany and Microbiology, University College, London

As a plant biochemist seeking to extend information about the amino acid complement of plants, I have frequently wondered whether fluoroamino acids will ever be established as plant products. At present there appears to be no evidence in support of this idea, but it seems reasonable to believe that plants like *Dichapetalum cymosum* and *Acacia georginae*, which produce fluoroacetate, may also possess biogenetic systems elaborating fluoroamino acids. Undoubtedly, fluoroacetate is extensively metabolized by seedlings of *A. georginae*, and L. H. Weinstein and co-workers (personal communication) have concluded that the seedlings possess an enzyme system which splits the C—F bond and liberates fluorine-free carbon fragments into intermediary metabolic pathways. In one experiment, ^{14}C, supplied as [2-^{14}C]fluoroacetate, was distributed among several fractions as follows: 3% evolved as CO_2, 11% present in lipids, 7% in cationic and neutral substances, and 79% in an anionic fraction that included the unchanged fluoroacetate. However, refined gas chromatographic-mass spectrometric analysis gave no indication of fluoro-fatty acids in the lipid fraction, and any evidence for fluoroamino acids was equally lacking.

Both *D. cymosum* and *A. georginae* contain unusual amino acids. Some are still unidentified, but those characterized do not contain fluorine. The presence of *N*-methylalanine (Eloff and Grobbelaar 1967) and *N*-methylserine (Eloff and Grobbelaar 1969) in *Dichapetalum* shoots is so far unique among plants, and their synthesis may depend directly on availability of fluoroacetate. Possibly fluoroacetate acts as an *N*-methylating agent, for J. N. Eloff (personal communication) has shown that alanine and serine are direct precursors of the *N*-methyl derivatives; he has also observed that the concentration of the *N*-substituted amino acids decreases, as does the content of fluoroacetate, as leaf and stem material ages. Synthesis then may involve an elimination of a molecule of HF with the initial formation of *N*-carboxymethyl amino acids;

subsequent decarboxylation would yield the N-methyl derivatives. Alternatively, HF elimination and decarboxylation may occur simultaneously, leading directly to N-methylalanine and N-methylserine. A synthetic mechanism of this type shows some analogy with Sanger's method for the N-dinitrophenylation of amino acids using fluorodinitrobenzene.

The lack of naturally occurring fluoroamino acids necessarily means that chemically synthesized compounds have been employed in all studies concerned with the influence of fluoro derivatives on protein synthesizing mechanisms and, indirectly, on the many cellular processes that depend on the delicate regulation of protein synthesis. The substitution of a fluorine atom for one of hydrogen in an amino acid molecule frequently causes only a slight modification in the properties of the amino acid (see Richmond 1962); this is especially so if the amino acid molecule is fairly large, for then the presence of the slightly larger fluorine atom, in place of hydrogen, produces only a very small change in the relative sizes of the normal and substituted molecules, for example, p-fluorophenylalanine compared with phenylalanine. However, substitution of fluorine may influence the ionization characteristics of a molecule because the fluorine atom tends to withdraw electrons from the remainder of the molecule. The number of fluorine substitutions, and their positions relative to the ionizable amino and carboxyl groups, are reflected in altered pK values (see Table 1).

TABLE 1

Comparison of pK_a values of fluorinated and natural amino acids

Compound	pK_a value
Leucine	2.34*
5,5,5-Trifluoroleucine	2.05†
5,5,5,5′,5′,5′-Hexafluoroleucine	1.81*
Valine	2.29*
4,4,4-Trifluorovaline	1.54*
4,4,4,4′,4′,4′-Hexafluorovaline	1.21*
Methionine	2.16*
S-Trifluoromethylhomocysteine (Trifluoromethionine)	2.05*

*   Lazar and Sheppard 1968.
†   Walborsky and Lang 1956.

pK_a values were determined by the method of Walborsky and Lang 1956.

A similar degree of fluorine substitution causes a larger change in the pK value when substitution is at a position nearer to the α-amino and α-carboxyl groups (cf. trifluorovaline with trifluoroleucine, or hexafluorovaline with hexafluoroleucine). Whereas trifluoroleucine can behave as an effective analogue, replacing leucine under a number of conditions in the metabolism of *Escherichia coli* (Rennart and Anker 1963; Fenster and Anker 1969), trifluorovaline, hexa-fluorovaline and hexafluoroleucine have little or no effect on the growth or metabolism of the bacterium. Steric considerations alone indicate that the latter three fluoroamino acids do not differ markedly from their natural counterparts, but the significant changes in pK values prevent them from being taken up by the bacterial cells and transported to the site of protein synthesis and/or being accepted by the aminoacyl activating systems (Lazar and Sheppard 1968).

Fluorine-containing analogues of amino acids may affect protein synthesis in a variety of ways, other than by direct action as a competitive substrate in amino acid activation and incorporation into peptide chains. Antagonism of any process concerned with amino acid biosynthesis will ultimately restrict protein synthesis. For example, an analogue may function as a metabolic inhibitor because it is accepted as a substrate by a specific bacterial permease system, thereby limiting the uptake of the normal protein amino acid; if such competition occurs in a strain auxotrophic for the normal amino acid, an inhibition of growth must ensue because insufficient of the particular amino acid is then available to sustain the normal rate of protein synthesis. In a similar way, the regulatory mechanisms of amino acid biosynthesis may be disturbed by fluorinated analogues. The analogue may resemble the normal amino acid sufficiently well to allow it to mimic the amino acid in its role as a feed-back inhibitor or repressor of key enzymes of the biosynthetic pathway. As with permeation, the end result is a reduction in the concentration of the normal amino acid within the cells. A final way in which fluoroamino acids may exhibit metabolic antagonism is by specifically affecting hydroxylating systems implicated in the formation of tyrosine and hydroxyproline.

FLUOROAMINO ACIDS IN RELATION TO HYDROXYLATING AND DEGRADATIVE ENZYME SYSTEMS

Normally, the C—F bond is particularly stable in animal systems, but it is readily broken when suitable compounds form the substrates for particular amino acid hydroxylating enzymes. The hydroxylations encountered in the enzyme-catalysed formations of tyrosine from phenylalanine or of 4-hydroxy-proline from proline involve direct oxidative attack at the *p*- and C-4 positions of the benzene and pyrrolidine ring systems, respectively. If a fluorine substituent is attached at these carbon atoms, as in *p*-fluorophenylalanine and 4-

fluoroproline, the molecules still form substrates for the hydroxylating enzymes, tyrosine or hydroxyproline formation now being associated with the elimination of a fluoride ion. Kaufman (1961) demonstrated that phenylalanine hydroxylase from rat or sheep liver utilized *p*-fluorophenylalanine at about one sixth the rate of phenylalanine itself. An *in vivo* conversion of [3-^{14}C]-*p*-fluorophenyl-alanine into [^{14}C]tyrosine was shown to occur in rats without any ^{14}C-label becoming incorporated into phenylalanine; the [^{14}C]tyrosine formed was incorporated into both plasma and pancreatic proteins (Dolan and Godin 1966). Initial studies with *cis*- and *trans*-[4-^{3}H]-L-prolines revealed that the hydroxylating system converting proline into *trans*-4-hydroxyproline in chick embryos effected the displacement of a *trans*-proton at C-4 and completely retained the congfiuration about this ring carbon atom. This displacement mechanism is then comparable with other systems that directly use molecular oxygen to introduce the hydroxyl function (Fujita *et al.* 1964). The ability of the hydroxylase to use *trans*-4-fluoro-L-proline, but not the corresponding *cis*-compound, as a substrate for the production of *trans*-4-hydroxy-L-proline (Gottlieb *et al.* 1965) was then not unexpected.

    m-Fluorophenylalanine also acts as a substrate for phenylalanine hydroxylase, but in this instance the fluorine atom is retained and *m*-fluorotyrosine is pro-duced. When *m*-fluorophenylalanine, or *m*-fluorotyrosine, is given to rats, the animals quickly exhibit convulsions followed frequently by death (Koe and Weissman 1967). The toxicity associated with these two fluoroamino acids is attributed to the conversion of *m*-fluorotyrosine into fluoroacetate by enzymes of the normal degradative pathway for tyrosine, for *m*-fluorotyrosine intake results in a characteristic accumulation of citric acid in the brain and kidneys of the animal (Weissman and Koe 1967). As yet, there is no convincing explana-tion of why *m*-fluorotyrosine is more active than fluoroacetate in inducing convulsions, but differential permeability of cell membranes to the two com-pounds may be important. Catabolism of *p*-fluorophenylalanine by *E. coli* has been reported to proceed with certainty only as far as *p*-fluorophenylacetic acid, *p*-fluorophenylpyruvic acid being an intermediate product; *p*-fluorophe-nylacetic acid showed no inhibitory effect upon the growth of the bacterium (Bowman and Mallette 1966). Another deaminating system for phenylalanine, that is phenylalanine-ammonia lyase from either wheat seedlings (Young and Neish 1966) or the fungus *Ustilago hordei* (Subba Rao, Moore and Towers 1967), also utilizes either *m*- or *p*-fluorophenylalanine as a substrate. Fluorinated flavonoid pigments or fluorinated lignin could conceivably result after further metabolism of the initial products, *m*- or *p*-fluorocinnamic acids.

THE EFFECTS OF FLUORO DERIVATIVES ON THE UPTAKE AND BIOSYNTHESIS OF AMINO ACIDS

Permease studies

Permease enzymes control the uptake of amino acids across cell membranes and lead to their accumulation in the cytoplasm of cells. The systems so far studied in detail have mainly involved bacteria or yeasts. Most permeases function specifically for the uptake of a single amino acid, but certain others, such as the aromatic amino acid permease, may be responsible for the uptake of a structurally related group of compounds. Fluoro derivatives of certain protein amino acids also act as permease substrates and, if present in a culture medium, reduce the uptake of the normal amino acid in a competitive manner. p-Fluorophenylalanine was initially shown to restrict the entry of [^{14}C]phenyl-alanine into the internal metabolic pools of *E. coli* and *Candida utilis* (Kempner and Cowie 1960). My colleagues, H. Tristram and G. Whitehurst (unpublished), have since shown that the fluoro derivative inhibits both phenylalanine and tyrosine uptake by *E. coli* C4 to an exactly comparable extent, and all their experiments suggest the existence of only a general aromatic permease in this strain. In contrast, *Salmonella typhimurium* (Ames 1964) appears to possess not only a general aromatic permease (capable of effecting the simultaneous and mutually competitive uptake of phenylalanine, tyrosine and tryptophan), but also three individual permeases (specific for phenylalanine, tyrosine and tryptophan, respectively). Competition experiments involving p-fluorophenyl-alanine suggested that the general aromatic permease was responsible for almost 90% of the uptake of the fluoro analogue, whereas the specific phenyl-alanine permease transported the remainder.

In general, the behaviour of a fluoroamino acid as an inhibitor of cell growth and protein synthesis implies that the compound forms a substrate for some permease system, although resistance, shown by certain bacterial mutants, to the antagonistic effects of fluoroamino acids may be explained by invoking an altered permease system; for example, the mutant strain may produce a per-mease able to discriminate partially or wholly against the fluoro derivative.

Studies on the control of amino acid biosynthesis

It is now well established that amino acids, representing end-products of biosynthetic pathways, may prevent their own oversynthesis by (*i*) acting as an inhibitor of the enzyme catalysing the first specific step in the biosynthetic sequence and *(ii)* inhibiting or repressing the synthesis of all enzymes concerned in the pathway. In many instances fluoro derivatives have been shown to mimic either one or both of these actions of the parent amino acid, with the result that the organism produces an insufficient amount of the amino acid to maintain normal protein synthesis. Again, mutation in simple organisms may serve to alleviate or completely overcome such false feed-back action. Resistant mutants of this type are now attracting considerable attention; an understanding of the basis of resistance may help to elucidate further the complex genetic basis of metabolic control.

Shive and his co-workers, who have made many contributions to the subject of analogue behaviour, have provided a good example of the way in which fluoro derivatives may affect the synthesis of aromatic amino acids. Working with preparations of 3-deoxy-D-*arabino*-heptulosonic acid 7-phosphate (DAHP) synthetase (the enzyme catalysing the first reaction involved in the shikimate aromatic pathway) from *E. coli*, they showed that *o*-, *m*-, and *p*-fluorophenyl-alanines can imitate the action of phenylalanine as a *feed-back* inhibitor of the phenylalanine-sensitive isoenzyme. The activity of the enzyme was inhibited by 50% when 0.02-mM phenylalanine was added to reaction mixtures. A similar degree of inhibition was produced by 0.01 mM-*o*-fluorophenylalanine, or 0.03 mM-*m*-fluorophenylalanine, or 0.2 mM-*p*-fluorophenylalanine. The activity of the isoenzymic form of DAHP synthetase that responds to tyrosine was almost unaffected by the presence of either phenylalanine itself or its various fluoro derivatives; however, a 50% inhibition of this enzymic activity was recorded in the presence of either 0.02 mM-tyrosine or 0.08 mM-3-fluorotyrosine (Smith *et al.* 1964).

Turning next to the topic of end-product repression of enzyme formation, one can find a good example in the related biosyntheses of valine, isoleucine and leucine. The formation of enzymes catalysing the initial reactions leading to valine and isoleucine is subject to multivalent repression in *E. coli* strain W and *Salmonella typhimurium*; that is isoleucine, valine and leucine (and panto-thenic acid) must all be included in the medium in which the organism is growing before the full repressive effect on enzyme synthesis is recorded. The formation of those enzymes functioning in the later steps that yield leucine is regulated only by leucine. 5,5,5-Trifluoroleucine can replace leucine in its controlling role in respect of all these enzymes in *E. coli*, but as shown in Table 2 (modified from Freundlich and Trela 1969), the presence of the fluoro analogue in the growth medium of *S. typhimurium* had no repressive effect on β-isopropylmalate dehydrogenase activity, although it repressed the activity of threonine deaminase, dihydroxy acid dehydrase and acetohydroxy acid syn-thetase (three enzymes of the valine-isoleucine pathway). This differential

TABLE 2

Effect of 5,5,5-trifluoroleucine (FL) on the specific activities of individual enzymes involved in the biosynthesis of isoleucine, valine and leucine in *Salmonella typhimurium*

Growth limitation in medium*	Specific activity of enzyme (μmol product/mg protein/hr)			
	Threonine deaminase	Dihydroxy acid dehydrase	Acetohydroxy acid synthetase	β-IPM† dehydrogenase
None	7.0	2.8	0.8	1.0
Leucine (limited to 15 μg/ml)	35.2	15.6	30.0	11.5
Leucine (15 μg/ml) + FL (100 μg/ml)	12.4	4.8	4.1	11.8
Leucine (15 μg/ml) + FL (500 μg/ml)	11.8	4.6	4.4	13.0

* The normal medium contained L-isoleucine (50 μg/ml), L-valine (100 μg/ml) and L-leucine 50 μg/ml).
† β-IPM: β-isopropylmalate.

behaviour of fluoroleucine on enzyme repression in the two organisms was shown to be a real effect and not merely a reflection of competition between the analogue and the normal branched chain amino acids for entry into the bacterial cells. The same authors claim that leucine must be activated by leucyl-tRNA synthetase before it can participate in repression of the valine-isoleucine and leucine biosynthetic enzymes. [A general requirement for activation before an analogue can function in repression is not universally established because several tyrosine analogues cause a marked repression of tyrosine-sensitive DAHP synthetase, yet are not activated by tyrosyl-tRNA synthetase (Freundlich and Trela 1969; Ravel, White and Shive 1965).] Activation and transfer of the aminoacyl moiety to tRNA may utilize several species of tRNA^{Leu}, and different species may be involved in the repression of the groups of enzymes implicated in the biosynthesis of valine and isoleucine, and of leucine, respectively. Trifluoroleucine then may become attached to the tRNA^{Leu} species involved in repression of the valine-isoleucine enzymes, but the analogue may be unable to bind to the particular species implicated in repression of β-isopropylmalate dehydrogenase. This explanation is supported by the observation that trifluoroleucine could protect only 70% of the leucine-specific tRNA against periodate oxidation. If the analogue does exhibit such differential binding to tRNA^{Leu} species, then it would be expected to show ambivalent effects on other metabolic processes in which tRNA^{Leu} may be involved in *S. typhimurium*. In another publication, Trela and Freundlich (1969) showed that trifluoroleucine allowed continued protein synthesis in a leucine auxotroph deprived of leucine, although RNA synthesis was completely blocked under these conditions. In contrast, trifluoroleucine allowed both protein and RNA synthesis to continue in a leucine auxotroph of *E. coli* strain W.

FLUOROAMINO ACIDS AND PROTEIN SYNTHESIS

General considerations

There are numerous reports of situations in which fluorinated amino acids have been incorporated into protein molecules, where they specifically and stoichiometrically replace residues of the corresponding protein amino acids. Substitutions of this type may lead to protein molecules with impaired function; for example, as enzymes or as structural (membraneous) components of cells or organisms. The extent of replacement is governed by the ability of the analogue to compete with the normal amino acid during activation and transfer to the specific tRNA; in such competitive situations high analogue concentrations naturally favour the incorporation of analogue residues. If the analogue then also behaves as a false feed-back inhibitor or end-product repressor, the resulting restriction of endogenous production of the normal amino acid will further enhance the degree of analogue incorporation into protein molecules.

The production of altered protein molecules, resulting from an incorporation of fluoroamino acids, may cause, either directly or indirectly, phenomena of the following types:

(1) A reduced synthesis (or activity) of a number of enzymes or their cellular products: examples include an inhibition of RNA polymerase synthesis in *E. coli* by *p*-fluorophenylalanine (Hardy and Binkley 1967), and a reduced induction of phenylalanine-ammonia lyase in excised embryonic axes of beans (*Phaseolus vulgaris*) treated with *p*-fluorophenylalanine (Walton 1968). *p*-Fluorophenylalanine also retarded the induction of β-galactosidase activity in *E. coli* cells, although the final concentration of β-galactosidase reached after 3 hours was higher than that of control cultures (Bowman *et al.* 1964). The initial effect of the fluoro analogue probably could be attributed to a retarded formation of an active permease for β-galactosides. *p*-Fluorophenylalanine (1 μmol/ml) when present in the culture medium of *E. coli* strain KB, caused a marked depression in glycogen accumulated by the bacterial cells. In contrast, glycogen production was unaffected in a fluorophenylalanine resistant strain (PFP-10), which does not incorporate the analogue into its protein (Cattenćo *et al.* 1966).

(2) An alteration in the shape and size of cells and blocking of cell division. When *E. coli* was grown in the presence of monofluorotryptophans, the cells

developed odd shapes, had a generally swollen appearance and exhibited irregular contrast under the microscope (Browne, Kenyon and Hegeman 1970): the DL-5-fluoro derivative (at 10^{-4} M) was more potent in inducing abnormal protein synthesis and cellular abnormalities than a similar concentration of either racemic 6- or 4-fluorotryptophan. Very similar cytological changes are induced in *E. coli* cultures subjected to trifluoroleucine (Rennart and Anker 1963). Cell division in wild-type cultures exposed to fluorotryptophans was quickly terminated, growth ceasing after some 40–50 % of the tryptophan of the bacterial protein had been replaced by fluorinated residues. *p*-Fluorophenyl-alanine similarly restricts cell division in *E. coli* (Hardy and Binkley 1967); it also blocks nuclear replication and division in HeLa cells (Mueller and Kajiwara 1966) and stops multiplication in *Paramecium aurelia* and *Tetrahymena pyriformis* (Rasmussen 1967). In all these instances, the *p*-fluorophenylalanine appeared not to inhibit the rapid early synthesis of DNA within the various cells; however, a special protein essential for the completion of cell division seems to be synthesized during this early phase of DNA synthesis and, if formed in the presence of *p*-fluorophenylalanine, this protein is presumably mal-functional.

(3) A delayed production of floral primordia in *Xanthium* is caused by *p*-fluorophenylalanine. Flowering in this genus is induced by subjecting the plant to conditions of short daylength. Induction is thought to depend upon the production of specific enzymes necessary to trigger the flowering process, which is consistent with the observed incorporation of *p*-fluorophenylalanine into the plant's proteins (Miller and Ross 1966). *p*-Fluorophenylalanine also interferes with the action of interferon. If this antiviral protein is to be effective, then protein and RNA synthesis must continue normally; thus the antagonistic action of *p*-fluorophenylalanine can be attributed once again to the associated synthesis of malfunctional protein molecules (Friedman and Sonnabend 1964).

(4) Numerous workers have reported that *p*-fluorophenylalanine brings about changes in the chromosomes of cells. The initiation of chromosomal replication in *E. coli* probably requires systems present in the cytoplasmic membrane of the cell. *p*-Fluorophenylalanine prevents replication (Friedman and Sonnabend 1964); it also induces the production of electron-dense meta-chromatic granules within the membranes of cells, and a marked change in their amino acid composition (Brostrom and Binkley 1969). This fluoroamino acid has been applied as a tool for obtaining haploids from diploid strains in *Aspergillus niger* and *A. nidulans* (Lhoas 1961; McCully and Forbes 1965), and in the yeast *Schizosaccharomyces pombe* (Gutz 1966), where linked genes segregated in parental combinations. A similar reduction in chromosome num-bers from 32 to between 8 and 24 is effected in a *Ribes* hybrid [*R. grossularia*

(gooseberry) × *R. nigrum* (blackcurrant)] by *p*-fluorophenylalanine (Knight, Hamilton and Keep 1963).

In the few instances in which the distribution of a fluoro analogue within a protein molecule has been studied, replacement of the normal constituent would seem to be random; that is, there is an equal likelihood of replacement of any residue, irrespective of its position in the molecule. A good example of this type of finding was provided by Richmond (1963) who made a tryptic digest of alkaline phosphatase, produced by *E. coli* under conditions of direct competition between phenylalanine and *p*-fluorophenylalanine, and established that the fluoro analogue was present in each of six peptides normally containing phenylalanine in amounts commensurate with their usual phenylalanine content. Evidence accumulating since Richmond's work has indicated that multiple (or degenerate) forms of tRNA^{Phe} may exist in many organisms. This is true for *E. coli*, where tRNA degeneracy may be invoked to explain a situation in which significantly more aminoacyl-tRNA is formed from phenylalanine than from *p*-fluorophenylalanine by cell-free extracts (Fangman and Niedhardt 1964), although the initial rates of transfer of the amino acids to tRNA were almost identical (Dunn and Leach 1967). It appears that phenylalanine can bind to tRNA species containing either the AAA or GAA anticodons, whereas *p*-fluorophenylalanine can only charge the AAA-containing tRNA (see Fig. 1).

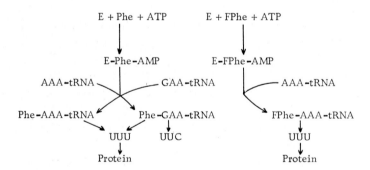

F ig. 1. Scheme illustrating the incorporation of phenylalanine (Phe) and *p*-fluorophenylalanine (FPhe) into protein as stimulated by AAA and GAA anticodons (after Dunn and Leach 1967).

These experiments support the idea that the specificity associated with amino acid incorporation into protein is more strictly determined at the stage of the transfer of the aminoacyl group to tRNA than in the initial step forming an aminoacyladenylate complex [see earlier reports by Loftfield 1963; Berg *et al.* 1961]. The *in vivo* consequences of tRNA degeneracy of this type are not evident, but the equivalence of phenylalanine residues may repay further examination in other *in vivo* situations.

Quantitative aspects of activation and incorporation into protein of fluoroamino acids

The ability of amino acids to act as substrates for aminoacyl-tRNA synthetase enzymes may be determined by several experimental techniques. The most reliable method measures the formation of aminoacyl-tRNA molecules by the combined reactions (*1*) and (*2*) (see below); because radioisotopically labelled amino acid substrates are indispensable for this type of assay, it has not found a general use in studies concerned with fluoro analogues, where labelled derivatives may be difficult to obtain. Alternative assay procedures are based on the ability of amino acid to stimulate (*a*) the exchange of ^{32}P between pyrophosphate and ATP, or (*b*) the formation of amino acid hydroxamate. Essentially, these two procedures measure only the ability of the amino acid substrate to participate in reaction (*1*), although synthetases are generally considered to show a greater discrimination between substrates in reaction (*2*) (see p. 150). Caution then must be exercised when evaluating data obtained from the different assay techniques.

(*1*) Amino acid + ATP + enzyme $\rightleftharpoons$ [Aminoacyladenylate-enzyme] + PP_i

(*2*) [Aminoacyladenylate-enzyme] + tRNA $\rightarrow$ Aminoacyl-tRNA + AMP + enzyme

Several reports generally agree that *p*-fluorophenylalanine is activated almost as readily as phenylalanine itself. Hydroxamate assay has been used in association with phenylalanyl-tRNA synthetase preparations from *E. coli*. Conway, Lansford and Shive (1962) found that *p*-fluorophenylalanine was activated at 95% of the rate determined for phenylalanine. Using a wild-type strain of *E. coli*, Fangman and Niedhardt (1964) noted that the *p*-fluoro analogue was activated more readily than phenylalanine when both were used at enzyme-saturating concentrations, but the K_m values (4×10^{-4} M and 4.4×10^{-4} M, respectively) indicated the slightly higher affinity of the enzyme for its normal substrate: *o*- and *m*-fluoro derivatives were included in this study (see Table 3). A mutant form (PFP-10), thought to result from a single genetic lesion in a structural gene, grew quite normally in the presence of *p*-fluorophenylalanine. The alteration in specificity was peculiarly discriminating, for *o*- and *m*-fluorophenyl-alanines were still well used as substrates. *m*-Fluorophenylalanine acted as a substrate (at 52% of the rate determined for phenylalanine as measured by ATP-$^{32}PP_i$ exchange) for the phenylalanyl-tRNA synthetase obtained from

TABLE 3

The substrate specificity of phenylalanyl-tRNA synthetase from wild-type and *p*-fluorophenyl-alanine-resistant strains of *Escherichia coli*

Amino acid substrate†	Hydroxamate formed*		
	Wild-type (A)	Resistant mutant (B)	Ratio B/A × 100
Phenylalanine	0.086	0.180	209
p-Fluorophenylalanine	0.170	0.006	3.5
m-Fluorophenylalanine	0.170	0.141	83
o-Fluorophenylalanine	0.061	0.072	118

* Expressed as μmoles formed per mg protein per hr.
† Each amino acid present in incubation mixture at a concentration of 3 μmol/ml.

horse chestnut (*Aesculus hippocastanum*) seed. With this enzyme, many types of structural alteration within the phenylalanine molecule were compatible with retention of substrate activity (Anderson and Fowden 1970).

Arnstein and Richmond (1964) examined the ability of cell-free preparations from rabbit reticulocytes to catalyse the transfer of phenylalanine and *p*-fluorophenylalanine to tRNA. When the amino acids were added singly to the reaction system, these workers observed that the ratio of phenylalanine- to *p*-fluorophenylalanine-transfer was about 1.3, a value in excellent agreement with the corresponding ratio of 1.4 determined for the amounts of the two amino acids incorporated into protein. However, competition experiments indicated that the active synthetase possessed a much higher affinity for phenylalanine than for *p*-fluorophenylalanine. Ezekiel (1965) reported experiments in which intracellular transfer of these two amino acids to tRNA^{Phe} was measured in four strains of *E. coli*; strain KB and a derived *p*-fluorophenyl-alanine-resistant mutant PFP-10 could synthesize phenylalanine, whereas strain C2 and its associated mutant PFP-10-Phe were auxotrophic for phenylalanine. Some of his data obtained with the C2 strain (see Table 4) confirms the view that phenylalanine is transferred to tRNA somewhat more effectively than

TABLE 4

The percentages of intracellular phenylalanine-specific tRNA molecules charged with amino acid after growth of various strains of *Escherichia coli*

Growth medium	%tRNA charged with amino acid			
	KB	PFP-10	C2	PFP-10-Phe
Minimal	57	50	1	5
Minimal + *p*-fluorophenylalanine	87	10	54	8
Minimal + phenylalanine	—	—	76	83

p-fluorophenylalanine. Little of the fluoro analogue became attached to tRNA in either *p*-fluorophenylalanine-resistant strain. Ezekiel noted a high correlation between the ability of particular strains to charge tRNA with amino acid and their capacity to sustain RNA synthesis; little RNA was synthesized when *p*-fluorophenylalanine was present in the growth medium of PFP-10 strains.

In an analogous way, 3-fluorotyrosine may act as an alternative substrate for preparations of tyrosyl-tRNA synthetase. At high substrate concentrations using the hydroxamate assay method, Schweet and Allen (1958) found that 3-fluorotyrosine was activated by a hog pancreatic enzyme at a rate approaching that measured for tyrosine itself; when ATP–^{32}PP$_i$ exchange was employed to determine activation, the reaction rate with 3-fluorotyrosine never exceeded 50% of that with the normal substrate. With a plant tyrosyl-tRNA synthetase (from *Phaseolus aureus* seed), the corresponding figure determined by ATP–^{32}PP$_i$ exchange was about 30% (Smith and Fowden 1968).

Prolyl-tRNA synthetase represents another widely studied activating enzyme that exhibits a fairly low imino-acid substrate specificity (see for example, Peterson and Fowden 1965; Papas and Mehler 1970). Using a prolyl enzyme purified from *E. coli*, Papas and Mehler (1970) reported that V_{max} for *trans*-4-fluoroproline was 66% of the corresponding value for proline. In another study with the same bacterium, *trans*-4-fluoroproline-^{3}H was shown to enter protein molecules much more rapidly than the diastereoisomeric *cis*-form. A similar finding was recorded when incorporation of the two fluoroprolines into protein of guinea-pig granuloma was followed. In experiments where ^{3}H-labelled *trans*-isomer was supplied, 20–30% of the radioactivity finally present in protein molecules was in the form of 4-hydroxyproline: negligible amounts of labelled hydroxyproline were formed from *cis*-4-fluoroproline-^{3}H (an expected result in view of the known stereospecificity of the oxidative attack at the C-4 atom, see p. 144) (Gottlieb *et al.* 1965). However, *cis*-4-fluoroproline strongly inhibited the incorporation of ^{14}C-proline into embryonic cartilage, and also restricted the hydroxylation of those proline residues that had become incorporated into protocollagen (Takeuchi and Prockop 1969). In these respects, *cis*-fluoroproline was quite as effective as the potent proline homologue, azetidine-2-carboxylic acid.

CONCLUSION

It is clear that fluoroamino acids may participate in many metabolic processes almost as readily as the related unsubstituted compounds. Fluoroamino acids then have been used to define more exactly the degree of substrate specificity associated with several types of enzyme, and to acquire a fuller understanding of genetic control mechanisms regulating amino acid biosynthetic pathways. Future studies could reveal that particular fluoroamino acids are natural products of plants—then some of the concepts of metabolic interaction discussed earlier may acquire an *in vivo* physiological significance.

SUMMARY

The possibility that fluoroamino acids occur naturally, especially in fluoro-acetate-containing plants, is considered first. However, in the absence of positive evidence, discussion of the ways in which fluoro derivatives may interact with amino acid- and protein-synthesizing systems cites investigations using chemically synthesized fluoro compounds. Introduction of one or more F atoms in place of protons in amino acid molecules causes only slight changes in molecular size and shape, that is fluoroamino acids are almost isosteric with their parent protein amino acid, but substitution of F for H can produce greater changes in pK values. These two factors are important in determining whether a fluoroamino acid forms a good analogue for metabolic investigations. Examples of the analogue function of fluoroamino acids are discussed at the level of (*i*) permease action controlling uptake of amino acids; (*ii*) the regulation of amino acid biosynthesis; and (*iii*) protein synthesis, especially in relation to the activities of aminoacyl-tRNA synthetases. In each of these ways fluoro-amino acids may directly influence the process of protein synthesis and in-directly cause anomalies in cellular differentiation and development.

References

AMES, G. F. (1964) *Arch. Biochem. & Biophys.* **104**, 1.
ANDERSON, J. W. and L. FOWDEN (1970) *Biochem. J.* **119**, 677.
ARNSTEIN, H. R. V. and M. H. RICHMOND (1964) *Biochem. J.* **91**, 340.
BERG, P., F. H. BERGMANN, E. J. OFENGAND and M. DIECKMANN (1961) *J. Biol. Chem.* **236**, 1726.
BOWMAN, W. H. and M. F. MALLETTE (1966) *Arch. Biochem. & Biophys.* **117**, 563.
BOWMAN, W. H., I. S. PALMER, C. O. CLAGETT and M. F. MALLETTE (1964) *Arch. Biochem. & Biophys.* **108**, 314.
BROSTROM, M. A. and S. B. BINKLEY (1969) *J. Bacteriol.* **98**, 1263.
BROWNE, D. T., G. L. KENYON and G. D. HEGEMAN (1970) *Biochem. & Biophys. Res. Commun.* **39**, 13.
CATTENĆO, J., N. SIGAL, A. FAVARD and I. H. SEGAL (1966) *Bull. Soc. Chim. Biol.* **48**, 441.
CONWAY, T. W., E. M. LANSFORD and W. SHIVE (1962) *J. Biol. Chem.* **237**, 2850.

DOLAN, G. and C. GODIN (1966) *Biochemistry* **5**, 922.
DUNN, T. F. and F. R. LEACH (1967) *J. Biol. Chem.* **242**, 2693.
ELOFF, J. N. and N. GROBBELAAR (1967) *J. S. Afr. Chem. Inst.* **20**, 190.
ELOFF, J. N. and N. GROBBELAAR (1969) *Phytochemistry* **8**, 2201.
EZEKIEL, D. H. (1965) *Biochim. & Biophys. Acta* **95**, 48.
FANGMAN, W. L. and F. C. NIEDHARDT (1964) *J. Biol. Chem.* **239**, 1839.
FENSTER, E. D. and H. S. ANKER (1969) *Biochemistry* **8**, 269.
FREUNDLICH, M. and J. M. TRELA (1969) *J. Bacteriol.* **99**, 101.
FRIEDMAN, R. M. and J. A. SONNABEND (1964) *Nature (Lond.)* **203**, 366.
FUJITA, Y., A. A. GOTTLIEB, S. PETERKOFSKY, S. UDENFRIEND and B. WITKOP (1964) *J. Am. Chem. Soc.* **86**, 4709.
GOTTLIEB, A. A., Y. FUJITA, S. UDENFRIEND and B. WITKOP (1965) *Biochemistry* **4**, 2507.
GUTZ, H. (1966) *J. Bacteriol.* **92**, 1567.
HARDY, C. and S. B. BINKLEY (1967) *Biochemistry* **6**, 1892.
KAUFMAN, S. (1961) *Biochim. & Biophys. Acta* **51**, 619.
KEMPNER, E. S. and D. B. COWIE (1960) *Biochim. & Biophys. Acta* **42**, 401.
KNIGHT, R. L., A. P. HAMILTON and E. KEEP (1963) *Nature (Lond.)* **200**, 1341.
KOE, B. K. and A. WEISSMAN (1967) *J. Pharmacol. & Exp. Ther.* **157**, 565.
LAZAR, J. and W. A. SHEPPARD (1968) *J. Med. Chem.* **11**, 138.
LHOAS, P. (1961) *Nature (Lond.)* **190**, 744.
LOFTFIELD, R. B. (1963) *Biochem. J.* **89**, 82.
McCULLY, K. S. and E. FORBES (1965) *Genet. Res.* **6**, 352.
MILLER, J. and C. ROSS (1966) *Plant Physiol.* **41**, 1185.
MUELLER, G. C. and K. KAJIWARA (1966) *Biochim. & Biophys. Acta* **119**, 557.
PAPAS, T. S. and A. H. MEHLER (1970) *J. Biol. Chem.* **245**, 1588.
PETERSON, P. J. and L. FOWDEN (1965) *Biochem. J.* **97**, 112.
RASMUSSEN, L. (1967) *Exp. Cell Res.* **45**, 501.
RAVEL, J. M., M. N. WHITE and W. SHIVE (1965) *Biochem. & Biophys. Res. Commun.* **20**, 352.
RENNART, O. M. and H. S. ANKER (1963) *Biochemistry* **2**, 471.
RICHMOND, M. H. (1962) *Bacteriol. Rev.* **26**, 398.
RICHMOND, M. H. (1963) *J. Mol. Biol.* **6**, 284.
SCHWEET, R. S. and E. H. ALLEN (1958) *J. Biol. Chem.* **233**, 1104.
SMITH, I. K. and L. FOWDEN (1968) *Phytochemistry* **7**, 1065.
SMITH, L. C., J. M. RAVEL, S. R. LAX and W. SHIVE (1964) *Arch. Biochem. & Biophys.* **105**, 424.
SUBBA RAO, P. V., K. MOORE and G. H. N. TOWERS (1967) *Can. J. Biochem.* **45**, 1863.
TAKEUCHI, T. and D. J. PROCKOP (1969) *Biochim. & Biophys. Acta* **175**, 142.
TRELA, J. M. and M. FREUNDLICH (1969) *J. Bacteriol.* **99**, 107.
WALBORSKY, H. M. and J. H. LANG (1956) *J. Am. chem. Soc.* **78**, 4314.
WALTON, D. C. (1968) *Plant Physiol.* **43**, 1120.
WEISSMAN, A. and B. K. KOE (1967) *J. Pharmacol. & Exp. Ther.* **155**, 135.
YOUNG, M. R. and A. C. NEISH (1966) *Phytochemistry* **5**, 1121.

Discussion

Peters: Professor Fowden thinks, and the rest of us would probably agree, that these fluoroamino acids will turn up somewhere in rather large quantities.

Fowden: Yes, I am sure some of the modern methods of analysis will ultimately provide such confirmation.

Gál: For the last three years, we have been trying to unravel the selective effect of irreversible inactivation of two mammalian enzymes by *p*-halophenylalanines. The two enzymes are phenylalanine 4-hydroxylase of the liver and tryptophan 5-hydroxylase of the central nervous system. Strangely enough, an enzyme system that also belongs to the same group, like tyrosine hydroxylase, is completely unaffected. Administration of *p*-fluorophenylalanine or *p*-chlorophenylalanine to animals will irreversibly inactivate these enzymes for 6–7 days. We have purified phenylalanine 4-hydroxylase to 100% homogeneity (Gál 1972), analysed the amino acids and demonstrated incorporation with ^{14}C-labelled *p*-halophenylalanines. Incorporation of 1 mole of that amino acid analogue into 1 mole of enzyme is essential for 100% inhibition. Cycloheximide and puromycin can prevent incorporation of these *p*-halophenylalanines. The site therefore is at the ribosomal site of protein synthesis. We also find that in the mammalian system there is a degeneracy of the codon, first described for *E. coli* by Dunn and Leach (1967), who demonstrated that a 3:1 ratio of the UUC to the UUU codon will minimize incorporation of *p*-fluorophenylalanine. Although *p*-fluorophenylalanine will incorporate into practically all the proteins that we have tested, only the two enzymes mentioned above are inactivated. Inactivation, we think, is tied to the incorporation of the analogue at an active centre, or at a site near the active centre. At present we are attempting, by peptide analysis and fingerprinting, to discover the exact position of *p*-halo analogues in the peptidic sequence.

Kent: Is there any demonstrable change in the immunological properties of these proteins containing *p*-fluorophenylalanine?

Goldman: Richmond (1960) has demonstrated some change in the immunological properties of the exopenicillinase synthesized by *Bacillus cereus* in the presence of *p*-fluorophenylalanine.

Gál: In phenylalanine 4-hydroxylase, for instance, no change in the circular dichroism and α-helix occurs as the result of *p*-halophenylalanine incorporation. There is a hint of change at 240 nm, and we might be able to detect the changes in the denatured enzyme more specifically. Similarly, there is no change in optical rotatory dispersion. However, there is some change in the mobility of the pure enzyme when it is inhibited on acrylamide gel.

Goldman: Professor Gál, you said that *p*-fluorophenylalanine is an irreversible inhibitor of phenylalanine hydroxylase from liver. How do you reconcile this with Kaufman's (1961) observation that it is also a substrate for the enzyme?

Gál: In an *in vitro* system, in the presence of mono-oxygenase, *p*-fluorophenylalanine undergoes a partial conversion to tyrosine more readily than either the chloro or bromo analogue. *In vivo* 50% of the fluorophenylalanine will show up as labelled tyrosine and the other 50% will incorporate and give a 50% inhibition of the enzyme. After administration of ^{14}C-labelled *p*-fluoro-phenylalanine, the ratio of radioactivity incorporated in protein as tyrosine to that incorporated as *p*-fluorophenylalanine is 0.31 to 0.49.

Armstrong: Professor Fowden, what is known about the properties of proteins which contain fluoroamino acid residues or about the fluoroamino acids you have synthesized, particularly as regards fluorophenylalanine and fluoroproline?

Fowden: Incorporation of an analogue affects the physical and chemical properties of proteins. In a protein exhibiting an enzymic role, such incorporation may lead to a marked loss of enzymic activity; alternatively, the effect may be minimal: much depends on the enzyme being considered.

Bergmann: In tissue culture experiments with *Acacia georginae* we never found a fluorinated amino acid if we used only *inorganic* fluoride (P.W. Preuss and R. Birkhahn, unpublished results). These tissue cultures synthesize fluoro-acetate from inorganic fluoride (Preuss, Calaisto and Weinstein 1970; Preuss, Birkhahn and Bergmann 1970). It might still be possible to obtain such amino acids if one gives organic fluoride, in the form of a suitable precursor for such synthesis.

I do not think one can expect fluoroacetic acid to be an alkylating or methylating agent for any amino acid. For methylation in this way, one should think of using fluoropyruvic acid.

Saunders: I agree, but we tried it and got no reaction.

Kun: Professor Bergmann, it was mentioned earlier (p. 91), that fluoro-pyruvate prepared by the decarboxylation of fluoro-oxaloacetate is not an alkylating agent. That form of fluoropyruvate is quite stable, so we suspect the fluoropyruvate prepared by other means contains impurities or that it may be another form.

Bergmann: Could it be the enol form?

Kun: No, the enol form is very unstable and goes to the keto form very quickly.

Cain: We have been interested in why not only *Acacia* but a lot of common species appear to have such high capacities for defluorinating added fluoro-acetic acid. We were originally interested in finding out whether Dr Goldman's defluorinating enzyme was responsible for this, but Dr Ward's comments about fluorine coming off in the lettuce and Professor Fowden's remarks about the putative HF-producing reaction leading also to the formation of an *N*-car-boxymethyl derivative, clearly indicate other possibilities. Professor Fowden, have you tested whether plant extracts can produce HF from a suitable amino acid and fluoroacetate, with the accompanying formation of an *N*-substituted compound?

Fowden: No, we have not performed that experiment. If fluoropyruvate is a natural product of some plants, the simplest route yielding a fluoroamino acid would involve a transamination reaction. However, when we used an active glutamate-pyruvate transaminase from plants, large amounts of alanine were formed from pyruvate, but no transamination product was observed when fluoropyruvate was used.

Kun: Fluoroalanine probably does not exist as the carboxylate ion. All carboxylic acid fluoro compounds with NH_2 on the neighbouring carbon atom defluorinate and decompose. Synthesis of the ester of fluoroalanine has been reported by Chinese workers (Yüan, Chang and Yeh 1959, 1960).

Fowden: If this happened, the final product formed after transamination of fluoropyruvate would be pyruvate.

Peters: Will fluoroalanine split off the fluoride?

Fowden: Do you mean β-fluoro-α-alanine?

Kun: Yes.

Heidelberger: But α-fluoro-β-alanine is stable.

Cain: Sir Rudolph, in the batch of *Acacia* homogenate that you fortified with pyruvate, ATP, magnesium ions and fluoride and then picked up small amounts of fluoroacetate (Peters, Shorthouse and Ward 1965), did you ever replace the ATP and the pyruvate with phosphoenylpyruvate? Perhaps HF adds on across that double bond. Is this chemically or energetically feasible?

Peters: We have not tried that. V. Desreux (personal communication) told me that when one makes fluorofumarate there is an internal splitting-off of HF. We could get no evidence of whether HF would add to a double bond.

Bergmann: Compounds like the β-fluoro-α-alanine are indeed unstable, but the compound has been made and described by Yüan, Chang and Yeh (1959) and later by Lettré and Wölcke (1967). 3-Fluoroglutamic acid is a very difficult compound to make but 4-fluoroglutamic acid is easily synthesized. It might also be possible to transaminate fluoropyruvic to β-fluoro-α-alanine.

Kun: We tried to do this chemically and enzymically and failed.

Bergmann: The C—F bond in fluoropyruvic acid is surprisingly reactive.

Kun: In the kind of fluoropyruvic acid we have it is quite stable.

Heidelberger: What is the structural difference between the non-reactive fluoropyruvic and the reactive fluoropyruvic acid?

Kun: I think it is a question of purity. The fluoropyruvic acid made by the usual methods may contain some trace compounds which perhaps facilitate the decomposition of fluoropyruvate in such a fashion that it will alkylate reactive groups. But if one prepares it from pure fluoro-oxaloacetic acid it is quite stable. We were surprised to observe this. It is not an alkylating agent at pH 7.4. It is a good substrate for lactic dehydrogenase.

Fowden: Is it a competitive inhibitor, rather than a substrate, of the glutamate-pyruvate transaminase?

Kun: I expect it is; I haven't studied that.

Kent: Chemical and electrolytic reduction of the purified hydrazines of pyruvic acid and esters of oxaloacetic acid leads almost invariably to loss of fluoride.

Kun: Yes, this is in agreement with our observations.

Taylor: Dr Kent and I found that 2-deoxy-2-fluoroglyceraldehyde also eliminates fluoride when treated with 2,4-dinitrophenyl hydrazine. The Schiff's base is formed first and then the fluorine is lost (Taylor and Kent 1956) and two molecules of the hydrazine react with one of the aldehyde.

Peters: Dr Saunders, didn't you have something to do with the fluorination reaction used initially by Sanger (1945)?

Saunders: Yes, we prepared 2,4-dinitrofluorobenzene and gave some to Dr Sanger. It was made initially as a vesicant.

References

DUNN, T. F. and F. R. LEACH (1967) *J. Biol. Chem.* **242**, 2693–2699.

GÁL, E. M. (1972) *Symposium on Biochemical Mechanism of Drug Activity in the CNS*, ed. R. Paoletti. Long Island: Raven Press. In press.

KAUFMAN, S. (1961) *Biochim. & Biophys. Acta* **51**, 619–621.

LETTRÉ, H. and U. WÖLCKE (1967) *Justus Liebigs Ann. Chem.* **708**, 75–85.

PETERS, R. A., M. SHORTHOUSE and P. F. V. WARD (1965) *Life Sci.* **4**, 749–752.

PREUSS, P. W., R. BIRKHAHN and E. D. BERGMANN (1970) *Israel J. Bot.* **19**, 609.

PREUSS, P. W., L. CALAISTO and L. H. WEINSTEIN (1970) *Experientia* **26**, 1059.

RICHMOND, M. H. (1960) *Biochem. J.* **77**, 112–121.

SANGER, F. (1945) *Biochem. J.* **39**, 507.

TAYLOR, N. F. and P. W. KENT (1956) *J. Chem. Soc.* 2150–2154.

YÜAN, C. Y., C. N. CHANG and I. F. YEH (1959) *Acta Pharm. Sin.* **7**, 237.

YÜAN, C. Y., C. N. CHANG and I. F. YEH (1960) *Chem. Abstr.* **54**, abstr. 10096.

General discussion II

Taves: There is a large discrepancy in the observed fluoride concentrations in human serum depending on whether or not the sample is ashed before fluoride analysis (Taves 1968*a*; Singer and Armstrong 1969). The diffusion recovery of added [^{18}F] fluoride from serum is $> 95\%$ (Taves 1968*b*). Thus the discrepancy could be accounted for either by a non-exchangeable form of fluoride, which is unlikely in the strong acid solution used in diffusion, or by an organic compound with an acid-stable fluorine bond, the fluorine atom being converted to inorganic fluoride on ashing. Electrophoresis of serum distinguish-ed two peaks of fluoride in ashed samples (Fig. 1) (Taves 1968*c*). One peak was coincident with added [^{18}F] fluoride and the other with albumin.

Dr Warren S. Guy has been continuing this investigation in our laboratory. The fluoro compound from human plasma has been purified 2000-fold. Plasma proteins are precipitated by the method of Micheal (1962) leaving the fluoro compound in a methanol supernatant solution. After evaporation of solvents the residue is fractionated in a liquid–liquid partition column with a dextran gel as the binder for water (Siakotos and Rouser 1965). The fluoro compound is neither part of the first lipid fraction containing neutral fats, glycolipids and phospholipids, nor of the last water-soluble fraction. It is eluted from the column in fractions of intermediate solubility, which contain compounds such as gangliosides, bile acids and urea. Subsequent silicic acid chromatography with a gradient elution of increasing amounts of methanol in chloroform has developed multiple peaks in at least one batch, but the experiment needs to be repeated. One of these peaks migrates to the positive pole on electrophoresis at pH 9.

This organic fluorine appears to be present only in human sera. In humans

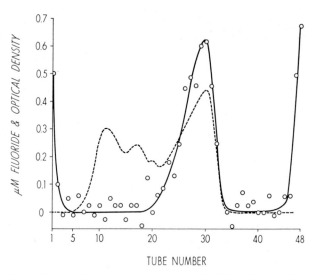

FIG. 1 (Taves). Curtain electrophoresis of a dialysed normal human serum. ———: concentration of fluoride in the electrophoresis samples after ashing; ------: optical density (280 nm) of a 1 in 10 dilution of the same sample; applied voltage: 2000 V; amperage: 105 ± 5 mA; solution in separating chamber: 0.12% $(NH_4)_2CO_3$; electrode rinse: 0.36% $(NH_4)_2CO_3$; flow rate: 100 ml per hour; samples: 12 ± 1 ml. Radioactive fluoride added to the dialysed serum only appeared at the two sides, not in the central peak.

we generally find 1 to 3 μM but in animals, if it is present, it is less than 1 μM. There is, of course, the possibility that in the ashing procedure we are losing organic fluorine so we shall have to check these conclusions with combustion methods in which all gases are retained, before we can be certain that these results represent all the organic fluorine present. But we feel quite certain from the results that there is some organic fluorine in human sera. We don't know the source of the fluorine, but we don't think it is Freon because a refrigerator repairman exposed to large amounts of Freon showed no increase in serum fluorine of either type (B. W. Fry and D. R. Taves, unpublished).

Gál: Is this a single compound or a complex of fluorinated compounds?

Taves: We find more than one peak on a silicic acid column, but it may represent breakdown during isolation.

Gál: What type of compound do you think it is?

Peters: I gather you think it is in the ganglioside.

Taves: There are several other things besides gangliosides in that fraction so any such conclusion is premature.

Armstrong: You referred to non-ionized fluoride. We find ultrafiltrates of plasma that contain inorganic fluoride that is responsive to the fluoride electrode and hence is ionic. However, we find additional amounts of fluoride in plasma ultrafiltrates after ashing. Therefore we believe that plasma contains some organically bound fluoride in compounds with molecular weights of less than 25 000.

NEW NATURALLY OCCURRING FLUORINE COMPOUNDS FROM DICHAPETALUM
TOXICARIUM

Harper: In association with J. A. Cooke and R. J. Hall at Newcastle, I have been investigating fluoro compounds in the seeds of *Dichapetalum toxicarium*. I have been looking at the non-saponifiable lipid fraction after separation from the fatty acids. The main fluorine-containing fraction of the fatty acids has already been identified by Peters and co-workers (Peters *et al.* 1960; Peters and Hall 1960; Ward, Hall and Peters 1964) as ω-fluoro-oleic acid and ω-fluoro-palmitic acid.

The seeds contain about 1500 p.p.m. organic fluorine and about 80 p.p.m. are located in the non-saponifiable fraction, distributed between three principal compounds. So far I have purified one of these, which accounts for about 20 p.p.m. organic fluorine in the seed, and have also isolated its non-fluorinated analogue. The fluorinated compound has a melting point of 88–90 °C, a molecular weight of 472, and an empirical formula (as determined from high resolution mass spectrometry) of $C_{27}H_{49}O_5F$, although it normally exists as a hydrate in association with $1\frac{1}{2}H_2O$. The infrared spectrum indicates the presence of a carbonyl group, probably a saturated ketonic function, though it must be sterically hindered since it does not react with 2,4-dinitrophenyl-hydrazine. The non-ketonic oxygens appear to be hydroxyl groups: n.m.r. demonstrates that the fluorine is probably part of a CH_2FCH_2-group. Assuming that this is only one carbonyl group, the molecule must contain 2 rings, as there is no evidence of unsaturation. The n.m.r. data indicate that there are at least 16 methylene groups.

The non-fluorinated analogue (m.p. 93–94 °C, molecular weight 454) appears to exist as the monohydrate and shows very similar properties and spectra, except that the CH_2F group is replaced by a methyl group. An interesting feature of both these compounds is that although practically neutral and highly insoluble in water, they are soluble in sodium hydroxide and cannot be extracted from it by organic solvents. However, both are readily extractable from sodium bicarbonate solution. This probably indicates that at least one of the hydroxyl groups is weakly acidic. We originally thought that the compounds might be sterols, but the high oxygen content and the presence of only two rings (they have an unsaturation equivalent of 3, accounted for by two rings and a carbonyl group) seems to rule this out unless they are sterol precursors. My feeling at present is that these compounds probably do not fall into any of the major groups of natural products.

Saunders: Does the $1\frac{1}{2}H_2O$ come off on heating?

Harper: The presence of water is inferred from the carbon, hydrogen and fluorine analysis. Whether it will come off on heating I do not know.

Saunders: Is it water of crystallization?

Harper: With small amounts of high molecular weight compounds it is difficult to be sure that water is not being absorbed from the atmosphere, but I am inclined to believe that it is probably water of crystallization.

Peters: Have you got enough to find out whether it is toxic?

Harper: Yes, we have about 20 mg, but we have not investigated this yet. It is interesting to note that the non-fluorinated compound occurs in approximately the same proportion (i.e. about 10:1) to the fluorinated compound as oleic acid occurred with respect to fluoro-oleic acid in *Dichapetalum toxicarium* in your investigations, Sir Rudolph. So it looks as though fluoroacetate is incorporated in approximately the same proportion into all substances in which acetate is used for biosynthesis.

Peters: Phillips and Langdon (1955) reported that liver incorporated fluoroacetate into an intermediate on the pathway towards the cholesterols and fatty acids.

Taylor: Was it a terminal fluorinated methylene group?

Harper: Yes.

Peters: How do you extract this compound from the seeds?

Harper: With hot ethanol. After alkaline hydrolysis of the extract the fatty acids are removed by chromatography on silica gel. Any residual acidic compounds are then extracted with sodium bicarbonate solution and the resulting approximately neutral mixture of lipids is purified by repeated column chromatography on silica gel.

Hall: A new analytical procedure seems to be needed to help to determine how widespread fluorinated compounds with covalent C—F bonds are in nature, and n.m.r. spectroscopy is a very fine way of studying this. In the past the low sensitivity of the n.m.r. technique has restricted its use; one has needed more than 10 mg of the material in 1 ml of solution. However, recent developments in n.m.r. technology, involving the use of larger sample tubes and of Fourier transform techniques, have improved the signal-to-noise ratio of the technique by 10^3, so in a few years one will be able to work with a few milligrammes of a mixture and it might be possible to use n.m.r. to pick out the presence of covalent C—F compounds in very crude extracts.

FLUORIDE: A PROBABLE CAUSE OF METHOXYFLURANE NEPHROTOXICITY

Taves: Our interest in the fluorinated anaesthetics arose from our discovery of very high concentrations of fluoride in a haemodialysis patient of Dr Posen's in Ottawa, Canada, in 1968. The initial fluoride determination showed 150 μM, or three times the 50 μM concentration that was present in the dialysate from the fluoridated water. The magnitude of the concentrations made it seem unlikely that the fluoride was coming from the dialysis or from previously stored fluoride in the bone. In looking for other possible sources of fluoride we found that the patient had received both halothane, $CF_3CBrClH$, and methoxyflurane, $CH_3OCF_2CCl_2H$, a few days earlier. A further check on the analytical reliability of the values showed that only 100 μM were present as inorganic fluoride, as demonstrated by the fluoride electrode, until after the sample had been acidified. If the diffusion from an acidified solution was prolonged, more fluoride was recovered. If the sample was ashed there was even more. This made it clear that there were probably at least two forms of organic fluoride as well as increased levels of inorganic fluoride present.

Three forms of fluoride were clearly demonstrated by electrophoresis (Fig. 2). There were two distinct acid-labile fluoride components, only one of which was inorganic fluoride, and a third peak was found after we had ashed the residue from the analysis for the acid-labile forms. This case was obviously rather complicated, since two types of fluorinated anaesthetic were administered.

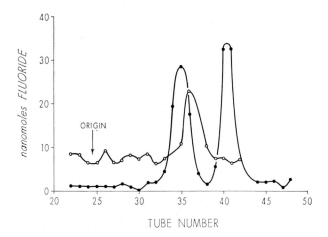

FIG. 2 (Taves). Peritoneal dialysate from a patient who had received methoxyflurane and halothane anaesthesia. The dialysate was diluted tenfold in 0.12% $(NH_4)_2CO$ buffer and run on the curtain electrophoresis apparatus. The solid circles are the results of analysis after diffusing fluoride from the acidified samples (Taves 1968*b*) and the open circles the results of analysis of the neutralized residue after ashing with $MgCl_2$.

Samples from other patients who received only one type of anaesthetic confirmed our suspicion that the acid-labile organic fluoride arose from the methoxyflurane and that the fraction available only after ashing arose from halothane.

Shortly after Dr R. B. Freeman of the Hemodialysis Unit in Rochester, New York, learned of the above findings, he saw a patient with high output renal failure after methoxyflurane anaesthesia. The concentration of inorganic fluoride (F) was 275 μM and the acid-labile organic fluoride (OALF), 3800 μM. Two patients who did not have evidence of renal failure after the use of methoxyflurane showed about 30 μM-F and less than 1600 μM-OALF (Taves *et al.* 1970). Dr Bill Fry, in our laboratory, has studied other patients with no complications, except a possible transient increase in urine volume, and shown that the peak concentrations average 35 μM-F and 1200 μM-OALF. The peak values occur 12 to 36 hours after the anaesthesia, when the urine volume and the renal fluoride clearance are also at their highest (Fry 1971). We think the sources of the inorganic fluoride are two metabolites of methoxyflurane, as outlined in Fig. 3.

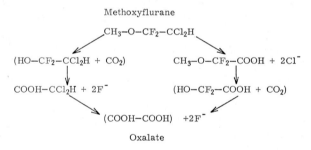

Methoxyflurane

$CH_3-O-CF_2-CCl_2H$

$(HO-CF_2-CCl_2H + CO_2)$ $CH_3-O-CF_2-COOH + 2Cl^-$

$COOH-CCl_2H + 2F^-$ $(HO-CF_2-COOH + CO_2)$

$(COOH-COOH)$ $+2F^-$

Oxalate

FIG. 3 (Taves). Metabolism of methoxyflurane to oxalate.

The metabolites without parentheses have been identified by Holaday, Rudofsky and Treuhaft (1970). The left-hand pathway is most important during anaesthesia, when the concentration of methoxyflurane is the highest, but the right-hand path becomes dominant the day after the anaesthesia, when the concentrations of the methoxydifluoroacetic acid, CH_3-O-CF_2-COOH (OALF), reach a maximum and the methoxyflurane concentration in the blood is only about 1% of the anaesthetic concentration. This change is reflected in a marked change in the ratio of OALF-to-F concentration that occurs in the 24 hours after anaesthesia (Fry, Taves and Merin 1972).

Dr Gary Whitford, in our laboratory, has found the serum fluoride concentration that produces a three-fold increase in urine flow rate in anaesthetized rats to be 275 μM. This is the same concentration as noted in the patient with severe nephrotoxicity, thus supporting the hypothesis that inorganic fluoride is the cause of the high output syndrome that occasionally follows methoxyflurane anaesthesia (Whitford and Taves 1971).

He has also found that the increased urine flow rate is probably caused by a loss of the sodium gradient in the loop of Henle. This loss is not accompanied by an increase in the loss of sodium or solute in the urine, so it is distinctly different from the usual diuretic (Whitford 1971).

Saunders: The oxalate in Fig. 3 is important. It ties up a little with what I was saying earlier (p. 18) about oxalate coming from fluorocitrate.

Taves: Yes. I was interested in your comment. If the methoxydifluoroacetic acid breaks down in acid to form oxalate, then all analyses of oxalate are inappropriately high after methoxyflurane anaesthesia. The problem is that all the oxalate procedures call for acidification before the oxalate is separated from the fluorometabolite. However, oxalate may be responsible for the lack of recovery of some patients with methoxyflurane nephrotoxicity since tremendous numbers of oxalate crystals have been found in the kidneys of such patients.

References

FRY, B. W. (1971) Ph.D. Thesis. University of Rochester.
FRY, B. W., D. R. TAVES and R. C. MERIN (1972) In prep.
HOLADAY, D. A., S. RUDOFSKY and P. S. TREUHAFT (1970) *Anesthesiology* 33, 579–593.
MICHEAL, S. E. (1962) *Biochem. J.* 82, 212.
PETERS, R. A. and R. J. HALL (1960) *Nature (Lond.)* 187, 537.
PETERS, R. A., R. J. HALL, P. F. V. WARD and N. SHEPPARD (1960) *Biochem. J.* 77, 117.
PHILLIPS, A. H. and R. G. LANGDON (1955) *Arch. Biochem. & Biophys.* 58, 247–249.
SIAKOTOS, A. N. and G. ROUSER (1965) *J. Am. Oil Chem. Soc.* 42, 913.
SINGER, L. and W. D. ARMSTRONG (1969) *Arch. Oral Biol.* 14, 1343–1347.
TAVES, D. R. (1968*a*) *Nature (Lond.)* 217, 1050–1051.
TAVES, D. R. (1968*b*) *Talanta* 15, 1015–1023.
TAVES, D. R. (1968*c*) *Nature (Lond.)* 220, 582.
TAVES, D. R., B. W. FRY, R. B. FREEMAN and A. J. GILLIES (1970) *J. Am. Med. Assoc.* 214, 91–95.
WARD, P. F. V., R. J. HALL and R. A. PETERS (1964) *Nature (Lond.)* 201, 611.
WHITFORD, G. M. (1971) Ph.D. Thesis. University of Rochester.
WHITFORD, G. M. and D. R. TAVES (1971) *Proc. Soc. Exp. Biol. Med.* 137, 458–460.

Synthesis and reactivity of fluorocarbohydrates

P. W. KENT

Department of Biochemistry, University of Oxford

The introduction of fluorine into a metabolite, such as a carbohydrate, is a unique way of achieving distinctive modification with minimal disturbance of the overall stereochemistry. The modification may profoundly alter the bio-chemical behaviour of the fluorosugar in enzyme inhibition or in 'lethal synthesis', by which the fluorosugar is partly metabolized into an even more toxic substance. Kinetic, inhibition and nuclear magnetic resonance (n.m.r.) studies of such fluorine-substituted compounds give additional new information about ways in which binding to macromolecules can occur. Although the atomic size of fluorine places it between hydrogen and hydroxyl (and distinguishes it from other halogens), fluorine can only engage in hydrogen bonding as an acceptor. X-ray crystal analysis of fluorocarbohydrates shows how replacement can occur in three-dimensional structures with maximum hydrogen-bonding. However, in naturally occurring fluoro compounds such as fluoroacetic acid and ω-fluoro-oleic acid, the fluorine atom may be regarded as substituting for a hydrogen, as discussed by Bergmann (p. 210). A further interesting and important instance is the discovery of a fluorine-containing antibiotic, nucleo-cidin (Morton *et al.* 1969), in which F replaces H-4 of a ribofuranosyl ring.

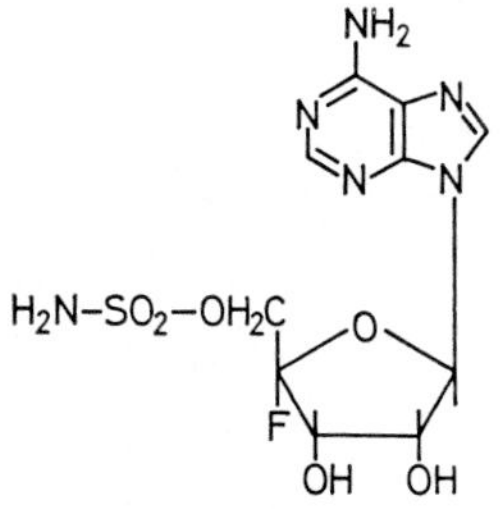

NUCLEOCIDIN

A number of specialist reviews pertinent to fluorosugars have been published. Attention is drawn to articles on glycosyl fluorides (Micheel and Klemer 1961), halogenated sugars (including 2-bromo and 2-iodoglycosyl fluorides; Barnett 1967), biochemical activities of fluorinated metabolite analogues (Goldman 1969) and earlier studies on fluorocarbohydrates (Kent 1969). Properties of available fluoromonosaccharides and their derivatives are set out in the Appendix to this paper (pp. 199–206).

SYNTHESIS OF FLUOROCARBOHYDRATES

Strategy of synthesis

The principal obstacle to the synthesis of fluorosugars rests in the fundamentally weak nucleophilicity in protic solvents of F^-, the weakest of all the halide anions. In the usual anionic exchange reactions, special solvent requirements are imposed for F^- (competing anions, e.g. OH^- and OMe^-, are avoided), and the acidity of the reaction is controlled (otherwise there may be elimination rather than substitution) (Parker 1962). In addition, unless reactive fluorinating reagents are employed, reactions involving F^- frequently require forcing conditions of temperature and concentration.

The strategy of synthetic design has been much influenced by progress in the steroid field, where conformation and mechanism can be more certainly ascertained.

In general, acetal or ketal protecting groups may be removed from a fluorosugar with ease by dilute aqueous acid at 50–80 °C or by polycationic exchange resins (H^+). Methanolic hydrogen chloride has been successfully used to deprotect 1,2-isopropylidene derivatives and simultaneously effect glycosidation.

Although *O*-acetyl groups have been widely used as blocking reagents, conditions for their removal from fluoro monosaccharides vary considerably. Peracetylated derivatives of methyl 6F-α-D-galactopyranoside and 2F-(±)-erythritol, for example, have been successfully deacetylated, and give good yields with catalytic amounts of sodium methoxide in anhydrous methanol at room temperature or at 40 °C. Glycosyl fluorides, similarly, can usually be obtained from their fully acetylated derivatives by this means, without loss of fluorine. The isolation and crystallization of 3-deoxy-3-fluoro-D-glucose from its di-isopropylidene derivatives has been achieved with dilute H_2SO_4 followed by barium carbonate for deprotection (Taylor *et al.* 1972).

2-Deoxy-2-fluoro-D-glucose and 2-deoxy-2-fluoro-α-D-glucosyl fluoride have proved somewhat more intractable. Deacetylation of the peracetates of these compounds, even under mild conditions, results for example in the formation of side-products, suggesting that competing elimination reactions are involved.

Early studies of the reaction between methyl-3:4-isopropylidene-2-*O*-toluene-*p*-sulphonyl (or methane sulphonyl) α-D-arabinoside with KF in a variety of solvents showed that although some exchange occurred that led to the 2-fluoro derivatives, exposure to aqueous or methanolic acids in attempts to remove the ketal blocking group resulted in extensive liberation of F^- and the formation of 2:5-anhydro compounds and unsaturated substances.

The use of benzyl ethers as blocking groups (Taylor and Riggs 1963) in the synthesis of fluorosugars has proved effective. These are removable by hydrogenation in the presence of platinum without displacement of F. Substituted benzyl ethers, such as *p*-nitro, also offer promise in this connection, with the further advantage of greater crystallinity (Kent and Young 1971). Gentle means of de-etherification, by the use of compounds such as BCl_3, has enabled *O*-methyl derivatives to be used as non-migrating protecting groups in fluorocarbohydrate synthesis (Johansson and Lindberg 1966). The development of new methods of de-etherification, for example ethanethiol and boron trifluoride etherate (Fletcher and Diehl 1971), extends the scope of this type of derivative for synthetic purposes.

Synthetic routes involving substitution reactions

(*a*) *Exchange*

Helferich and Gnütchel (1938) showed that methane sulphonyloxy esters of primary alcohols underwent ready exchange with NaI in boiling acetone to give an almost quantitative yield of the corresponding 6-iodo derivative. Exchange of toluene-*p*-sulphonyloxy ester proceeded with greater speed (Tipson 1953; Foster, Hems and Webber 1967). Exchange reactions of analogues with KF in methanol lead to the formation of 6-deoxy-6-fluoroglucose (Scheme 1) by treatment of 6-*O*-mesyl-3,5-benzylidene-1,2-*O*-isopropylidene-α-D-glucofuranose with KF in methanol at 100 °C (Helferich and Gnütchel 1941), and deprotection of the resulting crystalline fluorinated product. Under these strictly anhydrous conditions, 6-*O*-mesyl(or tosyl)-di-*O*-isopropylidene-α-D-galactose failed to react. Exchange occurred, however, in 17 hours at 160 °C to give a mixture of the 6-fluoro and the 6-*O*-methyl derivatives (Kent, Morris and Taylor 1960). The competing solvolysis was minimized by use of ethanediol at near-boiling point (Taylor and Kent 1958) for 75 minutes. This solvent is suitable for a wide

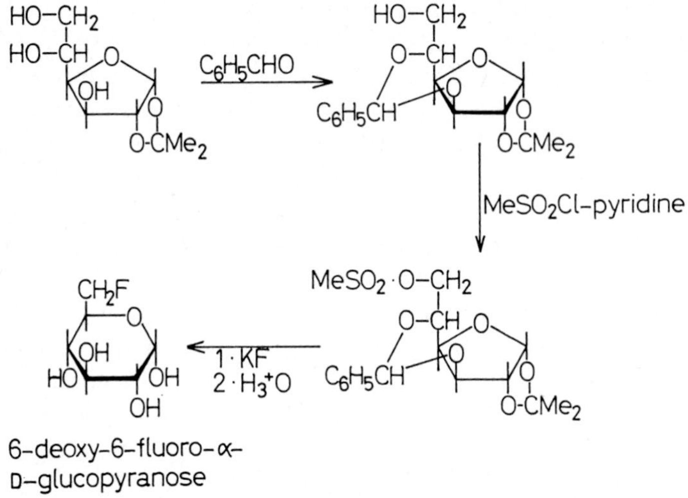

6-deoxy-6-fluoro-α-
D-glucopyranose

SCHEME 1: Synthesis of 6-deoxy-6-fluoro-D-glucose from 1,2-O-isopropylidene-D-glucose.

range of comparable carbohydrate reactions (Table 1), although some instances have been reported of $-OCH_2CH_2OH$ groups being introduced as side reactions (Barnett 1967; Foster and Hems 1969). Exchange reactions involving KF are strongly solvent-dependent and are hindered by the presence of impurities, such as pyridine or Cl^- (Pattison 1954; Pattison and Norman 1957; Ball and Parrish 1969).

TABLE 1

Treatment of 1,2:3,4-di-O-isopropylidene-6-O-toluene-p-sulphonyl-α-D-galactopyranose with tetrabutylammonium fluoride*

Solvent	Temp. (°C)	Reaction time	Molar excess of Bu₄NF	Ratio of fluoride: olefin	Yield of fluoride (%)
Ethylene glycol	100	22 h	5	No reaction	No reaction
Tetrahydrofuran	80	24 h	5	1.4:1	29
Acetone	56	16 h	9	1.6:1	†
Acetonitrile	81–83	10 h	6	2.2:1	69
	63–67	3.5 days	9	3.7:1	61
	37	9 days	6	5.5:1	39
	37	23 days	15	3.2:1	64
N,N-Dimethylformamide	50–60	3.5 days	6	3.2:1	53
	37	13.5 days	6	3.0:1	45

* From Buck et al. 1966.
† Additional products formed.

Other extra-cyclic sulphonyloxy esters, for example methyl 5-*O*-methan-sulphonyl-2,3-isopropylidene-β-D-ribofuranoside, also underwent ready exchange with anhydrous KF in boiling ethanediol in 1 hour, with a 60% yield (Taylor and Kent 1958).

The rate of fluoride exchange is profoundly influenced by the stereochemistry and dipolar contributions of substituents on atoms adjacent to that bearing the departing ester (Taylor 1958). Richardson (1969), in a detailed consideration of contributing polar effects, pointed out the conformational restrictions to exchange reactions that can result from accumulated electronegative substituents.

Elucidating the conditions for exchange of secondary sulphonyloxy esters proved more difficult until Cohen, Levy and Bergmann (1964) succeeded in isolating derivatives of 3-deoxy-3-fluoro-D-xylose from the products of the reaction between anhydrous KHF_2 and benzyl-2-*O*-tosyl-β-D-arabinopyrano-side in ethanediol or acetamide at 200 °C. The reaction probably proceeded via an epoxide and, also in 1964, Taylor, Childs and Brunt succeeded in obtaining the same fluorosugar from methyl 2,3-anhydro-4-*O*-benzyl-β-D-ribopyranoside with the same reagents. This method of epoxide scission appears to be more satisfactory than the use of HF in dioxane at 120 °C.

A notable advance was made in 1967 by Foster, Hems and Webber (1967) who, noting the extreme stability of 3-*O*-sulphonyloxy esters of di-isopropyl-idene-α-D-glucofuranose in exchange reactions, found that the corresponding esters of the 3-epimer (di-isopropylidene-α-D-allofuranose) readily exchanged with excess tetrabutylammonium fluoride (Henbest and Jackson 1962) in

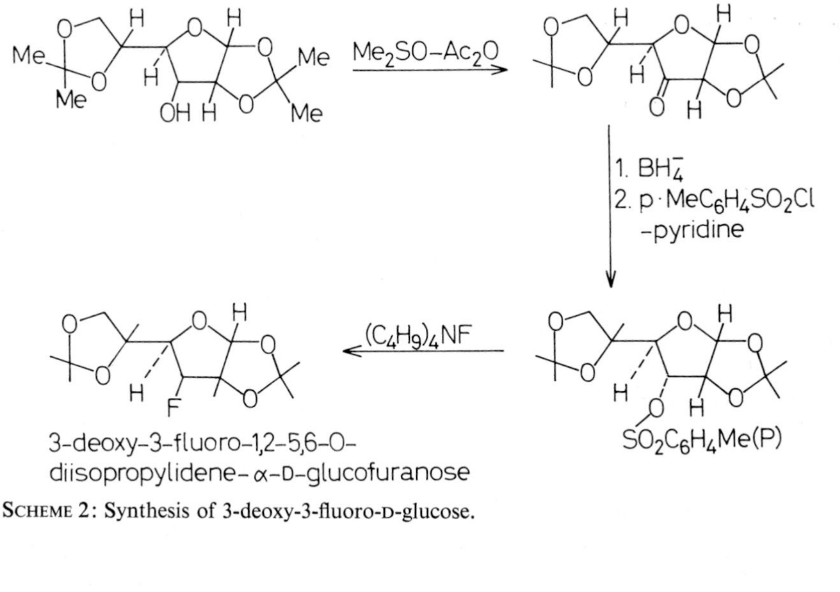

3-deoxy-3-fluoro-1,2-5,6-*O*-diisopropylidene-α-D-glucofuranose

SCHEME 2: Synthesis of 3-deoxy-3-fluoro-D-glucose.

anhydrous acetonitrile at 70–80 °C for $3\frac{1}{2}$ days (Scheme 2). Inversion, indicative of an SN_2 reaction, occurred and the corresponding 3-fluoroglucose was obtained in 71% yield (Buck *et al.* 1966). The product was characterized as the crystalline β-1,2,4,6-tetra-*O*-acetate and 1,2-*O*-cyclohexylidene derivatives (Foster, Hems and Webber 1967). The proof of the structure depended on the 3-fluorosugar being identified as the product obtained independently by epoxide cleavage by Johansson and Lindberg (1966) and on degradation by periodate oxidation to the known 3-deoxy-3-fluoro-D-xylose (Taylor, Childs and Brunt 1964; Cohen, Levy and Bergmann 1964). Subsequently, proton and ^{19}F-n.m.r. investigations added further confirmatory data.

Although this procedure is not universally applicable, nevertheless it opened the way to the synthesis of a number of other secondary fluorosugars including 3-deoxy-3-fluoro-D-galactose (Brimacombe *et al.* 1968). The assignments of structures were made on n.m.r. data.

An improved route to 4-deoxy-4-fluoro-D-glucose has been found (Foster, Hems and Westwood 1970) in which methyl 4-*O*-mesyl-2,3-di-*O*-methyl-6-*O*-trityl-α-D-galactopyranose reacts with calcium fluoride or tetrabutyl ammonium fluoride in boiling acetonitrile, the fluoro product being obtained in 71% yield.

A difluorosugar, 3,5-dideoxy-3,5-difluoro-D-xylose was obtained by exchange between 3-deoxy-3-fluoro-1,2-*O*-isopropylidene-5-toluene-*p*-sulphonyl-α-D-xylofuranose and anhydrous KF in boiling ethanediol (Foster and Hems 1969). The presence of a *p*-nitrobenzyl ether at C-3 of the xylofuranose ring substantially hindered exchange of 5-*O*-tosyl (or mesyl) groups with F^- (Kent and Young 1971), and no reaction occurred under the foregoing conditions. With tetrabutylammonium fluoride in boiling anhydrous CH_3CN, the reaction proceeded smoothly to 5-deoxy-5-fluoro-D-xylofuranose, one of the few crystalline reducing pentofuranoses.

As yet direct halogen exchange in the carbohydrate series has met with little success. Attempts to convert 6-deoxy-6-iodo-D-glucose by reaction with silver fluoride led principally to unsaturated compounds (Helferich and Himmen 1928).

There is clearly a continuing need for further fluorinating agents capable of reacting under gentle and neutral conditions. Attempts to introduce fluorine into monosaccharides with the Yaravenko reagent [*N*(2-chloro-1,1,2-trifluoro-ethyl)diethylamine] or by rearrangement of formyl fluorides have not proved successful so far (Wood, Fisher and Kent 1966; Welch and Kent 1962).

The replacement of sulphonyloxy groups attached to monosaccharide structures normally proceeds by an SN_2 mechanism. The customary transition state which leads to products of inverted configuration involves the concerted expulsion of the sulphonyloxy group by the attacking species. It has been pointed out that the formation of such a high polar transition state will be powerfully influenced by neighbouring polar constituents. Where an anionic attacking reagent is employed, the formation of the transition state is hindered, especially by electronegative neighbouring groups such as OMe and maximally so when the dipoles are antiparallel. Electropositive substituents (e.g. $\overset{+}{N}R_3$) result in a reverse effect, as do 'neutral' attacking species. Exchange reactions with hexopyranose-6-O-sulphonates are hindered markedly by the dipolar effect of the axial O-4 and the ring oxygen, in contrast to the greater reactivity of the corresponding glucopyranose derivatives. Similar considerations may apply to the resistance of hexopyranose 1-sulphonates to undergoing reaction, due to dipolar repulsion being generated from O-1 and O-2 in unfavourable configurations (Hill, Hough and Richardson 1968; Richardson 1969).

It is noted, however, that furanose sulphonates also show considerable differences in reactivity. 3,5-Benzylidene-1,2-isopropylidene-D-glucofuranose 6-mesylate undergoes ready replacement with F^- and other anions despite its tricyclic structure. The 5-mesylate (or tosylate) of 3-O-benzyl-1,2-isopropylidene-D-xylofuranose only undergoes exchange under vigorous conditions. This is consistent with the shielding of C-5 by O-3 arising from non-planarity of the furanose ring. In addition to dipolar effects on reactions of this type, the effect of solvent on stabilizing the transition state is significant, as is the conformational flexibility of the sugar ring structure when elevated temperatures are used.

Sulphonyloxy exchange reaction in the 3 and 4 positions of hexopyranosides is considered to be susceptible to the influence of β-*trans* axial substituents. This is shown in the greater ease of exchange of β-anomers compared with the α-anomer as in the case of 1,2,4,6-tetra-O-benzoyl-3-O-toluene-*p*-sulphonyl-D-glucopyranose. Steric factors may contribute largely in this type of effect. The effect of vicinal substituents is observed in cases where an axial group *cis* to a sulphonyloxy group hinders replacement.

These hypotheses have provided grounds for rationalizing many of the observed variations in reactivity of sulphonyloxy derivatives. More quantitative data are required on the temperature-dependence of pyranoid conformations, and especially of dihedral angles, to allow rigorous analysis of these reactions.

(b) Epoxide scission

In a pioneering study, Kunyants, Kil'desheva and Petrov (1949) showed that ethylene oxide reacted with HF in ether at 100 °C to give 2-fluoroethanol in high yield. Cleavage of unsymmetrical epoxides was reported to give single products with fluorine substituted terminally:

$$\text{HF} + \text{CH}_3{-}\overset{\displaystyle\overset{\text{O}}{\triangle}}{\text{CH}}{-}\text{CH}_2 \xrightarrow[\text{ether}]{} \text{CH}_3 - \overset{\displaystyle\overset{\text{OH}}{|}}{\text{CH}} - \text{CH}_2\text{F}$$

In the steroid series, Fried and Sabo (1954, 1955) successfully synthesized 9α-fluorohydrocortisone from the corresponding 9α, 11β oxide by HF in ethanol-free chloroform.

The first definitive syntheses (Scheme 3) of secondary fluorosugars accomplished by this route were the formation of 3-deoxy-3-fluoro-D and L-xyloses (Taylor, Childs and Brunt 1964; Cohen, Levy and Bergmann 1964), with either HF-dioxane or KHF_2 in ethanediol. Later studies (Wright and Taylor 1967) gave improved yields of fluoro product. Treatment of methyl 2,3-anhydro-5-O-benzyl-β-D-ribopyranoside with excess KHF_2 in boiling ethanediol for 40 min gave about 50% yield of the 3-fluoroxyloside. Correspondingly, methyl-2,3-anhydro-5-O-benzyl-α-D-arabinoside led, after deprotection, to 3-deoxy-3-fluoro-D-arabinose. In every case the expected *trans* scission of the epoxide was found.

In the hexose series studied by Johansson and Lindberg (1966), O-methyl (or O-acetyl) derivatives were used as protecting groups. Methyl 2,3-anhydro-4,6-di-O-methyl-α-D-allopyranoside with hydrogen tetrafluoridoborate in HF at −70 °C gave the corresponding 2-deoxy-2-fluoroaltrosyl α-fluoride in low yield (Scheme 4). It is interesting that the glycoside methoxy group was displaced by F^- in this reaction. In addition, 3-deoxy-3-fluoro-glucose derivatives were also obtained.

The location of the fluorine atoms in these products was assigned as a result of elimination reactions and oxidations by lead tetra-acetate. N.m.r. studies were reported to support the structures proposed. The use of locked conformations provided by β1:6-anhydrosugars has been investigated. Pacák, Točík and Černý (1969) formed 2-deoxy-2-fluoro-D-glucose by the sequence of reactions in which the essential fluorination step consisted of the reaction of 1:6, 2:3-dianhydro-4-O-benzyl-β-D-altropyranose with KHF_2 in boiling ethanediol giving *gluco* (chiefly) and some *altro* products. The 2-fluoroglucose was identical with that synthesized from CF_3OF and triacetylglucal (Adamson *et al.* 1969).

In this laboratory, A. S. Harrison found that methyl 3,4-anhydro-2,6-di-O-benzyl-α-D-galactoside, submitted to KHF_2 in boiling anhydrous ethanediol (45 min), gave a preponderance of the 3-fluoro*gulo* isomer and a small amount of the 4-fluoroglucoside (unpublished results).

A novel and important route for the direct synthesis of 2′-fluoronucleosides by reaction of 2,2′-anhydro-1-β-arabinofuranosyl uracil was devised by Fox and his co-workers (Codington, Doerr and Fox 1964, 1966) (Scheme 5).

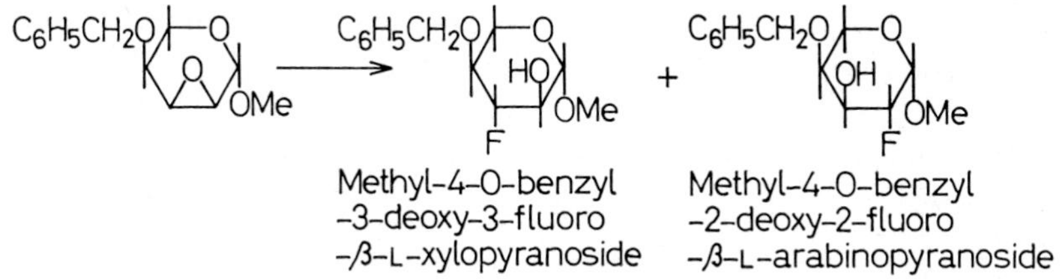

SCHEME 3: Synthesis of 3-deoxy-3-fluoro-β-L-xylopyranose and 2-deoxy-2-fluoro-β-L-arabino-pyranose by epoxide scission.

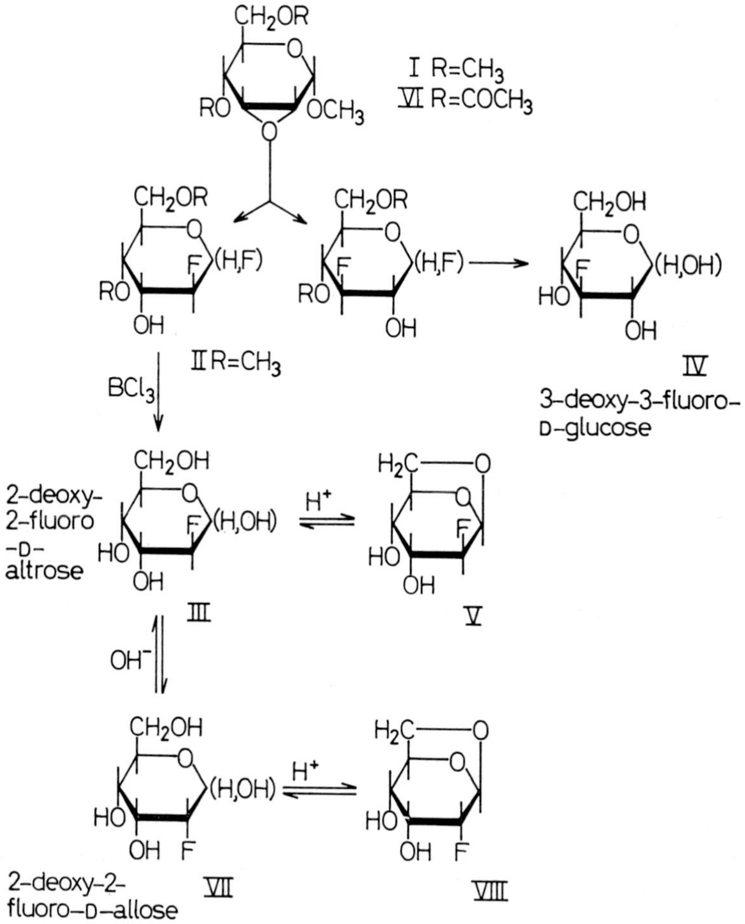

SCHEME 4: Action of hydrogen tetrafluoridoborate on methyl 2,3-anhydro-4,6-dimethyl-α-D-allopyranoside.

Reduction of 2'-deoxy-2'-fluorouridine to the 5,6-dehydro compound followed by glycoside cleavage of the sugar-pyrimidine bond allowed 2-deoxy-2-
fluoro-β-D-ribose to be obtained as the 1,3,4-tri-*O*-benzoate.

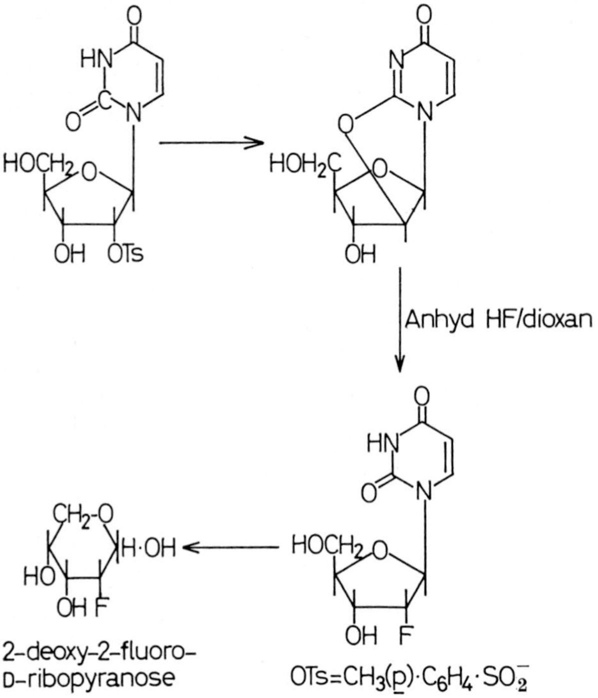

SCHEME 5: 2-deoxy-2-fluoro-D-ribose from uridine.

Synthetic routes involving addition reactions

Addition of halogens to unsaturated sugars has long been of interest (e.g.
Fischer, Bergmann and Schotte 1920), but only recently have techniques allowed
full exploration to be made. In an elegant series of investigations Lemieux and
his co-workers clarified the mechanism and course of addition reactions, for
example to acetylated 1,2 glycals (Lemieux and Frazer-Reid 1965). In general,
these reactions follow Markonikow predictions, the more cationic reactant
attacking C-2 and the anionic reactant C-1. The predominant products are
those in which the anionic substituent is located axially (α in the C-1 conformation). Anomerization equilibrium is considered to play a significant part in
determining the ratio of the products formed. This area has been extensively

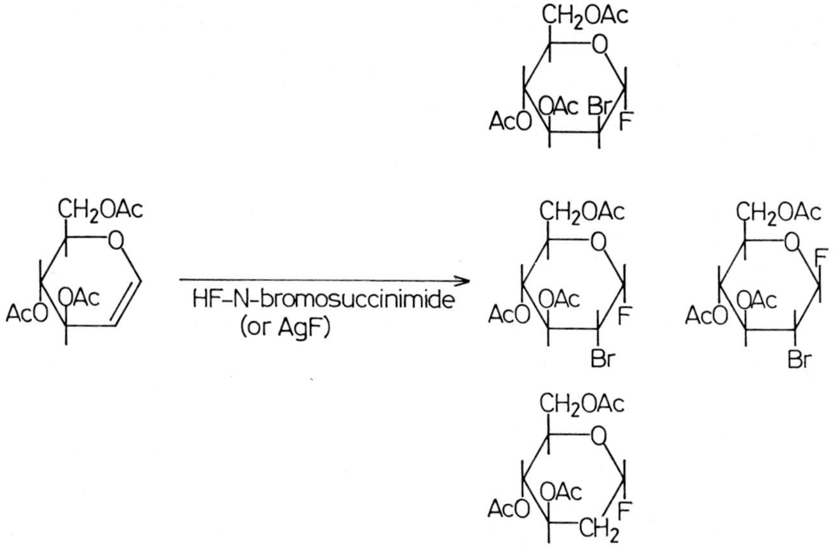

SCHEME 6: Products of bromofluorination of 3,6,4-tri-*O*-acetyl-D-glucal.

reviewed by Barnett (1967). Addition reactions involving fluorine, especially those of interhalogens ('BrF' generated from *N*-bromosuccinimide in liquid HF or with silver fluoride in CH_3CN) take a similar course (Scheme 6). Thus 2,3,4-tri-*O*-acetyl-D-glucal gave predominantly 2,3,4-tri-*O*-acetates of 3-bromo-2-deoxy-α-D-mannosyl and glucosyl fluorides (Kent, Robson and Welch 1963; Hall and Manville 1968; Campbell *et al.* 1968). The β-anomers were formed as minor products. This reaction has also been extended to the addition of 'ClF' and 'IF'. In no case, however, have any 2-fluorosugars yet been detected in the products.

A ready means for the synthesis of vicinal *cis*-fluorohydrin saccharides, nevertheless, has been ingeniously devised by additive use of fluoro-oxytrifluoromethane (CF_3OF) (Barton *et al.* 1969). Adamson *et al.* (1969) reported that triacetylglucal reacted smoothly with this reagent in $CFCl_3$ at $-78\,^\circ$C to give triacetyl 2-deoxy-2-fluoro-α-D-glycosyl fluoride (34%) and triacetyl 2-deoxy-2-fluoro-β-D-mannosyl fluoride (7.6%) (Scheme 7). In addition, the corresponding trifluoromethyl glycosides were found in similar proportions. The latter were ascribed to the dismutation, $CF_3OF \rightleftharpoons CF_2O^- + F^+ \rightleftharpoons CF_2 = C=O + F_2$.

Triacetyl-D-galactal, diacetyl-D-xylal and diacetyl-D-arabinal also react with CF_3OF giving corresponding glycosyl fluorides and trifluoromethyl glycosides (Dwek *et al.* 1970; Butchard and Kent 1971; Adamson and Marcus 1970).

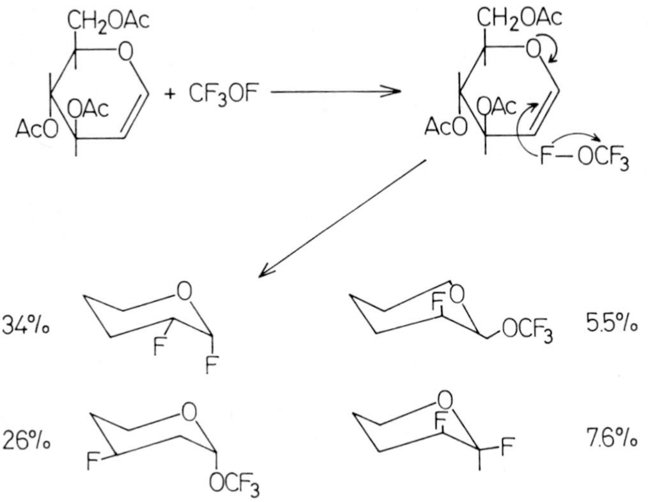

SCHEME 7: Addition of CF_3OF to 3,4,6-tri-*O*-acetyl-D-glucal.

Triacetyl 2-deoxy-2-fluoro-β-D-mannosyl fluoride has been found to undergo anomerization when treated with HF at −10 °C for 15 minutes (Hall *et al.* 1970).

Exploratory attempts in this laboratory to synthesize 2-fluorodisaccharides by this route indicate that peracetylated glycols, such as hepta-acetyl-lactal, react substantially less rapidly, and as yet only low yields of the 2-fluoro products have been obtained. This may be due to the conformational in-flexibility placed on the unsaturated ring by the second sugar at C-4.

In the steroid series, PbF_4 has been successfully employed in the synthesis of vicinal difluorides from alkenes. With triacetylglucal and diacetylarabinal, however, this reagent leads to 1,1-difluoro-2,5-anhydro sugars (Kent and Barnett 1966; Wood and Kent 1967) (Scheme 8).

The course of the reaction indicates bond migration with the preferential attack of the ring oxygen at C-2; the same has been found in other reactions involving an intermediate electron deficient at that carbon atom, for example, in the deamination of 2-deoxy-2-aminohexoses. Furthermore, rearrangements also occur when PbF_4 reacts with a 2,3-glycal (ethyl diacetyl-ψ-glucal) since the same 1,1-difluoro-2,5-anhydro-mannitol results as from triacetylglucal (Wood and Kent 1967). Parallel studies indicate that SeF_4 reacts in a similar fashion to SF_4, that is brings about conversion of $>C=O$ to $>CF_2$. The full potentialities of both these reagents in obtaining *gem*-difluorides in the sugar series await exploration.

Limited use has also been made of perchloryl fluoride as a means of forming secondary fluoro compounds by addition to unsaturated acetals or by replace-

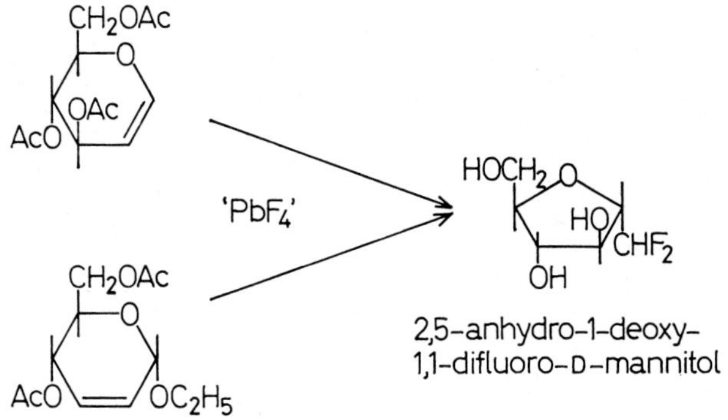

SCHEME 8: Addition of PbF$_4$ to glucal and ψ-glucal derivatives.

ment of active hydrogen (Pattison, Peters and Deane 1965), as in syntheses from fluoromalonic acid. The hazardous nature of this highly reactive fluorinating agent may restrict its applicability in the carbohydrate field.

Total synthesis

The first secondary fluorocarbohydrates to be synthesized were ($\pm$) 2-deoxy-2-deoxyerythritol and related compounds (Taylor and Kent 1954, 1956). These and 2-deoxy-2-fluoro-DL-ribitol (Kent and Barnett 1964) were obtained by total syntheses via Claisen condensations and selective reductions (Scheme 9). In these syntheses, fluorine was introduced as esters of fluoroacetic acid. Ethyl fluoroacetate readily condenses with diethyl oxalate in the presence of sodium methoxide or sodium hydride to give ethyl ethoxalylfluoroacetate as a yellow distillable oil. Partial reduction of the latter with sodium borohydride (at 0°C) led to dimethyl ($\pm$)-fluoromalate (Taylor and Kent 1954, 1956), a compound which, although not markedly toxic, possesses interesting biochemical properties as an enzyme inhibitor and inducer (see p. 236). More vigorous reduction of dimethyl oxalylfluoroacetate with lithium aluminium hydride or sodium borohydride in boiling ethanol gave a mixture of ($\pm$) 2-deoxy-2-fluoroerythritol (crystalline) and ($\pm$) 2-deoxy-2-fluorothreitol. Analogous 2,4-difluoro-tetritols were obtained from 2,4-difluoroacetoacetate, formed by self-condensation of methyl fluoroacetate.

Decarboxylation of dimethyl oxalylfluoroacetate in concentrated HCl gave fluoropyruvic acid.

In the course of these investigations, advantage was taken of the selective

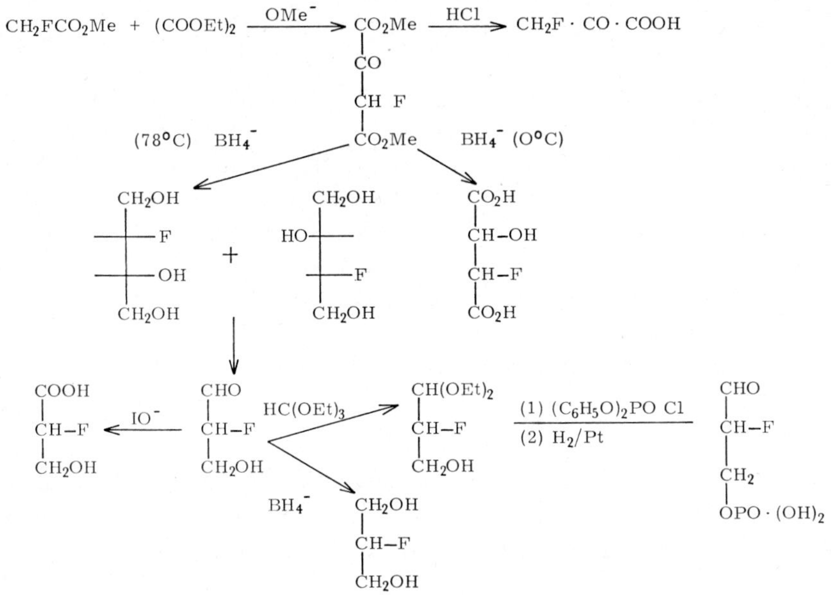

$$CH_2FCO_2Me + (COOEt)_2 \xrightarrow{OMe^-} \underset{\substack{CO\\ \mid \\ CH\ F \\ \mid \\ CO_2Me}}{CO_2Me} \xrightarrow{HCl} CH_2F \cdot CO \cdot COOH$$

(78°C) BH_4^- BH_4^- (0°C)

CH₂OH structures:

CH$_2$OH		CH$_2$OH	CO$_2$H
— F	+	HO—	CH–OH
— OH		— F	CH–F
CH$_2$OH		CH$_2$OH	CO$_2$H

$$\underset{\substack{COOH\\ \mid \\ CH-F \\ \mid \\ CH_2OH}}{} \xleftarrow{IO^-} \underset{\substack{CHO\\ \mid \\ CH-F \\ \mid \\ CH_2OH}}{} \xrightarrow{HC(OEt)_3} \underset{\substack{CH(OEt)_2\\ \mid \\ CH-F \\ \mid \\ CH_2OH}}{} \xrightarrow[(2)\ H_2/Pt]{(1)\ (C_6H_5O)_2PO\ Cl} \underset{\substack{CHO\\ \mid \\ CH-F \\ \mid \\ CH_2 \\ \mid \\ OPO\cdot(OH)_2}}{}$$

BH_4^- $\underset{\substack{CH_2OH\\ \mid \\ CH-F \\ \mid \\ CH_2OH}}{}$

SCHEME 9: Total synthesis of three and four-carbon fluorosugars. [All asymmetric products are racemic mixtures.]

reduction of hydroxy and keto carboxylic esters to alcohols by sodium borohydride at 68 °C in contrast to the corresponding unsubstituted esters which were themselves unreactive. Condensation of methyl fluoroacetate with methyl *O-iso*propylideneglycerate gave a mixture of fluoropentonates from which, after reduction and deprotection, 2-deoxy-2-fluoro-DL-ribitol was isolated (Kent and Barnett 1964). Though this route of synthesis has provided 3- and 4-carbon compounds of interest, its application to higher homologues is clearly severely limited by the accompanying creation of multiple asymmetric centres.

Other routes

In the search for fluoro-probes of enzyme structure, the introduction of CH_2FCO or CF_3CO- groups into metabolites offers a promising and fruitful approach. Considerable success has been achieved by the use of *N*-fluoroacetyl-D-glucosamine, synthesized by a carbodi-imide-catalysed reaction between 1,3,4,6-tetra-*O*-acetyl-β-D-glucosamine and sodium fluoroacetate (Dwek, Kent and Xavier 1971). An earlier report of the synthesis of this compound (Greig and Leaback 1963) was not supported by analysis of fluorine, and the constants are not wholly in agreement with later work. The *N*-fluoroacetates may also be synthesized directly from glucosamine, galactosamine and fluoroacetic acid by dicyclohexylcarbodi-imide in pyridine.

The fullest exploration of probe techniques requires multi-fluorinated monosaccharides and, more important, oligosaccharides. Those containing glucosamine residues are of particular interest and it is therefore appropriate to mention further reactions for introducing fluorine into aminosugars. *N*-Trifluoroacetylation has been used as a means of protecting the amine group of glucosamine during various syntheses (see Strachan *et al.* 1966; Wolfrom and Bhat 1967). The trifluoroacetyl group is notably more labile than the corresponding monofluoro compound under alkaline conditions, and may be removed along with *O*-acetyl groups during alkaline de-esterification.

Since reports of the use of *N*-fluoroacetylglucosamine as a molecular probe (Dwek *et al.* 1970; Dwek, Kent and Xavier 1971), *N*-trifluoroacetyl-D-glucosamine (see Dwek 1972) and *N*-trifluoroacetyltryptophan have also been employed for this purpose.

The α-anomeric fluoride of *N*-acetylglucosamine (Micheel and Wulff 1956) is formed when either α- or β-penta-acetyl-D-glucosamine is treated with anhydrous HF followed by careful deacylation with sodium methoxide at 35 °C for 1 hour. In contrast to other α-hexosyl fluorides, more vigorous alkaline treatment does not lead to the formation of a 1,6-anhydro sugar. The course of the reaction is, however, influenced by the *N*-substituent. The β-fluoride of *N*-tosylglucosamine was synthesized as a crystalline triacetate by halide exchange of 2-deoxy-2-(toluene-*p*-sulphonamide)-3,4,6-tri-*O*-acetyl-α-D-glucosyl bromide with anhydrous silver fluoride in CH_3CN. In contrast to the α-fluoride of *N*-tosylglucosamine, when the β-derivative was kept for 4 hours at room temperature, with 5% sodium in methanol, a mixture of 2-amino-2-deoxy *N*-tosyl-1,6-anhydroglucose (69%) and the corresponding methyl β-glycoside resulted. It is interesting that excess sodium methoxide favours the production of glycosides which retain the anomeric configuration. In hot 2M-NaOMe, the α-fluoro analogue was converted into the 1,6-anhydro derivative (60% yield) and the methyl β-glycoside (12%) (Micheel and Michaelis 1957). Micheel and Wulff (1956) propose the formation of an intermediate imine, for which the NH of the tosyl substituent is a particularly suitable participant, to account for these results. The mechanism is supported by the greatly increased resistance of F-1 to alkali which results in compounds in which the NH has been replaced by $N-CH_3$ (Micheel and Michaelis 1957).

Treatment of 1,3,4,6-tetra-*O*-acetyl-β-D-glucosamine with trifluoroacetic anhydride in pyridine readily gives the corresponding *N*-trifluoroacetate, from which a mixture of α- and β-glycosyl fluorides results on treatment with anhydrous HF. Unfortunately, selective removal of O-acetates from these products has proved difficult even under mild conditions, due to considerable loss of CF_3CO- and of the anomeric fluorine. Compounds of this type, bearing two types of fluorine atoms and found to be clearly distinguishable by ^{19}F-n.m.r., are of considerable importance as macromolecular probes. Multiple N-fluoroacetylation (or trifluoroacetylation) of N and N' amino groups in chitobiose and higher oligomers similarly can be expected to be a most fruitful avenue of approach to distinguishing between multiple specific binding sites in a single macromolecule such as lysozyme.

N-Acetyl-3-deoxy-3-fluoro-D-neuraminic acid, synthesized by Gantt, Millner and Binkley (1964) by condensation of 3-fluoropyruvic acid with N-acetyl-glucosamine or N-acetylmannosamine at pH 11 and purified chromatographic-ally on Dowex-1 and charcoal (overall yield 1.5 and 3.0%), was the first halogenated sialic acid to be obtained. The structure of the epimers at C-3 was not fully established. One epimer, however, irreversibly inhibited acyl-neuraminate aldolase. Recently, Schauer, Wirtz-Peitz and Faillard (1970) obtained N-fluoroacetylneuraminic acid (FNANA) in 29% yield, by reaction of the methyl β-glycoside of neuraminic acid with fluoroacetic anhydride in dioxane at 0°C, followed by removal of the glycosidic methoxyl group by Dowex-50 at 80°C. The N-fluoroacetyl group proved to be noticeably unstable in contrast to N-chloroacetyl and N-glycolyl analogues. The fluoroacetyl product (FNANA) is cleaved by acyl neuraminate-pyruvate-lyase (aldolase) from *Clostridium perfringens* (K_m 1×10^{-2}M), at a rate only 20% as fast as that with N-acetylneuraminic acid as substrate where K_m is 1.85×10^{-3}M. In addition, FNANA is a competitive inhibitor (K_i 0.69×10^{-4}M) of acylneuraminate-aldolase (Schauer et al. 1971).

CRITERIA OF PURITY

Chromatography

Routinely, paper chromatography has been employed to characterize and separate fluoromonosaccharides from non-fluorinated sugars. In general, the common solvent systems for carbohydrate separation prove effective. In exchange reactions where side products arise by solvolysis or elimination, such as in the formation of 6-O-methyl-D-galactose during the synthesis of 6-deoxy-6-fluorogalactose, the technique can be a useful guide to the complexity of the reaction products (Taylor, Kent and Morris 1960). The discrimination is frequently improved by removal of blocking groups before attempting chromatography. Different fluorosugar glycosides and reducing fluorosugars show variable responses but in general are detectable in the usual ways: with periodate-permanganate (for the former), ammonia-$AgNO_3$, aniline hydrogen oxalate or phthalate (for the latter). 2-Deoxy-2-fluoro-hexoses, for example 2-deoxy-2-fluoro-altrose and allose, react weakly on chromatograms with anisidine-HCl or silver nitrate-potassium hydroxide. Periodate-benzidine provides a more sensitive method for detecting these sugars. Separation of nearly related fluorosugars by paper chromatography is more difficult and, in general, gas–liquid chromatography is found to be a stringent analytical procedure. Fluorosugar phosphates, which as salts cannot be easily resolved by gas–liquid chromatography, have been found to separate by paper chromatography in solvents such as methanol: pyridine: ammonia: water (5:3:1:1 by vol.) or 2-methoxyethanol: pyridine: ammonia: water (5:2:1:2: by vol.) or 2-methoxyethanol:butanone:3M-ammonia (7:2:3 by vol.). These compounds are detected by ammonium molybdate reagents (Benson *et al.* 1950) down to approximately 0.2 µg phosphorus.

Preparative absorption chromatography is widely employed for the separation of reaction products. Adamson and co-workers (1969) separated the stereoadducts that result from the reaction of CF_3OF with triacetylglucal by chromatography on a column of Kieselgel-G (Merck) 7731, with light-petroleum: ether (1:1) as the elutant. Successful separations of other complex reaction products have also been reported for fluoropentoses (Dwek *et al.* 1970; Butchard and Kent 1971) and for other fluorohexoses (Adamson and Marcus 1970; Dwek *et al.* 1970).

Thin-layer chromatography has also been widely used. As a supporting medium, Kieselgel-G 7731 may be used with a variety of solvents; for example carbon tetrachloride:ether (9:1), ethylacetate:ethanol (2:1) or ethyl acetate: benzene (3:7). Fluorosugar derivatives can be detected by sulphuric acid, or by iodine vapour.

For quantitative and qualitative analysis, gas–liquid chromatography offers the most sensitivity. Compounds have been analysed either as glycoside acetates or as *O*-trimethylsilyl ethers on such supporting media as poly(ethyleneglycol adipate) on Celite (Foster, Hems and Webber 1967), SE30 (Adamson *et al.* 1969) or Diataport S80–100 mesh/3% SE30 (Dwek *et al.* 1970), at constant temperatures (e.g. 150°C or 170°C), or by the use of a temperature gradient, for example, 0.5°C/min (Dwek *et al.* 1970). Flame ionization is most commonly employed, though care has to be exercised in maintaining clean detectors as they become insensitive due to accumulation of fluorine decomposition products. The alternative procedure of reduction to fluoro-polyols and separation of the corresponding acetates by gas–liquid chromatography has yet to be fully explored.

Ionophoresis

Ionophoresis is valuable as a means of separating nearly related fluorosugars and as a means of providing configurational information by comparison with known monosaccharides. Johansson and Lindberg (1966) used borate and germanate buffers in establishing the structures of 2-deoxy-2-fluoro-D-altrose and 3-deoxy-3-fluoro-D-glucose, by comparison with 2-O-methylaltrose and 3-O-methylglucose, respectively (Table 2). 4-Deoxy-4-fluoro-D-glucose was found to have a mobility of 0.27 in borate buffer (pH 10) (Foster, Hems and Westwood 1970).

TABLE 2

Paper-electrophoretic separation of fluoromonosaccharides*

	Relative mobility	
	M_G *(borate buffer)*†	M_G *(germanate buffer)*†
2-deoxy-2-fluoro-D-altrose	0.65	2.58
2-deoxy-2-fluoro-D-allose	0.63	2.17
3-deoxy-3-fluoro-D-glucose	0.90	1.79
2-O-methyl-D-altrose	0.55	1.30
3-O-methyl-D-glucose	0.78	1.94

*   Johansson and Lindberg 1966.
†   (method of Lindberg and Swan 1960).

Isomorphism of fluorosugars

A considerable amount of evidence now supports the view that F may take the place of OH in, at least, monosaccharides with minimal disturbance of the interatomic disposition and intermolecular effects. Taylor has taken advantage of this concept in using isomorphous nucleation to induce crystallization of particular fluoroglycosides. Thus methyl 3-deoxy-3-fluoro-β-D-xyloside was crystallized from a reaction mixture resulting from the action of MeOH/HCl on the fluoropentose when a nucleating quantity of methyl β-D-xyloside was introduced (Wright and Taylor 1967).

The system of hydrogen bonds in the distinctive crystal structure common to erythritol and (±) 2-deoxy-2-fluoroerythritol led Bekoe and Powell (1959) to predict that other racemic fluorotritols would also adopt the same lattice, namely

$(\pm)$ 1-deoxy-1-fluoroerythritol
2,3-dideoxy-2,3-difluoroerythritol
1,4-dideoxy-1,4-difluoroerythritol.

Mass spectra

So far the application of mass spectral techniques to fluorosugars is in the exploratory stage, and though it may well be expected to provide valuable fingerprinting information for the characterization of main classes of these compounds, it is doubtful how far these techniques can discriminate between the configurational isomers. Foster, Hems and Westwood (1970) have used the technique in connection with the characterization of methyl 6-O-acetyl 4-deoxy-4-fluoro-2,3-di-O-methyl-α-D-glucoside. The fragmentation pattern (Fig. 1, Scheme 10) corresponded closely to that established previously for known acetylated and methylated hexopyranose derivatives, with accountable modi-

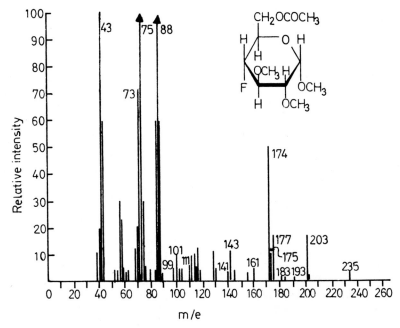

FIG. 1. Mass spectrum of methyl 6-O-acetyl-4-deoxy-4-fluoro-2,3-di-O-methyl-α-D-glucopyranoside.

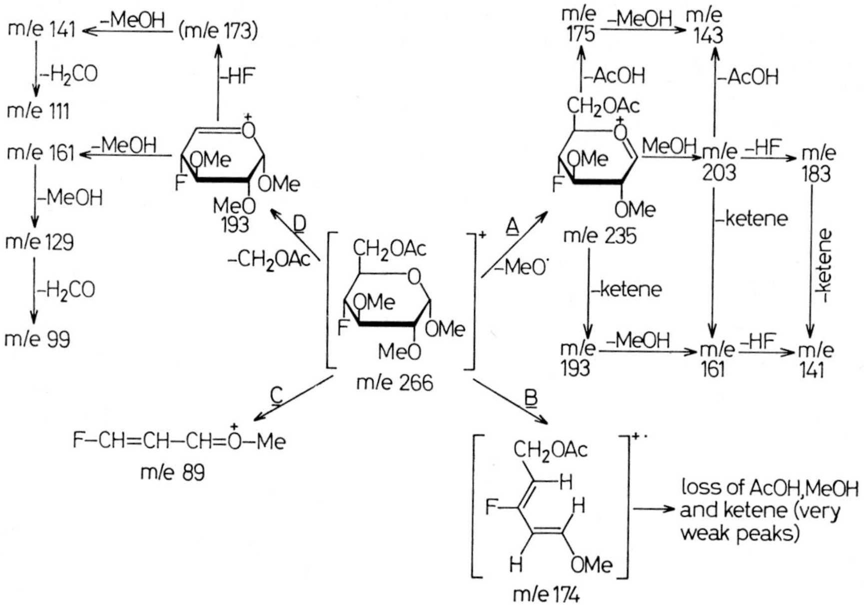

m/e 141
-MeOH
(m/e 173)
-H2CO
-HF
m/e 111
m/e 161
-MeOH
OMe
F
OMe
-MeOH
MeO
193
D
m/e 129
-CH2OAc
CH2OAc
-H2CO
OMe
F
OMe
MeO
m/e 99
m/e 266
C
F–CH=CH–CH=O–Me
m/e 89
m/e 175
-MeOH
m/e 143
-AcOH
-AcOH
CH2OAc
OMe
F
MeOH
m/e 203
-HF
m/e 183
OMe
m/e 235
A
-MeO·
-ketene
-ketene
-ketene
m/e 193
-MeOH
m/e 161
-HF
m/e 141
B
CH2OAc
H
F
H
H
OMe
m/e 174
loss of AcOH,MeOH and ketene (very weak peaks)

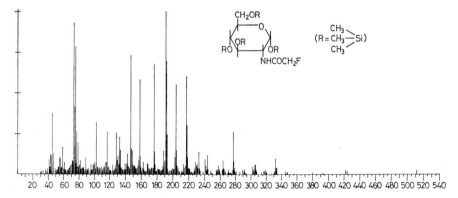

SCHEME 10: Mass spectrum (AEI, MS-12, 70eV, direct insertion, ion source temperature 80 °C) and fragmentation pattern of methyl 6-*O*-acetyl-4-deoxy-4-fluoro-2,3-di-*O*-methyl-α-D-glucopyranoside (from Foster, Hems and Westwood 1970).

fications resulting from the direct attachment of fluorine to the pyranose ring.

Mass-spectral characterization of *N*-fluoroacetyl-α-D-glucosamine has been achieved from the fragmentation pattern of the trimethylsilyl derivative (Fig. 2) (F. Wirtz-Peitz, unpublished results).

CH2OR
OR
RO
OR
NHCOCH2F
(R = CH3—Si)
CH3
CH3
20 40 60 80 100 120 140 160 180 200 220 240 260 280 300 320 340 360 380 400 420 440 460 480 500 520 540

FIG. 2. Mass spectrum and fragmentation pattern of per(trimethylsilyl) *N*-fluoroacetyl-α-D-glucosamine.

The principal fragments in the spectrum were closely related to those found in the corresponding pattern for N-acetyl-D-glucosamine with appropriate adjustment for the presence of fluorine in the former (Table 3).

TABLE 3

Mass-spectral assignments of per(trimethylsilyl) derivatives of N-fluoroacetyl-D-glucosamine, and N-acetyl-glucosamine

Fragments from per(trimethylsilyl) FNAG		Equivalent m/e of
m/e	Probable structure	N-acetyl- glucosamine
512	M^+- 15	494
424	M^+- 103	406
422	M^+- 105	404
332	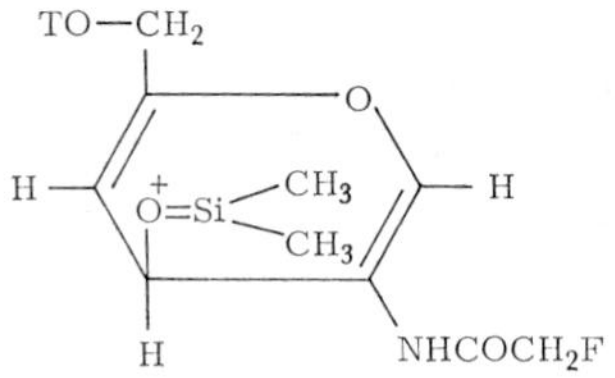 TO—CH$_2$ (cyclic structure)	314
277	TO—CH=C—CH=$\overset{+}{O}$T, with NH—COCH$_2$F	259
217	TO—CH=CH—CH=$\overset{+}{O}$T	217
204	[TO—CH=CH—OT]$^+$	204
191	[TO—CH=CH—NHCOCH$_2$F]$^+$	173
176	(CH$_3$)$_2$Si=$\overset{+}{O}$—CH=CH—NHCOCH$_2$F	158

STRUCTURE AND REACTIVITY

X-ray crystal analysis

With increased use of [19]F- n.m.r. data (see Dwek 1972), the structures of the majority of fluoromonosaccharides rest chiefly on nuclear magnetic resonance measurements of proton coupling constants and chemical shifts. Detailed investigation of two structures has been made by X-ray crystal analysis. In 1959 Bekoe and Powell reported details of the structure of ($\pm$) 2-deoxy-2-

fluoroerythritol, which was closely similar to that of erythritol in all respects.

Though a full analysis was not made, X-ray data were used in assigning the structure of 2-deoxy-2-fluoro-DL-ribitol obtained in crystalline form from a total synthesis (Kent and Barnett 1964). With a view to establishing the configuration of the halogen atoms, a full structural examination has been made of 3,4,6-tri-*O*-acetyl-2-bromo-2-deoxy-α-D-mannopyranosyl fluoride obtained by the addition of 'BrF' to triacetylglucal. The unit cell has the dimensions $a = 6.97 \pm 0.02$; $b = 15.47 \pm 0.03$ and $c = 14.72 \pm 0.03$ and the space group was $P2_1 P2_1 P2_1$. The C—F bond distance was 1.42 Å (E.S.D. 0.032). The results clearly establish the *trans* diaxial disposition of the Br and F atoms in this substance, details of which are shown in Fig. 3 (Campbell *et al.* 1969).

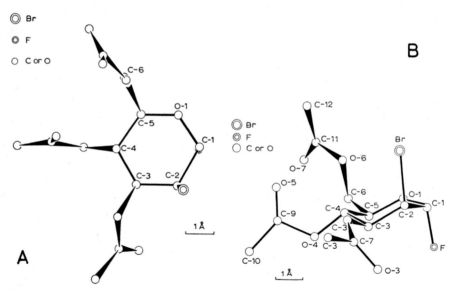

FIG. 3. The fluoride molecule projected (*A*) parallel and (*B*) perpendicular to the least-squares best plane through the atoms F, C-1, C-4, C-9, C-10, O-4 and O-5.

Glycoside formation

Monosaccharides carrying fluorine in an extra-cyclic position, as in 5-deoxy-5-fluoro-D-ribose, appear to undergo glycoside formation ·in the same manner and under conditions applicable to the unsubstituted sugar. Thus methyl α- and β-glycosides of 6-fluorogalactopyranose have been obtained by methanolic hydrogen chloride and by methanol/silver carbonate treatment of the galactosyl α-bromide, respectively. With fluorine in secondary positions, as in 3-deoxy-3-fluoroglucose, methanolic HCl leads to glycoside formation as indicated by loss of reducing properties and n.m.r. results, though separation of the anomeric glycosides has not been achieved by this route. It has been found by N.F. Taylor (unpublished results) that treatment of 3-fluoroglucose with benzyl alcohol and HCl gave an 80% yield of crystalline glycoside, purified by thin-layer chromatography, which had properties consistent with the β-compound. This benzyl β-glucoside and its tri-*O*-acetate each displayed a triplet with a chemical shift (in $CDCl_3$, 94 MHz) comparable with that displayed by 1,2,4,6-tetra-*O*-acetyl-3-deoxy-3-fluoro-β-D-glucose. The formation of a benzyl β-glycoside under these conditions has also been noted with D-arabinose (Ballou 1957).

Substitution reactions

Acylation (benzoylation and acetylation) of fluorocarbohydrates by anhydrides or acyl chlorides in anhydrous basic solvents (pyridine, *N*-methylaniline) follows a similar course to that of unsubstituted sugars, and peracylated derivatives are mainly obtained in good yield though difficulties may be encountered in their crystallization, particularly of the reducing sugar derivatives. Wright and Taylor (1968) found that 1,2,5-tri-*O*-acetates of 3-deoxy-3-fluoro-D-xylose and -arabinose reacted readily with HBr in acetic acid to give the corresponding glycosyl bromides from which adenyl nucleosides could be obtained, namely 9-(3′-deoxy-3′-fluoro-β-D-xylofuranosyl) adenine and its fluoroarabinose analogue. The peracetates of methyl fluoroglycosides appear to crystallize with greater ease and have been widely used as reference compounds for n.m.r. purposes.

Acetylation of 3-deoxy-3-fluoro-D-galactose with hot acetic anhydride and sodium acetate gave the crystalline β-tetra-*O*-acetate which was anomerized to the α-form in $ZnCl_2$ (Brimacombe *et al.* 1968).

Several examples of phosphorylated fluorosugars have been described (see also Taylor 1972). The first, 3-*O*-phosphate of 2-deoxy-2-fluoro-DL-glyceraldehyde, was obtained by treatment of the diethylacetal with diphenylphosphochloridate in pyridine, followed by hydrogenolysis. The product, obtained in moderate yield, was crystallized as the bis cyclohexylammonium salt (Kent, Marshall and Taylor 1966). Formation of 6-fluorogalactose α-1-phosphate was best carried out by a König-Knorr reaction from the corresponding triacetyl-fluorogalactosyl-α-bromide and silver diphenylphosphate in warm benzene (Kent and Wright 1972). The same compound resulted in an enzymic synthesis with the 6-fluorogalactose and galactokinase.

1-Deoxy-1-fluoro-DL-glycol 3-*O*-phosphate has been synthesized (Fondy, Ghangas and Reza 1970) from epifluorohydrin by reaction with excess phosphoric and/or dibenzylphosphoric acid. The compound was a substrate for NAD$^+$-linked glycerol 3-phosphate-dehydrogenase (rabbit muscle) with a K_m (approx.) of 8 mM at pH 7.5 and 6.5 mM at pH 9.0. The bromo and chloro analogues did not appear to be substrates for this enzyme except in the presence of 0.1 M-hydrazine.

Some further potentialities of this fluoro compound as an enzyme probe have been explored by J. R. Knowles and R. A. G. Smith (unpublished results), who confirm that it is a poor substrate for NAD$^+$-linked glycerophosphate dehydrogenase as shown by values for kinetic parameters and the equilibrium constant. The effect of hydrazine on the equilibrium suggests the occurrence of a carbonyl intermediate, possibly 1-fluoro-acetol-3-phosphate. The fluoroglycerol phosphate is a competitive inhibitor of this enzyme reaction with glycophosphate, dihydroxyacetone phosphate and NAD$^+$ as substrates. Triosephosphate isomerase is also competitively inhibited (K_i 5×10^{-3}M) by this fluoro compound.

As yet methylation has not been extensively employed in structural studies of fluorocarbohydrates. Such evidence as is available indicates that some fluorosugars, e.g. methyl 6-deoxy-6-fluoro-α-D-galactoside and isopropylidene-2-deoxy-2-fluoroerythritol, can be permethylated in good yields by the dimethyl sulphate–sodium hydroxide or methyl iodide–silver oxide methods, without loss of fluorine (Taylor, Kent and Morris 1960; Cherry and Kent 1962). Reagents such as BCl$_3$ are available for removal of stable protective groups in the synthesis of fluoromonosaccharides (see Scheme 4, p. 177).

N-Fluoroacylation, though not distinctive for the sugar series, nevertheless is worthy of mention in view of the use of such derivatives in the synthesis of probe molecules. *N*-Fluoroacetyl-β-D-glucosamine has been synthesized by treatment of 1,3,4,6-tetra-*O*-acetyl-β-D-glucosamine with sodium fluoroacetate and dicyclohexylcarbodi-imide in pyridine or in methylene dichloride. The derivative can also be synthesized with the α-anomer directly from glucosamine HCl in 20% yield with the same reagents (Dwek, Kent and Xavier 1971). *N*-Trifluoroacetyl-D-glucosamine (probably the α-form) has been obtained by Dr R. L. Foster by refluxing of the aminosugar hydrochloride with sodium trifluoroacetate and trifluoroacetic anhydride and de-*O*-acylation of the product in dry methanol (Foster and Capon 1971).

Elimination

Comparatively little work has yet been done on the relative stability of fluoro-atoms attached to the various carbon atoms of the pyranose ring. Evidence indicates that at C-6 and at the anomeric carbon, fluorine is substantially stable in alkaline reagents. Johansson and Lindberg (1966) reported that 3-deoxy-3-fluoro-D-glucose reacted with calcium hydroxide to give a mixture of 'α'- and 'β'-D-metasaccharinic acids, so supporting the evidence for the structure of the fluorosugar. It appears likely that difficulties encountered in alkaline deacetylation of derivatives of 2-deoxy-2-fluoroglucose may be due to eliminative loss of fluoride. Johansson and Lindberg (1966) also demonstrated that under mildly alkaline conditions (aqueous methanolic triethylamine overnight), acetylated 2-deoxy-2-fluoroaltrose was partially epimerized to 2-deoxy-2-fluoro-D-allose, in low yield. Higher yields (68 %) were obtained when epimerization was carried out at 60°C for 5 hours and conducted on a preparative scale. The route offers a useful means of obtaining some less accessible 2-fluorosugars.

Oxidation

In most cases so far studied, fluorosugars are oxidized smoothly and in the expected manner by sodium metaperiodate. In early studies, ($\pm$)2-deoxy-2-fluoro-erythritol was found to be quantitatively oxidized in 2 hours, 1 mol of IO_4^- being consumed. The reaction was subsequently used for the preparation of 2-deoxy-2-fluoro-DL-glyceraldehyde (2-fluoro-3-hydroxypropionaldehyde), the structure of which was confirmed by reduction to the corresponding 2-fluorodiol. Similarly, IO_4^- oxidation of methyl 6-deoxy-6-fluoro-α-D-galactoside proceeded rapidly, with the expected consumption of IO_4^-. So far, however, the accompanying product, 3-deoxy-3-fluoro-D-glyceraldehyde (3-fluoro-D-lactaldehyde) has evaded isolation, apparently through rapid polymerization.

Oxidation of epifluorohydrin with dimethyl sulphoxide in the presence of boron trifluoride provides a ready route to 3-fluorolactaldehyde (3-deoxy-3-fluoro-DL-glyceraldehyde) in the following way:

This fluoroaldehyde was characterized by ^{1}H- and ^{19}F-n.m.r. spectra. Attempts to obtain the 2,4-dinitrophenyl hydrazone led to loss of F$^-$ and the formation of the corresponding derivative of lactaldehyde. This fluoroaldehyde as well as 2-deoxy-2-fluoro-DL-glyceraldehyde acted as substrates for fructose-1,6-diphosphate aldolase from chicken muscle (J. R. Knowles and R. A. G. Smith, unpublished results).

Lead tetra-acetate, under the conditions used by Charlson and Perlin (1956), has also been used for investigating the structure of 2-deoxy-2-fluoro-D-altrose (Johansson and Lindberg 1966). Further aspects of the periodate oxidation of fluorosugars are discussed by Codington, Doerr and Fox (1966) in relation to their syntheses of 2-deoxy-2-fluoro-D-ribose from 2'-deoxy-2'-fluoro-D-uridine (see Scheme 5, p. 178). In none of these cases was gross over-oxidation encountered, nor did there appear to be formation of F$^-$. The Smith degradation (sequentially periodate oxidation, borohydride reduction and mild hydrolysis) was used successfully in the study of 1,1 *gem*-difluorides, namely 2,5-anhydro-1-deoxy-1,1-difluoro-D-mannitol (Kent, Barnett and Wood 1963). Vigorous oxidation of this compound with fuming nitric acid gave difluoroacetic acid, with negligible release of F$^-$.

Oxidation of aldehydic functional groups to carboxylic acids has been accomplished with bromine or sodium hypoiodite (Kent, Hebblethwaite and Taylor 1960), as in the formation of 2-deoxy-2-fluoroglyceric acid from the corresponding aldehyde. The hypoiodite reaction in that case could be used quantitatively for rapid estimation of the fluoroaldehyde in the range 0.01 to 0.08 mg/ml as well as in preparative reactions. The resulting DL-glyceric acid, characterized as the amide, was resolved into (+) and (−) quinine salts.

More vigorous oxidation of a 2-fluoro primary alcohol, as exemplified by 2-deoxy-2-fluoro-3,4-O-isopropylidene-erythritol, has been achieved without loss of fluorine with barium permanganate under neutral or slightly alkaline conditions (Cherry and Kent 1962). The selective oxidation of a -CHOH-CHF- group to -CO-CHF- proved more elusive since F is readily lost from the ketonic product. Investigation of this reaction in the case of 3,4,-di-O-acetyl (or benzoyl) -2-deoxy-2-fluoro-1-triphenylmethylerythritol by R. C. Young (unpublished results) gave only low yields of product with permanganate or Oppenauer reagents. Improved yields (60%) were obtained, however, when acetic anhydride-dimethyl sulphoxide mixtures were used. A significant by-product was identified as the 2-O-acetate, the formation of which tends to diminish the yields of the keto product. Owing to the lability of the product it has not yet been possible to remove the protecting groups to give the fluoro-tetrulose.

Recently, the synthesis of 3-deoxy-3-fluoro-D-gluconic acid has been achieved

(Taylor *et al.* 1972) by electrolytic oxidation of the 3-fluorosugar in dilute aqueous solution in the presence of $CaBr_2$ and $CaCO_3$ and with carbon electrodes. After passage of a steady current (40 mA) for 40 hours the product was isolated as a crystalline calcium salt in near-quantitative yield from the concentrated solution.

Reduction

As yet only metal hydrides (BH_4^- and AlH_4^-) have been used to any appreciable extent in the fluorocarbohydrate series. Treatment of fluoro-aldehyde and ketones with $NaBH_4$ or KBH_4 in ethanolic or methanolic solution at 0–5 °C results in rapid, smooth reduction. In this way, 2-fluoromalic acid was first synthesized from ethyl ethoxalylfluoroacetate (Taylor and Kent 1954, 1956). In a more detailed study of multifunctional compounds, Barnett and Kent (1963) found that fluoro-oxo esters underwent selective reduction (Scheme 9, p. 182) of oxo or ester function by KBH_4, depending on the conditions (Table 4).

Investigation of a number of 2-, 3- and 4-oxo-esters indicated that this unexpected reduction of the carboxylate function (not observed with unsubstituted esters) also occurred with the corresponding 2-, 3- and 4-hydroxy compounds. It is suggested that the reaction mechanism consists of a concerted complex involving both the oxo (or hydroxy) group and the carboxylate carbonyl group. These observations probably account for the much-used reduction of uronate esters to hexoses by BH_4^-, and may also account for the otherwise unprecedented reduction of aminoacrylate residues to (?DL) alanyl groups after elimination

TABLE 4

Reduction of oxo-esters by potassium borohydride.

	Conditions A		Conditions B		Conditions C		Conditions D	
Time (h) at 0°	2.5		1		1		1	
Time (h) at 20°	0		2		0		0	
Time (h) at 70°	0		0		2		5	
KBH_4 (mol)	1		1		1		2	
				Products				
	Hydroxy-	*Diol*	*Hydroxy-*		*Hydroxy-*		*Hydroxy-*	
Oxo-ester	*ester (%)*	*(%)*	*ester*	*Diol*	*ester*	*Diol*	*ester*	*Diol*
Methyl pyruvate	45	0	15	30	1	64	0	70
Ethyl acetoacetate	67	0	71	5	46	20	19	64
Methyl laevulate	77†	0	70†	0	65†	5	45†	20

† Isolated as pentano-1,4-lactone by heating at 70–75°/15 mm for 2 h.

of O-glycosyl derivatives of serine and threonine under alkaline conditions from certain polysaccharide–protein complexes.

Stereochemical dependence of BH_4^- reduction provided a key step in Foster's synthesis of 3-deoxy-3-fluoroglucose from 1,2:5,6-di-O-isopropylidene-α-D-allose. The latter was synthesized from the *gluco* epimer via the 3-oxo derivative and BH_4^- reduction, and resulted in the overall conversion of the OH configuration at C-3 (Foster, Hems and Webber 1967) (see Scheme 2, p. 173).

A little-explored route for the conversion of carboxylate esters into the corresponding aldehydes, under mild conditions suitable for the carbohydrate series, has been described by Kent and Barnett (1963). This takes advantage of the partial reduction of N-phenyl-N-methylimides by $LiAlH_4$ to -CHO and has been shown to be a means of converting methyl 2,3-O-isopropylidene-DL-glycerate into the correspondingly substituted glyceraldehyde.

Hydrogenolysis, with the Adams catalyst, is an efficient means of removing phenyl- or benzyl-ether or ester protecting groups of esters (Kent, Marshall and Taylor 1966; Taylor, Childs and Brunt 1964; Taylor and Riggs 1963), without disturbing fluorine substituents.

FLUOROCARBOHYDRATES AS PROBES OF SPECIFIC BINDING SITES IN MACROMOLECULAR SYSTEMS

Until now, virtually the only specific binding sites available for investigation in macromolecules have been those concerned with some distinctive chemical or biological activity. The now-extensive studies of those amino acid residues which comprise the catalytic sites and the binding sites of enzymes and antibodies in immunoglobulins illustrate the point. In a broader sense, the existence of allosterism, which provides a further means of controlling certain catalytic effects of enzymes, points to other types of specific binding sites outside the catalytic sites. Further evidence of specific interactions can be deduced from polymerase systems, such as glycogen synthetase, in which there is a specific dependence of the transferase on an acceptor (primer) for initiation of the polymerization. In principle, one may hypothesize that specific binding of small molecules by biological macromolecules may be a more general and widespread phenomenon than hitherto supposed, and that the above examples of enzymes and antibodies represent particular cases of the general phenomenon. Generally specific binding may be concerned with maintenance of tertiary structure and possibly with intermacromolecular interactions, for example glucose and α-crystallin.

The distinctive ^{19}F-n.m.r. properties of fluoromonosaccharides and their configurational distinctiveness for different positions of substitution in the ring (discussed by Dwek 1972) indicate that these sugars offer a promising route for the investigation of binding sites whether or not there is accompanying biological activity (Butchard *et al.* 1972).

Using conventional methods, Poretz and Goldstein (1970) showed that 3-deoxy-3-fluoroglucose was bound by concanavalin A 1–5 times less powerfully than D-glucose, whereas 6-deoxy-6-fluoro-D-glucose was 30 times less effective than glucose (So and Goldstein 1967), which points to the involvement of C-3 in the binding site and the probable exclusion of C-6. Introduction of fluorine into the aglycone, 2′,2′,2′-trifluoroethyl-β-D-glucopyranoside, appeared to enhance its binding capacity compared with the unfluorinated compound, ethyl β-D-glucoside. The lectins (plant haemagglutinins) offer a particularly interesting field for investigation by fluorocarbohydrate probes.

In view of the large amount of lysozyme in animal tissues, including human, a case may be advanced for this protein having other biological activities of a lectin-like nature and for its markedly basic properties as well as its hydrolytic action on chitin or *Micrococcus lysodeikticus* being incidental to its main function. The presence of 'mammalian lectins' of this sort is not unreasonable, and their possible function in masking selectivity, oligosaccharide side chains of glycolipids and glycoproteins, could play an important part in regulating the biosynthesis of such substances and in determining their final structure.

From studies of the effects of fluorosugars on ganglioside and glycoprotein biosynthesis in mouse cells either spontaneously transformed (T-AL/N) or transformed by the SV40 cancer virus (SV-AL/N), Kent and Mora (1971) have reported specific changes in aminosugar metabolism after the cells had been grown in 0.5 mM-*N*-fluoroacetylglucosamine or *N*-iodoacetylglucosamine. The overall effect of the modifiers may result in changes in the structures of oligosaccharides of membranes, hitherto produced by SV40 transformation (Brady and Mora 1970) and thought to arise from the absence or repression of UDP-*N*-acetylgalactosamine transferase. It appears that modification of aminosugar metabolism (at the intermediate level rather than at the level of uridyl transferases) is concerned in bringing about these effects, since neither sugar nor 6-deoxy-6-fluoro-D-galactose, 2-deoxy-2-fluoro-D-galactose nor 6-deoxy-6-fluoro-α-D-galactose -α-1-phosphate inhibit UDP-*N*-acetylgalactosamine transferase *in vitro* to ganglioside acceptors (F. A. Cumar, unpublished results).

CONCLUSION

Synthetic methods have now gone some way towards overcoming difficulties inherent in forming fluorosugars from their parent monosaccharides. In general these derivatives are stable and have many reactions in common with unsubstituted sugars. Detailed studies have shown a further close analogy between the molecular structures of these fluorinated and unsubstituted substances. ^{19}F-nuclear magnetic resonance investigations provide valuable means of determining the structure of fluorosugars and offer means of studying the binding of these sugars to macromolecules. The present results suggest that even more useful contributions to carbohydrate chemistry would be gained from studies of fluorine-substituted oligosaccharides, as well as providing further powerful molecular 'probes'.

Appendix

Fluoro-monosaccharides and their derivatives

(I)   Glyceraldehyde derivatives
(II) Tetrose derivatives
(III) Pentoses and derivatives
(IV) Hexoses and derivatives

Anomeric configurations are stated only in those cases where definite assignments are available.

n.c. = not crystalline
Ψ = λ 578 nm
*   = furanose form

I. FLUOROTRIOSES

	m.p. (°C)	$[\alpha]_D(°)$	t°C	Solvent	Reference
(±)2-Deoxy-2-fluoroglyceraldehyde (−)menthylhydrazone	150	—	—	—	Kent, Marshall and Taylor (1966)
diethyl acetal, 1,2-isopropylidene 3-*O*-phosphate, dicyclohexylammonium salt	104–105	—	—	—	Kent, Marshall and Taylor (1966)
2-Deoxy-2-fluoroglycerol 1,3-ditosyl-	109	—	—	—	Taylor and Kent (1956)
(±)1-Deoxy-1-fluoroglycerol	b.p. 42–45	—	—	—	Gryskiewicz-Trochimowski (1947)
(±)2-Deoxy-2-fluoroglyceric acid -amide	115	—	—	—	Kent, Hebblethwaite and Taylor (1960)
(−) quinine (+) salt	196	−151	21	H_2O	
(−) quinine (−) salt	179–180	−132	20	H_2O	

II. FLUOROTETROSES

	m.p. (°C)	$[\alpha]_D^t$(°)	t°C	Solvent	Reference
(±)**2-Deoxy-2-fluoroerythronic acid** -amide	103	—	—	—	Cherry and Kent (1962)
(±)**2-Deoxy-2-fluoro-erythritol**					
1,3,4-tri-*O*-tosyl-	85	—	—	—	Taylor and Kent (1956)
1,3,4-tri-*O*-mesyl	90	—	—	—	Barnett and Kent (1963)
1,3,4-tri-*O*-benzoyl	96–97	—	—	—	Barnett and Kent (1963)
2-Deoxy-2-fluoro-DL**-threitol**					
1,3,4-tri-*O*-tosyl-	115	—	—	—	Barnett and Kent (1963)
(±)**2,4-Difluoro-3-hydroxy-butanamide**					
(±)**2,4-Difluoro-1,3-butandiol**	70	—	—	—	Taylor and Kent (1956)
-bisphenylurethane	137–140	—	—	—	Barnett and Kent (1963)

III. FLUOROPENTOSES

	m.p. (°C)	$[\alpha]_D^t$(°)	t°C	Solvent	Reference
3-Deoxy-3-fluoro-D**-arabinose**					
free sugar (β)	120	−150→−105	20	c1.0, H₂O	Wright and Taylor (1967)
free sugar (α)	126–128	+75→+25.7	25	c1.7, H₂O	Wright and Taylor (1968)
-2,5-dichlorophenylhydrazone	120				Wright and Taylor (1967)
5-*O*-benzyl - (αβ)	n.c.	+107	20	c1.4, EtOH	Wright and Taylor (1967)
methyl - β-furanoside	n.c.	−62	20	c1,CHCl₃	Wright and Taylor (1968)
- phenylhydrazone	72				Wright and Taylor (1968)
Methyl glycoside (α)*	n.c.	+107	20	c1, EtOH	Wright and Taylor (1968)
- 2,5-diacetate*	n.c.				Wright and Taylor (1968)
- 2,5-dibenzoate*	81				Wright and Taylor (1968)
3-Deoxy-3-fluoro-L**-arabinose**					Taylor, Childs and Brunt (1964)

*   furanoside

	m.p. (°C)	$[\alpha]_D^t$ (°)	t°C	Solvent	Reference
2-Deoxy-2-fluoro-D-arabinose					
3,4-di-O-acetyl-β-fluoride trifluoromethyl	n.c.	−176	23	CHCl₃	Dwek et al. (1970)
3,4-di-O-acetyl-β-glycoside	n.c.	−226	26	c1.06, CHCl₃	Dwek et al. (1970)
2-Deoxy-2-fluoro-D-xylose					
3,4-di-O-acetyl-α-D-fluoride	109–111	−114	24	CHCl₃	Dwek et al. (1970)
trifluoromethyl 3,4-di-O-acetyl-β-glycoside	n.c.	−120	24	CHCl₃	Dwek et al. (1970)
2-Deoxy-2-fluoro-D-ribose					
free sugar -	106–112	−37	24	—	Codington, Doerr and Fox (1966)
methyl 3,4-di-O-benzoyl-glycoside	n.c.	−156	24	c0.4,CHCl₃	Codington, Doerr and Fox (1966)
methyl 3,5-di-O-benzoyl-(glycoside)	80–88	—	—	—	Codington, Doerr and Fox (1966)
methyl-(furanoside)	74–78	—	—	—	Codington, Doerr and Fox (1966)
2,5-anhydro-1-deoxy-1,1-difluoro-D-ribitol	75	+25	20	MeOH	Kent and Barnett (1966)
5-deoxy-5-fluoro-D-ribose					
- 2,5-dichlorophenylhydrazone	130	—	—	—	Taylor and Kent (1958)
methyl 2,3-isopropylidene-ß-glycoside	b.p. 32/0.025mm	−91.9	20	CHCl₃	Taylor and Kent (1958)
1-deoxy-1-fluoro-L-ribitol	89–90	—	—	—	Kent and Barnett (1966)
2-Deoxy-2-fluoro - (±)ribitol	90–91	—	—	—	Kent and Barnett (1964)
3-Deoxy-3-fluoro-D-xylose	125–7	+70 / +38	25	H₂O	Cohen, Levy and Bergmann (1964)
free sugar α	126–8	+75.3→ / +25	25	c1.7,H₂O	Wright and Taylor (1968)
	134–36	+37	20	c0.3,H₂O	Foster, Hems and Webber (1967)
1,2-O-isopropylidene (α)*	n.c.	−20	20	c1,CHCl₃	Foster, Hems and Webber (1967)
-p-toluenesulphonate	60–61	−27	—	c2,CHCl₃	Wright and Taylor (1967)
- 2,5-dichlorophenylhydrazone	75	—	—	—	

	m.p. (°C)	$[\alpha]_D(°)$	t °C	Solvent	Reference
-5-O-benzyl (αβ)	n.c.	-2.5	20	c1.1,EtOH	Wright and Taylor (1968)
Methyl pyranoside (β)	105	—	—	—	Wright and Taylor (1967)
- 2,4-ditosyl	138	-35	20	c0.57CHCl₃	Wright and Taylor (1967)
- 4-O-benzyl	84	-50.5	20	c1.35,CHCl₃	Wright and Taylor (1967)
Methyl furanoside (β)	n.c.	-63	20	c1.6,EtOH	Wright and Taylor (1968)
- 5-O-benzyl (β)	n.c.	-62	20	c1,CHCl₃	Wright and Taylor (1968)
- 2,5-di-O-benzoate	67	-22	22	c1.1,EtOH	Wright and Taylor (1968)
3-Deoxy-3-fluoro-D-xylose					
methyl 4-O-benzyl (β)					Taylor, Childs and Brunt (1964)
2-Deoxy-2-fluoro-D-xylose					
3,4-di-O-acetyl-α-fluoride	n.c.	—	—	—	Dwek et al. (1970)
trifluoromethyl 3,4-di-O-acetylglycoside (α)	150	+130	24	CHCl₃	Dwek et al. (1970)
5-Deoxy-5-fluoro-D-xylose					
free sugar	77–78	+52.7	24	c1.7,H₂O	Kent and Young (1971)
2,5-dichlorophenylhydrazone	100	—	—	—	Kent and Young (1971)
3,5-dideoxy-3,5-difluoro-D-xylose					
free sugar	n.c.	-9	22	c1-2,EtOH	Foster and Hems (1969)
- 2,5-dichlorophenylhydrazone	136.5–137	—	—	—	Foster and Hems (1969)
- 1,2-isopropylidene (α)	n.c.	-23	20	c2,CHCl₃	Foster and Hems (1969)
5-O-(2-p-phenylazobenzoyloxyethyl) (α)	76–77	—	—	—	Foster and Hems (1969)

IV. FLUOROHEXOSES

	m.p. (°C)	$[\alpha]_D(°)$	t °C	Solvent	Reference
2-Deoxy-2-fluoro-D-allose					
free sugar	165–167	+18 ψ →+28	22	c0.5,H₂O	Johansson and Lindberg (1966)
1,6-anhydro -	137–139	-79 ψ	22	c1,H₂O	Johansson and Lindberg (1966)

	m.p. (°C)	$[\alpha]_D(°)$	t °C	Solvent	Reference
2-Deoxy-2-fluoro-D-altrose					
free sugar	149–151	+42 → 53 ψ	22	c1, H₂O	Johansson and Lindberg (1966)
1,6-anhydro -	85–87	−270 ψ	22	c1, CHCl₃	Johansson and Lindberg (1966)
1,3,4,6-tetra-O-acetate (α)	112–114	+100 ψ	20	c1, CHCl₃	Johansson and Lindberg (1966)
1,3,4,6-tetra-O-acetate (β)	n.c.	+15 ψ	22	c1, CHCl₃	Johansson and Lindberg (1966)
6-Deoxy-6-fluoro-D-galactose					
free sugar	160	+135→ +76.5	20	H₂O	Taylor and Kent (1958)
- 2,5-dichlorophenylhydrazone	182	—	—	—	Taylor and Kent (1958)
Methyl α-pyranoside	139	+194	20	H₂O	Taylor and Kent (1958)
Methyl β-pyranoside	118	−1.3	21	H₂O	Kent and Wright (1972)
Methyl β-pyranoside tetra acetate	108	−11.0	22	CHCl₃	Kent and Wright (1972)
2,3,4-tri-O-acetyl-α-bromide	145–146	+231	20	CHCl₃	Kent and Wright (1972)
2,3,4-tri-O-methyl-α-glycoside	110	+151	20	CHCl₃	Taylor, Kent and Morris (1960)
2,3,4-trimethyl-1,5-lactone	b.p. 50/0.05	+71 → +11 (6 hr)	22	H₂O	Taylor, Kent and Morris (1960)
- phenylhydrazide	151–152	—	—	—	Taylor, Kent and Morris (1960)
α-1-phosphate (Mg salt)	decomp	+81	18	H₂O	Kent and Wright (1972)
1-Deoxy-1-fluoro-L-galactitol					
	173 −174	+4.2	21	H₂O	Kent and Wright (1972)
2-Deoxy-2-fluoro-D-glucose					
free sugar	170–176	+37° (2 min) +62 (3 h)	—	H₂O	Pacák, Točík and Černý (1969)
free sugar	160–165	+56	—	H₂O	Adamson et al. (1969)

	m.p. (°C)	$[\alpha]_D^t$ (°)	t°C	Solvent	Reference
Trifluoromethyl 3,4,6-tri-*O*-acetyl α	84–85	+158	—	CHCl₃	Adamson *et al.* (1969)
3,4,6-tri-*O*-acetyl-α-fluoride	91–92	+138	—	CHCl₃	Adamson *et al.* (1969)
3,4,6-tri-*O*-acetyl-α-fluoride	99–101	+75	25	CHCl₃	Hall *et al.* (1970)
3-Deoxy-3-fluoro-D-glucose					
free sugar	n.c.	+36Ψ	22	c0.5,H₂O	Johansson and Lindberg (1966)
free sugar	n.c.	+47	—	H₂O	Foster, Hems and Webber (1967)
free sugar	108	+64	22	c1,H₂O	Taylor *et al.* (1972)
2,5-dichlorophenylhydrazone	135	—	—	—	Taylor *et al.* (1972)
1,2,5,6-di-isopropylidene (α)	n.c.	−22	30	c1,CHCl₃	Foster, Hems and Webber (1967)
1,2,4,6-tetra-*O*-acetate (β)	119–120	−12	20	c0.9,CHCl₃	Foster, Hems and Webber (1967)
1,2,4,6-tetra-*O*-acetate (α)	n.c.	+15	22	c2,CHCl₃	Johansson and Lindberg (1966)
1,2-cyclohexylidene (α)	121–122	−12	25	c0.7,CHCl₃	Johansson and Lindberg (1966)
1,2-isopropylidene (α)	50–52	−18	—	c0.8,CHCl₃	Johansson and Lindberg (1966)
1,2-isopropylidene-5,6-benzeneboronate (α)	115–116	−2	—	c2.1,CHCl₃	Johansson and Lindberg (1966)
1,2-isopropylidene-5,6-carbonate (α)	84–85	−27	25	c1.0,CHCl₃	Johansson and Lindberg (1966)
3-Deoxy-3-fluoro-D-gluconic acid	119	−6	21	c1,H₂O	Taylor *et al.* (1972)
4-Deoxy-4-fluoro-D-glucose					
free sugar	182–186	+45	22	c0.5,H₂O	Barford *et al.* (1969)
free sugar					Foster, Hems and Westwood (1970)
Methyl 2,3-di-*O*-methyl, α-glucoside	b.p. 115–120 (0.2 mm)	+133	22	—	Foster, Hems and Westwood (1970)

	m.p. (°C)	$[\alpha]_D(°)$	t°C	Solvent	Reference
... 6-trityl (α)	129–122	+85	22	—	Foster, Hems and Westwood (1970)
... 6-O-acetate (α)	b.p. 120–130 (0.25 mm)	+115	22	—	Foster, Hems and Westwood (1970)
6-Deoxy-6-fluoro-D-glucose					
free sugar	155	+85.8 →+46.8	19	H_2O	Helferich and Gnüchtel (1941)
tetra-acetate	125–126	+20.1	19	pyridine	Helferich and Gnüchtel (1941)
1,2-isopropylidene-3,5-benzylidene-α-	104–105	+14.2	21	benzene	Helferich and Gnüchtel (1941)
triacetyl α-bromide	127–128	+234	21	$CHCl_3$	Helferich and Gnüchtel (1941)
methyl α-pyranoside	113–115	+135	25	cl,H_2O	So and Goldstein (1967)
... trimesyl	133–134	+93.1	22	pyridine	Helferich and Gnüchtel (1941)
phenyl β-pyranoside	148–149	−79	21	H_2O	Helferich and Gnüchtel (1941)
... triacetate	167–168	−8.2	19	$CHCl_3$	Helferich and Gnüchtel (1941)
vanillin β-pyranoside	181–182	−48.6	19	pyridine	Helferich and Gnüchtel (1941)
... triacetate	166–167	−35.7	20	$CHCl_3$	Helferich and Gnüchtel (1941)
2-Deoxy-2-fluoro-D-galactose					
free sugar (α)	131–135	+78.5 (5 min) →+92 (12 hr)	—	H_2O (c2.3)	Adamson and Marcus (1970)
3,4,6-tri-O-acetyl-α-fluoride	68–70	+130	20	$CHCl_3$	Dwek et al. (1970)
	67–68	+151	—	$CHCl_3$ (c3.2)	Adamson and Marcus (1970)

	m.p. (°C)	[α]$_D$ (°)	t °C	Solvent	Reference
trifluoromethyl 3,4,6-tri-O-acetyl-α-glycoside	53–55	+147	25	CHCl$_3$	Dwek et al. (1970)
	71–72	+136	—	CHCl$_3$ (c2.8)	Adamson and Marcus (1970)
3-deoxy-3-fluoro-D-galactose					
free sugar	114–116	+76 (equil)	—	c0.9,H$_2$O	Brimacombe et al. (1968)
1,2:5,6-di-O-isopropylidene (α)	48–49	−30	—	c1.0	Brimacombe et al. (1968)
β-tetra-O-acetate	126–127	+35	—	c0.8,CHCl$_3$	Brimacombe et al. (1968)
α-tetra-O-acetate	97–99	+26	—	c1.1,CHCl$_3$	Brimacombe et al. (1968)
3-Deoxy-3-fluoro-D-gulose					
methyl α-pyranoside	122–123	+146	28	MeOH	Harrison and Kent (unpublished results)
2,6-di-O-benzyl	61	+64	22	CHCl$_3$	Harrison and Kent (unpublished results)
2-Deoxy-2-fluoromannose					
Trifluoromethyl 3,4,6-tri-O-acetyl-β-	96–97	−21	—	CHCl$_3$	Adamson et al. (1969)
3,4,6-tri-O-acetyl-β-fluoride	113–114	−3.5	—	CHCl$_3$	Adamson et al. (1969)
3,4,6-tri-O-acetyl-α-fluoride	89–90	+27	25	CHCl$_3$	Hall et al. (1970)
2,5-anhydro-1-deoxy-1,1-difluoro-D-mannitol.					
- tri-O-tosyl	95	+16	25	CHCl$_3$	Wood and Kent (1967)
Aminosugar derivatives					
N-Fluoroacetyl-β-D-glucosamine	161–163	+22.8	25	c0.3,H$_2$O	Dwek, Kent and Xavier (1971)
tetra-O-acetate	185	+1.2	20	c1,CHCl$_3$	Dwek, Kent and Xavier (1971)
N-trifluoroacetyl-α-D-glucosamine	119–202	+13.6 (equil)	30	c1,H$_2$O	Foster and Capon (1971)
N-Fluoroacetyl-D-neuraminic acid	155–157	−32.5	25	H$_2$O	Schauer, Wirtz-Peitz and Faillard (1970)

References

ADAMSON, J., A. B. FOSTER, L. D. HALL and R. H. HESSE (1969) *Chem. Commun.* 309–310.

ADAMSON, J. and D. M. MARCUS (1970) *Carbohydr. Res.* **13**, 314–316.

BALL, D. H. and F. W. PARRISH (1969) *Adv. Carbohydr. Chem.* **24**, 139–197.

BALLOU, C. E. (1957) *J. Am. Chem. Soc.* **79**, 165–166.

BARFORD, A. D., A. B. FOSTER, J. H. WESTWOOD and L. D. HALL (1969) *Carbohydr. Res.* **11**, 287–288.

BARNETT, J. E. G. (1967) *Adv. Carbohydr. Chem.* **22**, 177–227.

BARNETT, J. E. G. and P. W. KENT (1963) *J. Chem. Soc.* 2743–2747.

BARTON, D. H. R., L. J. DANKO, A. K. GANGULY, R. H. HESSE, G. TARZIA and M. M. PECKET (1969) *Chem. Commun.* 227.

BEKOE, A. and H. M. POWELL (1959) *Proc. R. Soc. A.* **250**, 301–315.

BENSON, A. A., J. A. BASSHAM, M. CALVIN, T. C. GOODALE, V. A. HAAS and W. STEPKA (1950) *J. Am. Chem. Soc.* **72**, 1710–1718.

BRADY, R. O. and P. T. MORA (1970) *Biochim & Biophys. Acta* **218**, 308–319.

BRIMACOMBE, J. S., A. B. FOSTER, R. HEMS and L. D. HALL (1968) *Carbohydr. Res.* **8**, 249–250.

BUCK, K. W., A. B. FOSTER, R. HEMS and J. M. WEBBER (1966) *Carbohydr. Res.* **3**, 137–138.

BUTCHARD, G. C., R. A. DWEK, P. W. KENT, R. J. P. WILLIAMS and A. XAVIER (1972) *Eur. J. Biochem.* In press.

BUTCHARD, G. C. and P. W. KENT (1971) *Tetrahedron* **27**, 3457–3463.

CAMPBELL, J. C., R. A. DWEK, P. W. KENT and C. K. PROUT (1968) *Chem. Commun.* 34–35.

CAMPBELL, J. C., R. A. DWEK, P. W. KENT and C. K. PROUT (1969) *Carbohydr. Res.* **10**, 71–77.

CHARLSON, A. J. and A. S. PERLIN (1956) *Can. J. Chem.* **34**, 1200–1208.

CHERRY, R. C. and P. W. KENT (1962) *J. Chem. Soc.* 2507–2509.

CODINGTON, J. F., I. L. DOERR and J. J. FOX (1964) *J. Org. Chem.* **29**, 558–564.

CODINGTON, J. F., I. L. DOERR and J. J. FOX (1966) *Carbohydr. Res.* **1**, 455–466.

COHEN, S., D. LEVY and E. D. BERGMANN (1964) *Chem. & Ind.* 1802–1803.

DWEK, R. A. (1972) *This volume*, pp. 239–271.

DWEK, R. A., P. W. KENT, P. T. KIRBY and A. S. HARRISON (1970) *Tetrahedron Lett.* 2987–2990.

DWEK, R. A., P. W. KENT and A. XAVIER (1971) *Eur. J. Biochem.* **23**, 343–348.

FISCHER, E., M. BERGMANN and H. SCHOTTE (1920) *Chem. Ber.* **53B**, 509–547.

FLETCHER, H. G., JR. and H. W. DIEHL (1971) *Carbohydr. Res.* **17**, 383–391.

FONDY, T. P., G. S. GHANGAS and M. J. REZA (1970) *Biochemistry* **9**, 3272–3280.

FOSTER, A. B. and R. HEMS (1969) *Carbohydr. Res.* **10**, 168–171.

FOSTER, A. B., R. HEMS and J. M. WEBBER (1967) *Carbohydr. Res.* **5**, 292–301.

FOSTER, A. B., R. HEMS and J. H. WESTWOOD (1970) *Carbohydr. Res.* **15**, 41–49.

FOSTER, R. L. and B. CAPON (1971) *Chem. Commun.* 512–513.

FRIED, J. and E. F. SABO (1954) *J. Am. Chem. Soc.* **76**, 1455–1456.

FRIED, J. and E. F. SABO (1955) *J. Am. Chem. Soc.* **77**, 3181, 3166, 1068.

GANTT, R., S. MILLNER and S. B. BINKLEY (1964) *Biochemistry* **3**, 1952–1960.

GOLDMAN, P. (1969) *Science* **164**, 1123–1130.

GREIG, C. G. and D. H. LEABACK (1963) *J. Chem. Soc.* 2644–2647.

GRYSZKIEWICZ-TROCHIMOWSKI, E. (1947) *Recl. Trav. Chim. Pays-Bas Belg.* **66**, 427–429.

HALL, L. D., R. N. JOHNSON, J. ADAMSON and A. B. FOSTER (1970) *Chem. Commun.* 463–464.

HALL, L. D. and J. F. MANVILLE (1968) *Chem. Commun.* 35–36, 37–38.

HELFERICH, B. and A. GNÜCHTEL (1938) *Chem. Ber.* **71**, 712–718.

HELFERICH, B. and A. GNÜCHTEL (1941) *Chem. Ber.* **74**, 1035–1039.

HELFERICH, B. and E. HIMMEN (1928) *Chem. Ber.* **61B**, 712,1825–1835.

HENBEST, H. B. and W. R. JACKSON (1962) *J. Chem. Soc.* 954–959.

HILL, J., L. HOUGH and A. C. RICHARDSON (1968) *Carbohydr. Res.* **8**, 7–8, 19–28.

JOHANSSON, I. and B. LINDBERG (1966) *Carbohydr. Res.* **1**, 467–473.
KENT, P. W. (1969) *Chem. & Ind.* 1128–1132.
KENT, P. W. and J. E. G. BARNETT (1963) *Nature (Lond.)* **197**, 492–493.
KENT, P. W. and J. E. G. BARNETT (1964) *J. Chem. Soc.* 2497–2500.
KENT, P. W. and J. E. G. BARNETT (1966) *Tetrahedron* **7**, suppl. 69–74.
KENT, P. W., J. E. G. BARNETT and K. R. WOOD (1963) *Tetrahedron Lett.* 1345–1348.
KENT, P. W., G. HEBBLETHWAITE and N. F. TAYLOR (1960) *J. Chem. Soc.* 106–108.
KENT, P. W., D. R. MARSHALL and N. F. TAYLOR (1966) *J. Chem. Soc.* 1281–1282.
KENT, P. W. and P. T. MORA (1971) *Abstr. Meet. Am. Chem. Soc.* Washington, abstr. 254.
KENT, P. W., A. MORRIS and N. F. TAYLOR (1960) *J. Chem. Soc.*, 298–303.
KENT, P. W., F. O. ROBSON and V. A. WELCH (1963) *J. Chem. Soc.* 3273–3276.
KENT, P. W. and J. R. WRIGHT (1972) *Carbohydr. Res.* In press.
KENT, P. W. and R. C. YOUNG (1971) *Tetrahedron* **27**, 4057–4064.
KUNYANTS, I. L., O. V. KIL'DESHEVA and I. P. PETROV (1949) *Zh. Obshch. Khim.* **19**, 95–109.
LEMIEUX, R. U. and B. FRAZER-REID, B. (1965) *Can. J. Chem.* **43**, 1460–1475.
LINDBERG, B. and B. SWAN (1960) *Acta Chem. Scand.* **14**, 1043–1050.
MICHEEL, F. and A. KLEMER (1961) *Adv. Carbohydr. Chem.* **16**, 85–103.
MICHEEL, F. and E. MICHAELIS (1957) *Chem. Ber.* **91**, 188–194.
MICHEEL, F. and H. WULFF (1956) *Chem. Ber.* **89**, 1521–1530.
MORTON, G. O., J. E. LANCASTER, G. E. VAN LEAR, W. FULMOR and W. E. MEYER (1969) *J. Am. Chem. Soc.* **91**, 1535–1537.
PACÁK, J., Z. TOČÍK and M. ČERNÝ (1969) *Chem. Commun.* 77.
PARKER, A. J. (1962) *Q. Rev. Chem. Soc.* **16**, 163–187.
PATTISON, F. L. M. (1954) *Nature (Lond.)* **174**, 737–741.
PATTISON, F. L. M. and J. J. NORMAN (1957) *J. Am. Chem. Soc.* **79**, 2311–2316.
PATTISON, F. L. M., D. A. V. PETERS and F. H. DEANE (1965) *Can. J. Chem.* **43**, 1689.
PORETZ, R. D. and I. J. GOLDSTEIN (1970) *Biochemistry* **9**, 2890–2896.
RICHARDSON, A. C. (1969) *Carbohydr. Res.* **10**, 395–402.
SCHAUER, R., M. WEMBER, F. WIRTZ-PEITZ and C. FERREIRA DO AMARAB (1971) *Hoppe-Seyler's Z. Physiol. Chem.* **352**, 1073–1080.
SCHAUER, R., F. WIRTZ-PEITZ and H. FAILLARD (1970) *Hoppe-Seyler's Z. Physiol. Chem.* **351**, 359–364.
SO, L. L. and I. J. GOLDSTEIN (1967) *J. Immunol.* **99**, 158–163.
STRACHAN, R. G., W. V. RUYLE, T. Y. SHEN and R. HIRSCHMANN (1966) *J. Org. Chem.* **31**, 507–509.
TAYLOR, N. F. (1958) *Nature (Lond.)* **182**, 660–661.
TAYLOR, N. F. (1972) *This volume*, pp. 215–235.
TAYLOR, N. F., R. F. CHILDS and R. V. BRUNT (1964) *Chem. & Ind.* 928–929.
TAYLOR, N. F., B. HUNT, P. W. KENT and R. A. DWEK (1972) *Carbohydr. Res.* In press.
TAYLOR, N. F. and P. W. KENT (1954) *Nature (Lond.)* **174**, 40.
TAYLOR, N. F. and P. W. KENT (1956) *J. Chem. Soc.* 2150–2154.
TAYLOR, N. F. and P. W. KENT (1958) *J. Chem. Soc.* 872–875.
TAYLOR, N. F., P. W. KENT and A. MORRIS (1960) *J. Chem. Soc.* 298–303.
TAYLOR, N. F. and G. M. RIGGS (1963) *J. Chem. Soc.* 5600–5603.
TIPSON, R. S. (1953) *Adv. Carbohydr. Chem.* **8**, 107–215.
WELCH, V. A. and P. W. KENT (1962) *J. Chem. Soc.* 2266–2270.
WOLFROM, M. L. and H. B. BHAT (1967) *J. Org. Chem.* **32**, 1821–1832.
WOOD, K. R., D. FISHER and P. W. KENT (1966) *J. Chem. Soc.* 1994–1997.
WOOD, K. R. and P. W. KENT (1967) *J. Chem. Soc.* 2422–2425.
WRIGHT, J. A. and N. F. TAYLOR (1967) *Carbohydr. Res.* **3**, 333–339.
WRIGHT, J. A. and N. F. TAYLOR (1968) *Carbohydr. Res.* **6**, 347–354.

Discussion

Foster: During the last few years synthetic methods have developed to the point where fluorinated carbohydrates are now relatively accessible (Foster and Westwood 1972) and many of them are available in quantities which ought to stimulate interest in their biological activity. One can study the coupling between protons and fluorine and the manifestations of this in the n.m.r. pattern in a wide range of fluorinated carbohydrates. As a result of the accumulation of this data, especially on long-range (^4J, ^5J) F—H couplings, one can use ^{19}F resonances to probe the configuration and conformation of carbohydrates in conditions where proton resonances are not easily observed. Two ^{19}F signals are observed for an aqueous solution of 3-deoxy-3-fluoroglucose compound and these can be assigned to the α- and β-pyranose forms on the basis of the splitting pattern which reflects coupling between the fluorine in each anomer and the various protons. Fig. 1 illustrates this situation for 4-fluoro-D-glucose.

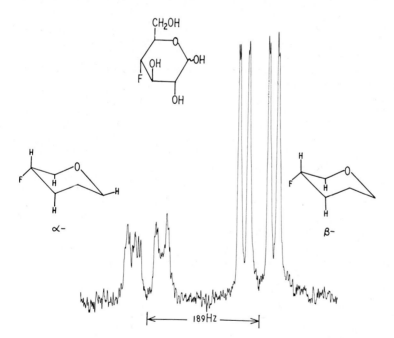

FIG. 1 (Foster). Structure and ^{19}F-n.m.r. pattern for 4-fluoro-D-glucose.

From the fluorine resonance one can determine the configuration at a carbon atom at the other end of the molecule. In the β-anomer three hydrogens are coupled to fluorine to give a relatively simple splitting pattern: in the α-anomer four are involved thereby giving a more complex signal. Thus one can use a fluorine substituent to investigate anomeric equilibria (Barford *et al.* 1971). We have recently obtained results in the 3-fluoro-L-idose series in which not only are the two pyranose forms present as in glucose but also the two furanose forms and one can identify each anomer quite easily.

Taylor: Is 4-fluoroglucose obtained as a crystalline conglomerate of equal mixtures of the α- and β-anomers? This is what one gets with 3-fluoroglucose, and it strikes me as rather unusual that one should end up with an equal conglomerate of the anomers in a crystalline form.

Foster: We have crystallized 3-fluoroglucose under a variety of conditions and we have always obtained the near-equimolar mixture of anomers which is, I think, similar to the anomeric equilibrium mixture in aqueous solution. 4-Fluoroglucose, however, can be crystallized in a form which is essentially the β-anomer which, on dissolution in water, gives a normal mutarotational equilibrium which can be followed by conventional procedures. However, crystallizing conditions may exist which will give the α- and β- forms of 3-fluoroglucose.

Taylor: This is true. In solution our crystalline conglomerate of 3-fluoro-glucose does not rotate. Is this phenomenon rare in carbohydrate chemistry?

Barnett: I have made all the 3-deoxy-3-halogeno-D-glucoses, and each one crystallizes as an equilibrium mixture of α- and β-anomers. There is no n.m.r. evidence, but the crystalline forms have a sharpish melting point, and do not mutarotate.

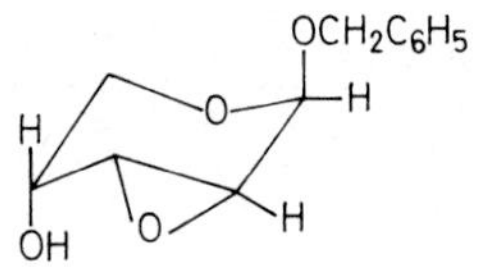

(I) 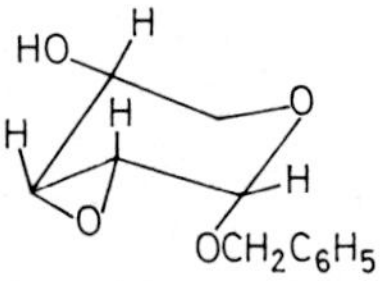   (II)

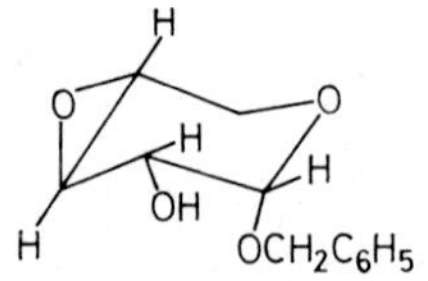

(III) 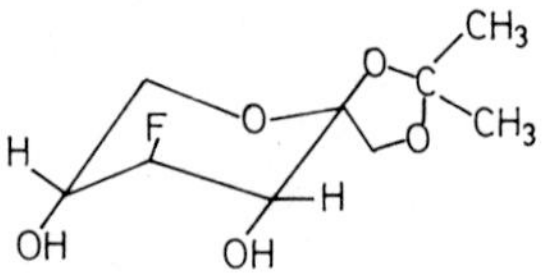   (IV)

Bergmann: We had previously shown (Cohen, Levy and Bergmann 1964) that benzyl 3-*O*-tosyl-β,D-arabinopyranoside reacts with potassium fluoride in acetamide to give benzyl 3-deoxy-3-fluoro-β,D-xylopyranoside via the intermediate 2,3-anhydro-compound (I). The same treatment transforms benzyl

3-*O*-tosyl-α,D-xylopyranoside in 25% yield into benzyl 3-deoxy-3-fluoro-α,D-xylopyranoside (S. Cohen, N. Friedman and E. D. Bergmann, unpublished). The formula of this compound is established by its inertness towards periodate and by the infrared spectrum. Also in this case, an anhydro-sugar (II) is the intermediate in the reaction. In fact, when benzyl 3-*O*-tosyl-α,D-xylopyrano-side was treated with sodium methoxide in benzene, two anhydrosugars were obtained, an oily one for which formula II was established and which gave the same final product on treatment with potassium fluoride, and a crystalline one which has formula III. The formulae are based on the infrared spectra and the application of Hudson's rule. We have also recently prepared the first fluoro-deoxyketo sugar, 4-fluoro-4-deoxy-D-sorbopyranose, as its acetonide (IV), starting from the known di-isopropylidene-fructose (M. Sarel and E. D. Berg-mann, unpublished).

Kent: Is this compound stable?

Bergmann: Yes, it crystallizes beautifully.

Kent: We tried to make 3-deoxy-3-fluoro-DL-erythrulose but the free keto sugar was not obtained owing to the difficulty in removing the blocking groups.

Bergmann: But did you have the fluorine in the α-position to the carbonyl?

Kent: Yes.

Bergmann: It is interesting to note that in the sugar series one gets iso-morphism if a fluorine atom is substituted for a hydroxyl group, whereas in the aromatic series there appears to be isomorphism if a hydrogen atom is replaced by a fluorine atom. 2-Fluoronaphthalene, for example, is isomorphous with naphthalene or 4-fluorobiphenyl with biphenyl but not with the corre-sponding hydroxyl compound.

Kent: Could one of the major contributors to the lattice energy in the sugar be the multiplicity of hydrogen bonding?

Bergmann: Yes.

Heidelberger: Could you elaborate on what you feel are the implications of these studies of the transformed cells relative to carcinogenesis?

Kent: Any speculation on the implications for carcinogenesis would be premature. I am interested in seeing whether one can alter the confluence-growth properties by biochemical manipulation. If contact phenomena depend on a new sort of membrane-bound information, mediated by the oligosacchar-ides of the surface membranes, then it is important to know whether im-munogenicity and tumorigenicity are changed as a result of deliberate interven-tion with enzymes synthesizing these oligosaccharides.

Wettstein: Are these reactions with fluoroxy-trifluoromethane your work?

Kent: We have done some and Professor Foster has too.

Heidelberger: I am puzzled because you are comparing one malignant cell with another maligant cell rather than a non-malignant cell with a malignant one.

Kent: One has the normal untransformed cell from which T-AL/N is spontaneously derived and produces the other by SV40 infection. So these are kindred cells.

Heidelberger: Do you have data on the normal cells?

Kent: The normal cells closely resemble the spontaneously transformed cells (T-AL/N) in many respects such as ganglioside composition.

Heidelberger: When you treated these SV40 transformed cells with N-fluoroacetylglucosamine did you get a diminution in the quantity of these higher gangliosides?

Kent: There was also a substantial reduction in the neuraminidase hydrolysable sialic acid residues, which would seem to implicate glycoproteins and gangliosides (Kent and Mora 1971).

Heidelberger: What does this do to the contact inhibition or malignancy of these cells?

Kent: We cannot make any statement about that yet.

Foster: If the fluorinated sugars replace the ordinary ones in the cell surface this could have an effect on the function of the cell surface. Is there any information on this?

Kent: At present we have no data, but the possiblity exists. At the moment we do not know whether the N-fluoroacetylglucosamine or 6-fluorogalactose are incorporated. One of the sad things about this field is the lack of a good isotope of fluorine.

Foster: Both of these sugars are of the type that appear in cell surfaces.

Kent: That is why we tried them. Professor Faillard's group in Bochum have been interested in N-fluoroacetylneuraminic acid which is a marked inhibitor of the acylneuraminate aldolase by which a synthesis can occur (Schauer, Wirtz-Peitz and Faillard 1970). If one can achieve, by rational biochemical means, a change not only in the general oligosaccharide structure but in the sialic acid content, then one may vary extensively what is generally thought to be the 'foreignness' of some of these cells.

References

ADAMSON, J., A. B. FOSTER, L. D. HALL, R. N. JOHNSON and R. H. HESSE (1970) *Carbohydr. Res.* **15**, 351.

BARFORD, A. D., A. B. FOSTER, J. H. WESTWOOD, L. D. HALL and R. N. JOHNSON (1971) *Carbohydr. Res.* **19**, 49.

COHEN, S., D. LEVY and E. D. BERGMANN (1964) *Chem. & Ind.* 1802.

FOSTER, A. B., R. HEMS and J. M. WEBBER (1967) *Carbohydr. Res.* **5**, 292.

FOSTER, A. B. and J. H. WESTWOOD (1972) *Angew. Chem.* In press.

HENBEST, H. B. and W. R. JACKSON (1962) *J. Chem. Soc.* 954.

KENT, P. W. and P. T. MORA (1971) *Abstr. Meet. Am. Chem. Soc.* Washington, abstr. 254.

SCHAUER, R., F. WIRTZ-PEITZ and H. FAILLARD (1970) *Hoppe-Seyler's Z. Physiol. Chem.* **351**, 359–364.

The metabolism and enzymology of fluorocarbohydrates and related compounds

N. F. TAYLOR

Biochemistry Group, School of Biological Sciences, University of Bath, Somerset

Interest in the natural occurrence of fluorinated organic compounds probably arose as a result of the discovery of fluoroacetic acid in the leaves of the South African shrub *Dichapetalum cymosum* by Marais (1944). Since then fluoroacetic acid has been identified in either the seeds or leaves of a large number of plants (Peters 1972). Other naturally occurring compounds reported to contain the carbon-fluorine (C—F) bond are ω-fluoro-oleic acid (Peters *et al.* 1960), ω-fluorocaproic, ω-fluoromyristic and ω-fluoropalmitic acids (Ward, Hall and Peters 1964). Recently nucleocidin, an anti-trypanosomal antibiotic, has been assigned the structure of a nucleoside (1) in which the fluorine atom is attached to the carbohydrate moiety (Morton *et al.* 1969), and the volatile constituent

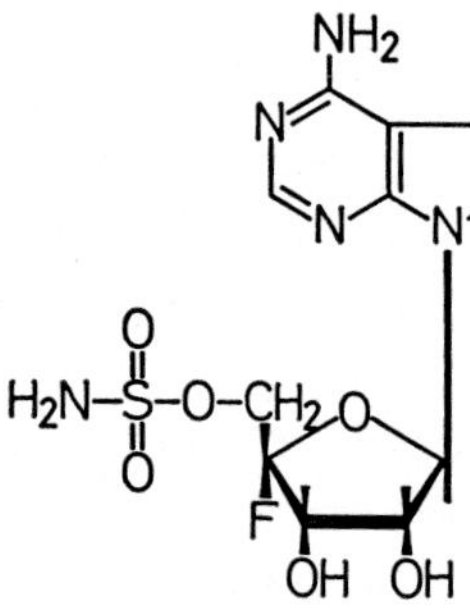

(1)

formed by homogenates of *Acacia georginae* has now been identified as fluoro-acetone (Peters and Shorthouse 1971). Evidence is accumulating, therefore, which suggests that the presence of the C—F bond in nature is more common than was originally thought. Further interest in fluorinated compounds was undoubtedly stimulated by the elucidation of the biochemical mode of toxic

action of fluoroacetate by Sir Rudolph Peters and co-workers (1953). One of
the many important aspects of this work resided in the demonstration that a
naturally occurring fluorinated substance, which in itself was non-toxic, was
enzymically transformed into a toxic product which retained the C—F bond.
To those interested in the structural requirements of metabolic transformations
in the cell such a 'lethal synthesis' shattered any previous misconceptions
about enzyme specificity. This was undoubtedly a factor which led to the
development of new techniques for the synthesis of a wide range of fluorinated
compounds for their biological activity although, as pointed out by Smith
(1970), the biochemical rationale for many of these synthetic substances is not
always apparent.

In many of the examples in which biological activity has been observed it is
a hydrogen atom of the natural substrate that has been replaced by fluorine,
so producing a molecule capable of 'deceiving' one or more enzymes and
thereby leading to inhibition, lethal synthesis or incorporation. Bergmann
(1961) has argued that the similarity in the Van der Waals' radii between the
elements hydrogen (1.20 Å) and fluorine (1.35 Å) is sufficiently close to account
for pseudosubstrate activity. The physicochemical similarity between C—F and
C—OH, however, may be considered closer in a number of respects (e.g. bond
length, force constants, atomic refractivity and electronegativity) and this
suggested that the fluorine atom might act as a hydroxyl or oxy analogue in
biochemical transformations. Support for this idea was also found in the close
similarity in crystal lattice structures between erythritol and 2-deoxy-2-fluoro-
erythritol (Bekoe and Powell 1959). As a consequence of this basic idea,
considerable effort was devoted between 1953 and 1965 to the synthesis of
monosaccharides and related compounds in which a hydroxyl group had been
stereospecifically replaced by fluorine (Kent 1969, 1972). Since hydrogen
bonding is often involved with enzyme-substrate interaction and fluorine and
hydroxyl can hydrogen bond in the same direction, it is possible that fluorinated
substrates and/or metabolites may bind especially to allosteric enzymes and
exert some interesting metabolic effects. As indicated elsewhere (Dwek 1972),
the extent and nature of hydrogen bonding of fluorine with an enzyme may be
detectable by the chemical shifts observed in the ^{19}F-n.m.r. spectrum, and this
will probably prove to be a most useful tool in future studies. A wide range of
fluorinated carbohydrates may now be synthesized and as these become available a rapid growth in their application to biochemical systems might be anticipated.

The first reported enzyme studies with fluorinated sugars were those by
Helferich, Grünler and Gnüchtel (1937). The aim of this work was to examine
the enzyme specificity of β-glucosidase. They demonstrated that both phenyl

(2) and vanillyl 6-deoxy-6-fluoro-β-D-glucopyranosides (3) would act as substrates for this enzyme (as measured by K_m), which suggested that the nature of the aglycone was not critical providing it had the β-configuration, and that position C-6 was probably not involved directly with the active site. Further,

(2) R=C$_6$H$_5$
(3) R=C$_8$H$_7$O$_2$

substitution of fluorine at the C-6 position of (2) and (3) by chlorine, bromine and iodine did not prevent hydrolysis, and Helferich established that the enzyme activity was inversely proportional to the atomic volume of the substituent at C-6. Various modifications of the hydroxyl groups at C-2, C-3 and C-4 in (2) and (3), however, resulted in complete loss of activity and hence it was concluded that the enzyme has an absolute specificity at these positions of the substrate. In contrast to these results, Barnett (1967) has argued that the C-6 hydroxyl group of α-D-glucosides is important in the binding to α-D-glucosidase. It would be interesting to re-examine the α- and β-glucosidase specificities with especially the 2-deoxy-2-fluoro-, 3-deoxy-3-fluoro- and 4-deoxy-4-fluoro-α- and β-D-glucopyranosides, all of which may now be synthesized from the corresponding free fluorosugars.

Only a few isolated enzyme studies have been done with the terminally fluorinated sugars. 6-Deoxy-6-fluoro-D-glucose (6FG) has been found to give the following percentage activities compared with glucose: glucose dehydrogenase, 100 (Metzger, Wilcox and Wick 1964); glucose oxidase, 3 (Blakley and Boyer 1955); and maltose phosphorylase, 80 (Morgan and Whelan 1962). A cell-free extract of *Aerobacter aerogenes* has been reported to oxidize 6FG to a 6-deoxy-6-fluoro-D-arabinohexulosonic acid (Blakley and Ciferri 1961). Recently phenyl 6-deoxy-6-fluoro-α-D-maltoside has been shown to be a poor substrate compared with phenyl α-D-maltoside for Taka-amylase A (Arita and Matsushima 1971), which suggests that position C-6 of the substrate is involved with this enzyme. Kissman and Weiss (1958) utilized 5-deoxy-5-fluoro-D-ribose in their attempts to provide suitably modified purine and pyrimidine nucleosides as potential antitumour agents. Although none of their 5′-fluoronucleosides showed antitumour activity, the inactivity of 6-mercapto-9-(5-deoxy-5-fluoro-β-D-ribofuranosyl)-purine (4, X=F) indicated the possibility that the known antitumour activity of 6-mercaptopurine riboside (4, X=OH) (Skipper *et al.* 1957) involves phosphorylation at the 5′-position of the sugar.

The effect of 6-deoxy-6-fluoro-D-glucose on intact yeast cells has been studied by Blakley and Boyer (1955). These authors showed that at molar concentra-

(4) X=F
   X=OH

tions, comparable to those of the glucose and fructose used, 6FG inhibited the rate of fermentation of yeast cells. The effect of 6FG on the fermentation of cell-free extracts was negligible, nor was there any significant effect on yeast hexokinase. In whole cells the inhibition of fermentation was competitive with the normal substrates and it was suggested, therefore, that 6FG probably influences a specific transport process, not hexokinase limited, which controls the rate of entry of glucose and fructose into the cell. Since 6FG cannot undergo enzymic phosphorylation it might be anticipated that this compound will affect sugar uptake, particularly in those tissues where the cell entry mechanism is rate limiting (Crane, Field and Cori 1957). Support for this contention has been demonstrated by a number of workers. Serif and Wick (1958) have reported the competitive inhibition of glucose and fructose oxidation in rat kidney slices; using specifically labelled glucose they showed that the inhibition occurred before glucose-6-phosphate formation. Similar observations have been made with rat epididymal adipose tissue and rat diaphragm muscle (Serif *et al.* 1958; Stewart and Wick 1960). A number of interesting studies on carbohydrate transport have been carried out with intestinal membranes; these involve the concept of both active and facilitated transport mechanisms (Crane 1960). Thus in the series D-galactose, 6-deoxy-6-fluoro-D-galactose, 6-deoxy-6-chloro-D-galactose and 6-deoxy-6-iodo-D-galactose, active transport occurs in the following order of substituent at C-6: OH $>$ F $\gg$ Cl, H $>$ I. This suggests that hydrogen bonding occurs at C-6 in the active transport process since only oxygen and fluorine form such bonds readily. Similar studies show that 6FG competes with glucose for the active transport site in mammalian intestine (Wilson and Landau 1960) as well as for facilitative transport sites (Wick and

Serif 1959). The use of 6-deoxy-6-fluorohexoses and other fluorinated sugars (e.g. 3-deoxy-3-fluoro-D-glucose) to provide more detailed information about the possible bonding sites of carrier and substrate in transport systems is dealt with more extensively by Barnett (1972).

Although no biological data have yet been reported, 6-deoxy-6-fluoromuramic acid has been synthesized (Diana 1970) as a potential inhibitor of bacterial wall synthesis. Recently *N*-fluoroacetyl-α-D-glucosamine has been used as a molecular probe for lysozyme structure, and the possibility exists that this analogue will inhibit glycoside transport and bacterial growth (Kent and Dwek 1971).

Most of the studies so far considered have been confined to extracyclic or terminally substituted fluorohexoses and pentoses. In recent years, however, a number of carbohydrates and related compounds in which a secondary hydroxyl group has been stereospecifically replaced by fluorine, have become available. Such fluorinated substrates might be expected to exhibit interesting metabolic activity since the terminal or primary hydroxyl groups are now available for attack by kinases and dehydrogenases. Various biological studies have been reported with the *rac*-1-deoxy-1-fluoroglycerols, which still have free primary hydroxyl groups. Pattison and Norman (1957) showed that *rac*-1-deoxy-1-fluoroglycerol is toxic to mice, the animal dying after a rather long interval of time but with the usual signs of fluoroacetate poisoning (convulsions and citrate accumulations in various organs, especially the heart). O'Brien and Peters (1958*a*) demonstrated, however, that the conversion of *rac*-1-deoxy-1-fluoroglycerol does not occur in the heart since in a simultaneous perfusion of the heart and liver there is no poisoning of the heart unless the liver is perfused first. This suggests that the liver is the major site for the conversion of *rac*-1-deoxy-1-fluoroglycerol into metabolites utilized by kidney and heart. Although chromatographic evidence for the formation of fluorocitrate in the heart-liver perfusate was found, the precise metabolic pathway for its formation has not yet been established. Elucidation of this pathway will be assisted when the enantiomorphic modifications of ^{14}C-labelled 1-deoxy-1-fluoroglycerols become available. Already both 1-deoxy-1-fluoro-*sn*-glycerol* (5) and its enantiomer 3-deoxy-3-fluoro-*sn*-glycerol (6) have been unambiguously synthesized (Eisenthal *et al.* 1970; Lloyd and Harrison 1971), and the synthesis of 2-deoxy-2-fluoroglycerol (7) has been improved to obtain more information about the substrate specificity model of glycerol kinase proposed by Gancedo, Gancedo and Sols (1968). Although (5) and (7) are substrates for yeast glycerol kinase (K_m values 120 mM and 150 mM respectively), (6) is not a substrate, presumably because the fluorine atom occupies the phosphorylation site (*A*) (Fig. 1) of

---

* Stereospecific numbering (sn) as recommended by the I.U.P.A.C.-I.U.B. Commission 1967.

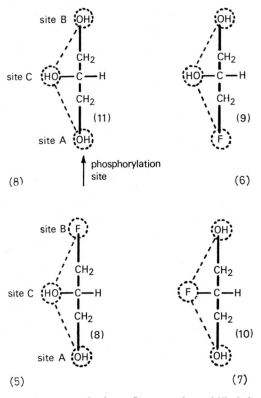

FIG. 1. Structures of 1-deoxy-fluoro-*sn*-glycerol (5); 3-deoxy-3-fluoro-*sn*-glycerol (6); 2-deoxy-fluoroglycerol (7); and *sn*-glycerol (8).

sn-glycerol (8). However, (6) is an effective inhibitor of either dihydroxyacetone or glycerol (K_i, 5 mм). We have also recently shown (Eisenthal, R., Harrison, R., Lloyd, J. and Taylor, N. F. 1971, unpublished) that (+)-1,2-propanediol [hydrogen analogue of (6)] is not a substrate for glycerol kinase whereas (−)-1,2-propanediol and 1,3-propanediol [hydrogen analogues of (5) and (7)] are. It is hoped to determine by the isolation and optical properties of the enzymic phosphorylated product whether, for example, the fluorine atom at C-2 in (7) is binding to site *C* (Fig. 1) in a similar manner to the OH group of *sn*-glycerol. Also recent studies with *rac*-1-deoxy-1-fluoroglycerol-3-phosphate by Fondy, Ghangas and Reza (1970) have shown that *rac*-1-deoxy-fluoroglyc-erol-3-phosphate can act as a substrate for L-α-glycerol phosphate dehydro-genase. They have argued that it might act, therefore, as a selective toxic agent to certain human cancer cells lacking the NAD-linked glycerol-3-phosphate dehydrogenase but possessing an active phosphatase, thereby allowing the formation of the toxic 1-deoxy-1-fluoro-*sn*-glycerol. The substrate activity and toxicity of 1-deoxy-1-fluoro-*sn*-glycerol-3-phosphate (and its enantiomer) and the 1-deoxy-1-fluoro analogue of dihydroxyacetone-3-phosphate are now awaited with interest.

2-Deoxy-2-fluoroglycerol (7) and *rac*-2-deoxy-2-fluoroglyceraldehyde also cause citrate accumulation in the heart and kidney of the rat (O'Brien and Peters 1958*b*). Since 2-deoxy-2-fluoroglycerol and *rac*-2-deoxy-2-fluoroglyceraldehyde are substrates for liver alcohol and aldehyde dehydrogenases respectively and L-fluoroglyceric acid is more toxic than the D-isomer, Treble and Peters (1962) suggested that fluoroacetate is produced from 2-deoxy-2-fluoroglycerate via the enzyme serine hydroxymethyl transferase. The suggestion by Pattison (1959) that the fluoroglycerate is oxidized to fluoromalonate with subsequent decarboxylation to fluoroacetate, however, still remains a possibility. We have recently synthesized [2-^{14}C]-2-deoxy-2-fluoroglycerol (Harrison, R. 1971, unpublished) to determine the *in vivo* metabolic fate of this compound. In the tetritol series, *rac*-2-deoxy-2-fluoroerythritol (Taylor and Kent 1956) has been shown to be a potent growth inhibitor of *Brucella abortus* (Smith *et al.* 1965) for which erythritol is a growth factor (Anderson and Smith 1965). The biochemical mode of inhibition has not been established but the competitive nature of the inhibition was illustrated by its reversal on addition of erythritol. In recent years a number of irreversible enzyme inhibitors have been found that combine with a functional group at the active site of the enzyme, for example di-isopropylphosphorofluoridate (DFP) and its reaction with the serine hydroxyl group of esterases and proteases (Oosterbaan and Cohen 1964). A compound in the carbohydrate series which may react in a similar manner is 3-fluoro-*N*-acetylneuraminic acid which inhibits the *in vitro* cleavage of *N*-acetylneuraminic acid to *N*-acetylmannosamine and pyruvic acid by *N*-acetylneuraminic acid aldolase (Gantt, Millner and Binkley 1964). No attempt was made, however, to show either that the inhibitor is covalently bound to the enzyme or that fluoride is produced during the process.

A number of *in vivo* and *in vitro* studies with 3-deoxy-3-fluoro-αβ-D-glucose (3FG) have now been carried out in our laboratories. Preliminary respirometric results with resting cells of *Saccharomyces cerevisiae* (Brunt and Taylor 1967) were based on experiments involving the use of syrupy 3FG (Foster, Hems and Webber 1967), which we subsequently showed to be due, in part, to ethanol contamination (Woodward, Taylor and Brunt 1969). All our subsequent experiments have been carried out with pure 3FG, which can now be obtained as a crystalline conglomerate of an equal mixture of α- and β-anomers (Taylor *et al.* 1972). It is now established that there is a small uptake of 3FG by resting cell suspensions of *S. cerevisiae* and that the intracellular 3FG or its metabolites affect subsequent glucose and galactose metabolism. Thus with glucose, although the respiratory rate was only transiently affected, both glucose uptake and total polysaccharide synthesis were inhibited; whereas with galactose long term respiration was inhibited, sugar uptake was not affected and total polysaccharide synthesis was stimulated (Table 1). These results suggest an inhibitory form of 3FG acting on phosphoglucomutase or uridine diphosphate glucose (UDPG) phosphorylase in a manner somewhat similar to that proposed for

TABLE 1

Effect of 3FG treatment on subsequent glucose and galactose metabolism in *S. cerevisiae*†

	Glucose			Galactose		
	Control	*3FG*	*% change*	*Control*	*3FG*	*% change*
Sugar uptake (μmoles of hexose/min/ 10 mg dry wt) $n=8$	0.32 ± 0.03	0.18 ± 0.03	$-44*$	0.19 ± 0.03	0.17 ± 0.04	-10
Oxygen uptake (μg-atom of O/ min/10 mg dry wt) $n=4$	0.63 ± 0.01	0.61 ± 0.03	-3.2	0.18 ± 0.02	0.12 ± 0.02	$-33*$
Total polysaccharide accumulation (μmoles of hexose equivalent/10 mg dry wt) $n=4$	4.38 ± 0.39	1.77 ± 0.33	$-60*$	2.63 ± 0.55	3.63 ± 0.37	$+38*$

Washed cells suspended in 0.04 M-phosphate buffer, pH 6.8, at the equivalent of 10 mg dry wt/3 ml were incubated with shaking at 30 °C with 5 mM-3FG for 90 min. At this point glucose or galactose was added to a final concentration of 3 mM. Control cells incubated without 3FG also received an equivalent amount of sugar at this time. Results are given as means ± S.E.M.; n=no. of estimations. The data were analysed by the method of paired comparisons.
* Significant difference at P=0.05.
† Data from Woodward, Taylor and Brunt (1969).

2-deoxy-D-glucose (Heredia, Dela Fuente and Sols 1964). Although as yet no intracellular 3FG metabolites (e.g. 3FG-6-phosphate or UDP3FG) have been isolated, we have measured the amounts of certain glycolytic intermediates and adenine nucleotides after treatment of *S. cerevisiae* with 3FG (Woodward, Taylor and Brunt 1971). It may be significant that the only hexosemonophosphate concentration to increase after treatment of the cells with 3FG is glucose-1-phosphate (G-1-P), although concurrently there are decreases in UDPG, 2- and 3-phosphoglyceric acids (2 and 3PGA) and phosphoenolpyruvic acid

(PEP). In connection with the decreased amounts of UDPG there is a corresponding decrease in the concentration of adenosine triphosphate (ATP) and an increase in that of adenosine diphosphate (ADP) (Fig. 2). Apart from the possible effect of the concentrations of ATP-ADP on the activity of regulatory

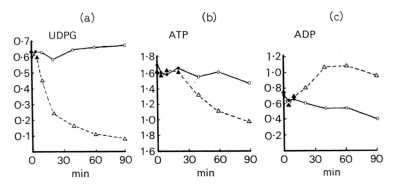

FIG. 2. The effect of 5 mM-3FG on the concentrations of (*a*) UDPG; (*b*) ATP; and (*c*) ADP. ●–○: control; ▲–△: 3FG; open symbols: significant difference, t-test, n=3. Units µM/g wet wt. (Data from Woodward, Taylor and Brunt 1971).

enzymes such as yeast phosphofructokinase (Sols 1968), the decrease in ATP and the increase in ADP correlate with our finding that there is a 20% decrease in the inorganic phosphate found in 3FG-treated cells (Table 2). Thus it can be argued that 3FG or its metabolites are binding inorganic phosphate and that this could affect the energy balance in the cell to produce the observed metabolic responses. Maitra and Estabrook (1964) have reported the same ATP-ADP changes in yeast cells treated with 2-deoxy-D-glucose and they also correlated these changes with the binding of inorganic phosphate. Whether the binding of phosphate by 3FG is sufficiently strong to continue to operate during the metabolism of external glucose or whether, like 2-deoxy-D-glucose, metabolites directly inhibitory to enzymes are produced has not yet been established.

TABLE 2

The effect of 3FG on the levels of total and inorganic phosphate*

	Total P		*Inorganic P*	
Time (min)	0	90	0	90
Control	172.0 ± 3.8	172 ± 4.7	25.4 ± 1.1	25.3 ± 1.0
3FG	171.0 ± 3.8	172 ± 6.2	25.9 ± 2.2	20.3 ± 1.2

Data given as µmoles/g wet wt of cells. ±S.E.M.; n=8.

* Data from Woodward, Taylor and Brunt (1971).

To enable us to examine the possible effects of 3FG-phosphates on glycolytic enzymes and UDPG pyrophosphorylase we recently synthesized 3FG-1-α-phosphate (9) and 3FG-6-phosphate (10) (Wright, J. A., Taylor, N. F. and Brunt, R. V. 1971, unpublished). The synthetic route (Fig. 3) to (9) is based on direct treatment of 1,2,4,6-tetra-*O*-acetyl-3-deoxy-3-fluoro-β-D-glucose (11) with

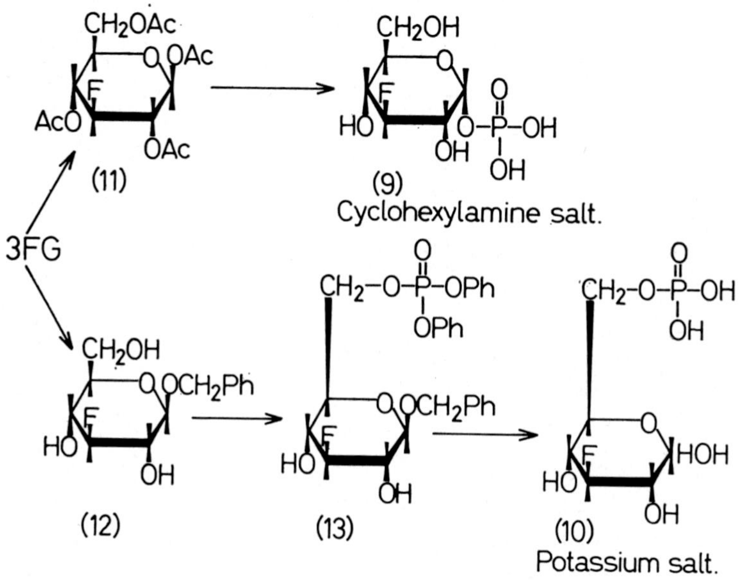

Fig. 3. Synthesis of 3FG-1α-phosphate and 3FG-6-phosphate (Wright, J. A., N. F. Taylor and R. V. Brunt 1971, unpublished).

anhydrous phosphoric acid and isolation of the product as the bis cyclohexylamine salt. Synthesis of (10) was based on the preparation of crystalline benzyl 3-deoxy-3-fluoro-β-D-glucopyranoside (12) by the action of benzyl alcohol and hydrochloric acid on 3FG, and the selective phosphorylation of the glycoside (12) by dibenzyl phosphorochloridate. Hydrogenolysis (by palladium on charcoal) of the product (13) yielded (10) which was isolated as the potassium salt. Some preliminary enzyme studies indicate that 3FG-1-α-phosphate can form the 3FG-1,6-diphosphate with glucomutase but the enzyme does not appear to convert this to the 3FG-6-phosphate. At the concentrations used in the assay procedure 3FG-6-phosphate is a substrate for glucose-6-phosphate dehydrogenase, but at higher concentrations the 3FG-6-phosphate inhibits the normal substrate conversion of G-6-P to gluconic acid-6-phosphate. It has not yet been possible to examine the effect of 3FG-1-α-phosphate on UDPG pyrophosphorylase, but we have confirmed our previous observations that

3FG is a poor substrate for yeast hexokinase (Brunt and Taylor 1967) and that at suitable concentrations 3FG-6-phosphate will also act as a substrate for phosphoglucoisomerase with the presumed formation of 3FG-fructose-1,6-diphosphate.

The poor substrate activity of 3FG for yeast hexokinase led us to examine the effect of this substrate on the glucokinase- and hexokinase-free system of *Pseudomonas fluorescens* (Taylor, White and Eisenthal 1972). In 0.67 M-buffer, washed resting-cell suspensions of this organism oxidize various concentrations of 3FG with the consumption of 1 g-atom of oxygen/mole of substrate (Table 3). This result is consistent with the production of a fluorinated aldonic acid,

TABLE 3

Oxidation of 0.01–0.05 M-3FG by washed suspensions of *Pseudomonas fluorescens* in 0.67 M-buffer*

Substrate oxidized (*μmoles*)	Period of incubation (*h*)	Rate of oxidation (*moles oxygen/ mg dry wt/h*)	Net O_2 uptake ml (*endogenous subtracted*)	Moles O_2/ mole substrate oxidized	Viable bacteria per ml (*×10^{-7}*)
—	0	—	—	—	35
—	5	—	—	—	5
3FG					
20	5	0.27	230	0.51	18
40	5	0.53	432	0.48	7
60	5	0.87	675	0.50	5
80	5	1.13	922	0.51	29
100	5	1.57	1332	0.60	25

Cells, grown for 20 h with aeration at 30 °C in glucose/mineral salts medium were collected at 4040 g for 10 min, washed twice in 0.67 M-phosphate buffer, pH 7, and resuspended in buffer of the same molarity and pH to 20 mg dry wt/ml. Their ability to oxidize 3FG was determined manometrically.
* Data from White and Taylor (1970).

and this was confirmed when large-scale incubations of 3FG allowed isolation of an acid from washed cells and supernatant which had chromatographic and electrophoretic mobilities identical with chemically synthesized (Taylor *et al.* 1972) 3-deoxy-3-fluoro-D-gluconic acid (3FGA). Fluoride determinations by the fluoride electrode on both cell supernatants and cell extracts (Woodward, Taylor and Brunt 1970) indicated that no detectable free fluoride anion was present at any time during the oxidation of 3FG. Although the oxidation of 3FG was limited to the consumption of 1 g-atom of oxygen/mole of substrate, provision of the oxidation product (3FGA) as an exogenous substrate allowed further oxidation to a product to which we have tentatively assigned the structure 3-deoxy-3-fluoro-2-keto-D-gluconic acid (3F2KGA) (Fig. 4, scheme *a*). These results are consistent with the accepted glucose catabolism by *Ps. fluorescens* (Wood and Schwerdt 1954) by a cytochrome oxidase system within the cytoplasmic membrane. To obtain more information about the location and specificity of the enzyme system involved with these metabolic transformations, we have recently examined the action of a cell-free enzyme preparation from *Ps. fluorescens* on 3FG and 3FGA (Taylor, White and Eisenthal 1972).

Whole cell: 3FG ——→ (δ-lactone?) ——→ 3FGA
                                      3FGA ——→ 3F2KGA   Scheme a

Cell extract: 3FG ————————————————(3FGA)—→ 3F2KGA   Scheme b

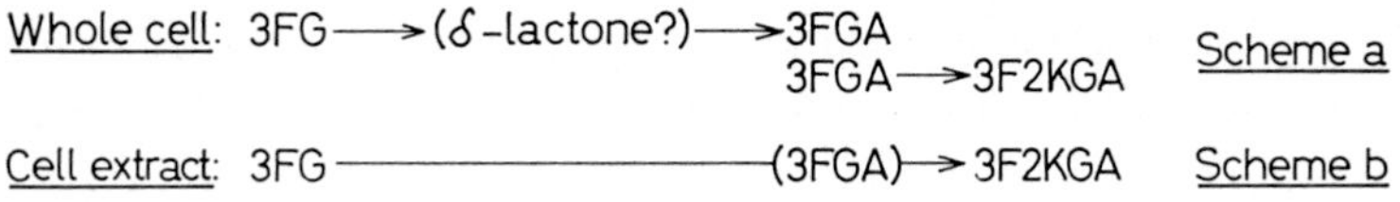

FIG. 4. Oxidation of 3FG by whole cells (scheme *a*) and cell-free preparations (scheme *b*) to 3F2KGA.

Tested manometrically, the results demonstrate that the supernatant/particulate enzyme preparation oxidizes glucose, gluconic acid, 3FG and 3FGA rapidly (Table 4) and that the net oxygen consumption per mole of substrate added was approximately 2 g-atoms/mole for glucose and 3FG and 1 g-atom/mole for gluconic acid and 3FGA. These are the theoretical values for the conversion

TABLE 4

Oxidation of 3FG and 3FGA by cell-free extracts of glucose-grown *Pseudomonas fluorescens**

Substrate oxidized	Initial rate of oxidation (μmoles O_2/mg protein/h	Net oxygen consumption (μl) (endogenous subtracted)	Moles oxygen/mole substrate oxidized
20 μmoles glucose	0.52	410	0.92
20 μmoles 3FG	0.29	410	0.92
10 μmoles glucose	0.36	235	1.05
10 μmoles 3FG	0.20	260	1.16
20 μmoles gluconate	0.57	225	0.50
20 μmoles 3FGA	0.45	235	0.53
10 μmoles gluconate	0.34	145	0.65
10 μmoles 3FGA	0.31	155	0.69

Cells from a 20 h growth were harvested, washed twice in 0.067 M-phosphate buffer, pH 7.0, and resuspended in the same buffer to form a thick cell paste which was ruptured ultrasonically. The combined particulate and soluble fraction, obtained after removal of whole cells by centrifugation (20 000 × g for 15 min) was used. Oxygen uptake was followed in Warburg manometers at 30 °C. Each flask contained 1.0 ml extract (48.5 mg protein), 1 μmole NAD, 0.67 M-phosphate buffer, pH 7.0, to 1.8 ml in main well; 0.2 ml KOH in centre well. Reaction was initiated by tipping in 10 or 20 μmoles of substrate from side-arm.

*   Data from Taylor, White and Eisenthal (1972).

of hexoses and aldonic acids to ketaldonic acids. Tests on the suspending fluid from Warburg flasks at the end of the 3FGA oxidations revealed the presence of reducing material with chromatographic properties similar to those produced by whole cells (White and Taylor 1970) and to which we have assigned the structure 3F2KGA. There is an important difference, however; whereas whole cells will only oxidize 3FG as a primary substrate to 3FGA (Fig. 4, scheme *a*), cell-free preparations will oxidize 3FG in a two-step process to a fluoroketo-aldonic acid (Fig. 4, scheme *b*). This limited oxidation of 3FG by whole cells may be related to the difference in binding and stability of the 'porter protein' for the 3FGA substrate. However, in the light of the studies on the specificity of the glucono-δ-lactonase and hydrolyase of *Ps. fluorescens* by Jermyn (1960), an alternative explanation for the limited oxidation by whole cells may be that the formation of the 3FGA proceeds via the 3-deoxy-3-fluoro-D-gluconic acid lactone (Fig. 4, scheme *a*) and that the specificity of the normal hydrolyase prevents the formation of the free 3FGA. In due course we hope to synthesize the 3FGA-δ-lactone to test this hypothesis.

The fact that the cell-free preparations from *Ps. fluorescens* oxidize 3FG to 3F2KGA suggests that either (*i*) the same enzymes that oxidize glucose to 2-ketogluconic acid do not possess the necessary specificity at C-3 of glucose and gluconic acid, or that (*ii*) there are separate enzymes for the 3FG and 3FGA substrates. Evidence has been presented that a single particulate enzyme of *Ps. fragi* (Weimberg 1963) oxidizes various sugars in the furanose form with the hydroxyl group at C-2 in the D-configuration to γ-lactones. In contrast, the existence of separate enzymes for the oxidation of 2-deoxy-D-glucose and D-glucose has been demonstrated with cell extracts of *Ps. aeruginosa* (Williams and Eagon 1959). Also, Blakley and Ciferri (1961) have postulated separate enzymes for the oxidation of 2-deoxy-D-glucose and 6-deoxy-6-fluoro-D-glucose with an enzyme preparation from *A. aerogenes*. If separate enzymes are involved in the oxidation of 3FG and 3FGA, then it might be expected that 3FG would increase reaction rates after the enzymes responsible for oxidizing glucose had been saturated with this substrate. In our experiments we were unable to demonstrate increased oxidation rates with glucose/3FG mixtures (Table 5). This suggests that the same enzymes are involved, but until further fractionation of the enzyme system is achieved the evidence is equivocal. The particulate nature of the 3FG and 3FGA oxidizing system was demonstrated by its ability to be separated from the soluble proteins by high speed centrifugation. Spectroscopic studies of the respiratory catalysts have given evidence implicating the cytochrome pigments as carriers in the 3FG oxidation. Thus addition of 3FG to a protein extract, made between 500 and 600 nm, established the cytochrome *c* α-peak at 555 nm and another component at 560 nm corresponding to the α-peak of cytochrome b_1. The β-absorption peaks of reduced cytochromes b_1 and c at 520–530 nm could not be resolved (Fig. 5). For these

TABLE 5

Manometric data on the oxidation of 3FG/glucose mixtures by cell-free extracts of *Pseudomonas fluorescens**

Substrate oxidized	Initial rate of oxidation (μmoles O_2/mg protein/h	Net oxygen consumption (μl) (endogenous subtracted)	Moles oxygen/mole substrate oxidized
10 μmoles glucose	0.41	250	1.12
10 μmoles 3FG	0.13	245	1.09
10 μmoles glucose 10 μmoles 3FG	0.36	450	1.01
10 μmoles glucose 20 μmoles 3FG	0.35	680	1.01
10 μmoles glucose 30 μmoles 3FG	0.43	880	0.94
10 μmoles glucose 40 μmoles 3FG	0.38	1080	0.96

The cells were grown and crude cell-free extracts prepared as in Table 4.

* Data from Taylor, White and Eisenthal (1972).

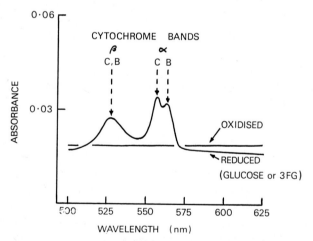

Fig. 5. Cytochrome absorption spectra of cell-free extracts of *Pseudomonas fluorescens.* Cuvettes contain 0.5 ml extract (15 mg protein), together with 2.5 ml of 0.067 m-phosphate buffer, pH 7.0. The extracts were oxidized either by flushing out the cuvettes with oxygen or by the addition of 0.05 ml of 1 % (w/v) potassium ferricyanide. A few crystals of glucose or 3FG were added as required and the absorption was measured against a blank containing oxidized extract.

measurements the instrument was set with both test and blank cuvettes containing aliquots of the same oxygenated preparation to obtain a horizontal base line; thus subsequently observed changes in absorption were due only to the formation of reduced components resulting from substrate additions to the test cuvette. The cytochrome spectra observed after addition of 3FG agree essentially with those recorded by Wood and Schwerdt (1953) for the glucose oxidizing system in *Ps. fluorescens*. Moreover, the absorption curve observed when glucose was added to the test cuvette could be cancelled by the addition of 3FG to the blank and vice-versa.

Although strain A 3.12 of *Ps. fluorescens* oxidizes 3FG it cannot utilize it when it is present as the sole source of carbon. Similar results have been found with halogenonicotinic acids (Hughes 1955) and halogenobenzoic acids (Smith and Cain 1965; Hughes 1965; Tranter and Cain 1967). The effect of 3FG on the lag and log phase of growth of the organism requires detailed study, but the addition of 3FG to a medium containing glucose as the carbon source reduces growth, which suggests that 3FG is acting as a competitive inhibitor of glucose in these circumstances. The inability of *Ps. fluorescens* to grow on 3FG is probably a consequence of the fact that the oxidation of the analogue is not carried to a stage at which the carbon may be assimilated in the cytoplasm by the action of kinases and into the Entner-Doudoroff or hexose monophosphate shunt system which is known to operate in this organism (De Ley 1960). There is no direct evidence that any 3FGA or 3F2KGA is transported from the cell membrane to the cytoplasm, but we have recently demonstrated that with a cell-free extract, 3FG, 3FGA or the postulated 3F2KGA are not substrates for the kinases present. Moreover, using preparations of kinase purified 2–3 times and the assay method of Colowick and Kalckar (1943), we have been able to demonstrate by a Dixon plot that both 3FG and 3FGA are competitive inhibitors of the enzyme for gluconic acid (K_i, 47.5 mM and K_i, 14.8 mM respectively) (Taylor, Hill and Eisenthal 1972). These results might partly explain the inhibition of growth of *Ps. fluorescens* on glucose by 3FG and might also partly account for the inability to identify the presence of fluorinated sugar phosphates or Entner-Doudoroff intermediates in whole-cell or cell-free extracts.

The oxidation of 3FG or 3FGA by either whole cells or cell-free extracts, proceeds without enzymic cleavage of the C—F bond. However, the adaptive formation of specific C—F bond cleaving enzymes in pseudomonad species has been demonstrated by Goldman (1965) and is dealt with elsewhere (Goldman 1972). Our interest in the potential use of fluorocarbohydrates as growth inhibitors of pathogenic microorganisms has led us to examine the effect of 3FG on the growth of *Ps. aeruginosa* (NCTC 6750). Preliminary growth studies, however, indicated that the rate of growth of this organism on a mixture of 3FG and glucose was parallel to that on glucose alone. A mutant strain has now been isolated which grows on 3FG and 3FGA with the release of fluoride anion during the log phase of growth (Fig. 6). During the lag phase there is no release of fluoride by 3FG or 3FGA and during this phase of growth the substrates consume 2 g-atoms and 1 g-atom of oxygen/mole of substrate respectively. This would indicate that 3FG is metabolized to 3F2KGA (or the 6-phosphate) before C—F cleavage occurs. The localization of the dehalogenase enzyme and identification of the fluorinated metabolites has yet to be established. It will also be of interest to determine the concentration of fluoride anion released by 3FG necessary to inhibit the growth of this *Ps. aeruginosa* mutant.

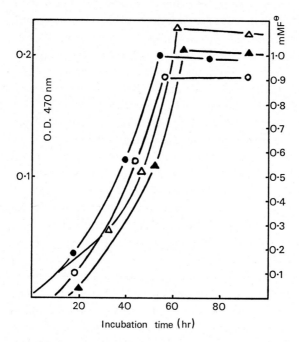

FIG. 6. Growth of *Pseudomonas aeruginosa* on 3FG and 3FGA.
Cells were grown on 0.5 mM-3FG/0.5 mM-glucose mixture, spun down (3000 r.p.m., 30 min) on reaching the stationary phase and resuspended in Melling medium. A 1 ml aliquot was added to fresh media containing either 1 mM-3FG or 1 mM-3FGA and incubated at 30 °C. Fluoride release was determined by the fluoride electrode. ●–●: 3FG; ○–○: Fluoride released from 3FG; △–△: 3FGA; ▲–▲: Fluoride released from 3FGA. (From Scott-Foster, J., N. F. Taylor and G. Whitfield 1971, unpublished).

Reference has already been made to the use of terminal fluorinated ribonucleosides in antitumour studies (p. 217) and the biological activities of fluorinated pyrimidines are dealt with elsewhere by Heidelberger (1972). Some interesting results have now been reported, however, with secondary fluorosugar analogues of arabinosyl and xylosylcytosine (Wright, Wilson and Fox 1970).

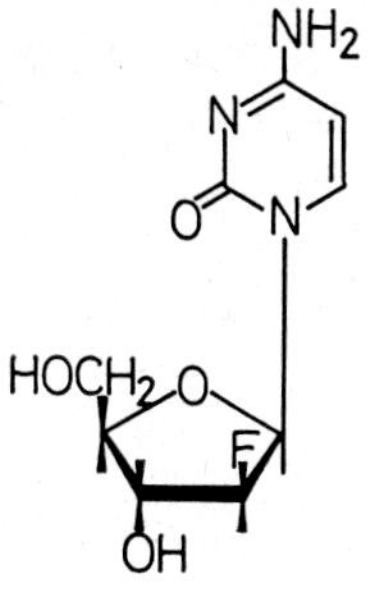

(15)

It is known that the antitumour activity of 1-β-D-arabinofuranosylcytosine (14) (Cohen 1966) is lost by its conversion into an inactive metabolite, 1-β-D-arabinosyluracil, through the action of a deaminase (Dollinger *et al.* 1967). The susceptibility of 1-(β-D-2-deoxy-2-fluoroarabinofuranosyl)-cytosine (15), which may be considered a hydroxyl analogue of (14), to a partially purified

TABLE 6*

Susceptibility of 1-β-D-pentofuranosylcytosines to cytidine deaminase

Compound	Configuration at 3′,2′	Relative velocity
Cytidine	HO — OH	1
2′-Deoxycytidine	HO — (OH below)	0.41
ara-C (14)	HO — (F below)	0.15
2′-F-*ara*-C (15)	(HO above) HO — (F above)	0.13
xylo-C	HO — (OH below)	0
2′-F-*xylo*-C	HO — (F below)	0
3′-F-*xylo*-C	(F above) — OH (below)	0

* From Wright, Wilson and Fox (1970).

cytidine deaminase from pig kidney is of particular interest. As can be seen
(Table 6), the results indicate that (14) and the 2'-fluoroanalogue (15) are
deaminated more slowly than cytidine and that the replacement of -OH by F
at the 2'-position of (14) does not change the relative rate of deamination.
2'-Deoxycytosine, however, in which F or –OH has been replaced by hydrogen,
increases the rate of deamination threefold. Configurational changes at position
C–3' yield *xylosyl* analogues of cytosine that are not substrates for the
deaminase. 1-(β-D-2-Deoxy-2-fluoroarabinofuranosyl)-cytosine (15) has a
growth inhibitory effect on L1210 mouse leukaemia suspension culture comparable to (14) (5.0 ng/ml for 50% inhibition), which can be reversed by the
addition of 2'-deoxycytosine. Although the biochemical level of inhibition is
not known, this latter observation would suggest that the fluoroanalogue (15)
interferes with the metabolic conversion of ribonucleotides to deoxyribonucleo-
tides and hence DNA synthesis.

CONCLUSION

There is now sufficient evidence to support the contention that suitably
substituted fluorinated carbohydrates and related compounds can act as
pseudosubstrates in biological systems. Perhaps the analogy between F and
−OH should not be emphasized too strongly until more precise information
is available about the nature of binding and bonding between fluoro substrate
and enzyme. Indeed, there are as yet few *in vivo* examples in which a fluorosugar
or related compound may be considered either to act competitively for the
normal substrate or to undergo metabolic transformation with retention of the
C—F bond. Apart from the studies reported with 3-deoxy-3-fluoro-D-glucose
and the monodeoxyfluoroglycerols, another interesting example has been
provided by Kay and Kornberg (1971) in their studies on the uptake of C_4-
dicarboxylic acids by *E. coli*. They isolated a fluoromalate-resistant mutant able
to grow in the presence of DL-fluoromalate as a sole carbon source (Kay and
Kornberg 1969) and also able to grow on L-(−)-tartrate. Enzyme specificity
studies in which deoxyfluoro- and the corresponding deoxysugars are systemat-
ically examined both kinetically and by ^{19}F-n.m.r. (especially at positions C-2,
C-3 and C-4 in hexoses), are undoubtedly required to pursue the analogy
between fluorine and hydroxyl further. The binding of fluorosugars to protein
(see Kent 1972; Dwek 1972; Barnett 1972) also suggest that certain fluorosugars
might act as either gratuitous inducers or inhibitors of induction. Some recent
preliminary studies (Taylor and White, unpublished results) indicate that
whereas 3FG will not induce the gluconic acid oxidizing enzyme in *Ps. fluor-*

escens grown in asparagine, it will inhibit the induction by glucose. Whether the analogy between F and —OH has any real validity or not, the concept has proved most useful for the design of potential metabolic deceptor molecules and has promoted a number of biochemical studies which would otherwise not have been undertaken.

Although the metabolic fate of 3FG (and other fluorosugars) in animals has not yet been established, we have shown no apparent toxic effects when it is administered (intraperitoneally or orally) to the rat (Taylor, N. F. and Riley, G., unpublished). Further investigations in this area will require ^{14}C or tritium specifically labelled fluorosugars. In view of the apparently non-toxic effects of 3FG on animals, the inhibition of polysaccharide synthesis in yeast cells and the inhibition of growth of *Ps. fluorescens*, future studies with 3FG and other fluorosugars may lead to the development of new synthetic antibiotics.

SUMMARY

After a brief review of naturally occurring fluorinated compounds, a rationale, based on physicochemical data, for the synthesis of C—F compounds as oxy or hydroxyl analogues is discussed. The biological effects of fluorinated monosaccharides and related compounds is reviewed. The metabolism and enzyme specificity of 3-deoxy-3-fluoro-D-glucose (3FG), with particular reference to the glycolytic enzymes of *Saccharomyces cerevisiae*, *Pseudomonas fluorescens* and *Pseudomonas aeruginosa*, is considered. Whole-cell and cell-free studies with the latter two organisms indicate that 3FG may act as a pseudo-substrate in which the C—F bond is retained (*Ps. fluorescens*) or broken (*Ps. aeruginosa*). Some recent studies with monofluoroglycerols and fluorinated sugar phosphates are described. Brief reference is made to the future requirements and likely developments in the metabolism of fluorinated sugars.

ACKNOWLEDGEMENTS

I am particularly grateful to Sir Rudolph Peters and Dr P. W. Kent who initiated my interest in the biochemistry of fluorinated compounds. The collaboration of my colleagues, Drs R. V. Brunt, R. Eisenthal, R. Harrison, J. Wright and J. Scott-Foster, as well as the expert technical assistance of Mrs. B. Hunt, is also very much appreciated.

References

ANDERSON, J. D. and H. SMITH (1965) *J. Gen. Microbiol.* **38**, 109–124.
ARITA, H. and Y. MATSUSHIMA (1971) *J. Biochem. (Tokyo)* **69**, 409–413.
BARNETT, J. E. G. (1967) *Adv. Carbohydr. Chem.* **22**, 177–227.
BARNETT, J. E. G. (1972) *This volume*, pp. 95–110.
BEKOE, A. and H. M. POWELL (1959) *Proc. R. Soc. A* **250**, 301–315.
BERGMANN, E. D. (1961) *Bull. Res. Counc. Israel* **10A**, 1–16.
BLAKLEY, E. R. and P. D. BOYER (1955) *Biochem. & Biophys. Acta* **16**, 576–582.
BLAKLEY, E. R. and O. CIFERRI (1961) *Can. J. Microbiol.* **7**, 61–73.
BRUNT, R. V. and N. F. TAYLOR (1967) *Biochem. J.* **105**, 41–43c.
COHEN, S. S. (1966) *Progr. Nucleic Acid Res. & Mol. Biol.* **5**, 1–88.
COLOWICK, S. P. and H. M. KALCKAR (1943) *J. Biol. Chem.* **147**, 117–126.
CRANE, R. K. (1960) *Physiol. Rev.* **40**, 789–825.
CRANE, R. K., R. A. FIELD and C. F. CORI (1957) *J. Biol. Chem.* **224**, 649–662.
DE LEY, J. (1960) *J. Appl. Bacteriol.* **23**, 400–441.
DIANA, G. D. (1970) *J. Org. Chem.* **35**, 1910–1912.
DOLLINGER, M. R., J. H. BURCHENAL, W. KREIS and J. J. FOX (1967) *Biochem. Pharmacol.* **16**, 689–706.
DWEK, R. A. (1972) *This volume*, pp. 239–271.
EISENTHAL, R., R. HARRISON, J. LLOYD and N. F. TAYLOR (1970) *Chem. Commun.* 1507–1508.
FONDY, T. P., J. S. GHANGAS and M. J. REZA (1970) *Biochemistry* **9**, 3272–3280.
FOSTER, A. B., R. HEMS and J. M. WEBBER (1967) *Carbohydr. Res.* **5**, 292–301.
GANCEDO, C., J. M. GANCEDO and A. SOLS (1968) *Eur. J. Biochem.* **5**, 165–172.
GANTT, R., S. MILLNER and S. B. BINKLEY (1964) *Biochemistry* **3**, 1952–1960.
GOLDMAN, P. (1965) *J. Biol. Chem.* **240**, 3434–3438.
GOLDMAN, P. (1972) *This volume*, pp. 335–349.
HEIDELBERGER, C. (1972) *This volume*, pp. 125–137.
HELFERICH, B., S. GRÜNLER and A. GNÜCHTEL (1937) *Hoppe-Seyler's Z. Physiol. Chem.* **248**, 85–95.
HEREDIA, C. F., G. DELA FUENTE and A. SOLS (1964) *Biochim. & Biophys. Acta* **86**, 216–223.
HUGHES, D. E. (1955) *Biochem. J.* **60**, 303–310.
HUGHES, D. E. (1965) *Biochem. J.* **96**, 181–188.
JERMYN, M. A. (1960) *Biochim. & Biophys. Acta* **37**, 78–92.
KAY, W. W. and H. L. KORNBERG (1969) *FEBS Lett.* **3**, 93–96.
KAY, W. W. and H. L. KORNBERG (1971) *Eur. J. Biochem.* **18**, 274–281.
KENT, P. W. (1969) *Chem. & Ind.* 1128–1132.
KENT, P. W. (1972) *This volume*, pp. 169–208.
KENT, P. W. and R. A. DWEK (1971) *Proc. Biochem. Soc.* **121**, 11p–12p.
KISSMAN, H. M. and M. J. WEISS (1958) *J. Am. Chem. Soc.* **80**, 5559–5564.
LLOYD, J. and R. HARRISON (1971) *Carbohydr. Res.* **20**, 133–139.
MAITRA, P. K. and R. W. ESTABROOK (1964) *Anal. Biochem.* **7**, 472–484.
MARAIS, J. S. C. (1944) *Onderstepoort J. Vet. Sci. Anim. Ind.* **20**, 67–73.
METZGER, R. P., S. S. WILCOX and A. N. WICK (1964) *J. Biol. Chem.* **239**, 1769–1772.
MORGAN, K. and W. J. WHELAN (1962) *Nature (Lond.)* **196**, 168–169.
MORTON, G. O., J. E. LANCASTER, G. E. VAN LEAR, F. FULMOR and W. E. MEYER (1969) *J. Am. Chem. Soc.* **91**, 1535–1537.
O'BRIEN, R. D. and R. A. PETERS (1958a) *Biochem. J.* **70**, 188–195.
O'BRIEN, R. D. and R. A. PETERS (1958b) *Biochem. Pharmacol.* **1**, 3–18.
OOSTERBAAN, R. A. and J. A. COHEN (1964) In *Structure and Activity of Enzymes*, pp. 87–95, ed. T. W. GOODWIN, J. I. HARRIS and B. S. HARTLEY. New York: Academic Press.
PATTISON, F.L.M. (1959) *Toxic Aliphatic Fluorine Compounds*, pp.129-130. Amsterdam: Elsevier.

PATTISON, F. L. M. and J. J. NORMAN (1957) *J. Am. Chem. Soc.* **79**, 2311–2316.
PETERS, R. A. (1972) *This volume*, pp. 55–70.
PETERS, R. A., R. J. HALL, P. E. V. WARD and N. SHEPPARD (1960) *Biochem. J.* **77**, 17–23.
PETERS, R. A. and M. SHORTHOUSE (1971) *Nature (Lond.)* **231**, 123–124.
PETERS, R. A., R. W. WAKELIN, P. BUFFA and L. C. THOMAS (1953) *Proc. R. Soc. B*, **140**, 497–506.
SERIF, G. S., C. J. STEWART, H. I. NAKADA and A. N. WICK (1958) *Proc. Soc. Exp. Biol. & Med.* **99**, 720–722.
SERIF, G. S. and A. N. WICK (1958) *J. Biol. Chem.* **233**, 559–561.
SKIPPER, H. E., J. R. THOMSON, D. J. HUTCHINSON, F. M. SCHABEL and J. A. JOHNSON (1957) *Proc. Soc. Exp. Biol. & Med.* **95**, 135–138.
SMITH, A. and R. B. CAIN (1965) *J. Gen. Microbiol.* **41**, xvi.
SMITH, F. A. (1970) *Handb. Exp. Pharmakol.* **20**, pt. 2, 253–408.
SMITH, H., J. D. ANDERSON, J. KEPPIE, P. W. KENT and G. M. TIMMIS (1965) *J. Gen. Microbiol.* **38**, 101–108.
SOLS, A. (1968) In *Aspects of Yeast Metabolism*, p. 47, ed. A. K. MILLS. Oxford: Blackwells.
STEWART, C. J. and A. N. WICK (1960) *Fed. Proc. Fed. Am. Soc. Exp. Biol.* **19**, 83.
TAYLOR, N. F., L. HILL and R. EISENTHAL (1972) In preparation.
TAYLOR, N. F., B. HUNT, P. W. KENT and R. A. DWEK (1972) *Carbohydr. Res.*, in press.
TAYLOR, N. F. and P. W. KENT (1956) *J. Chem. Soc.* 2150–2154.
TAYLOR, N. F., F. H. WHITE and R. EISENTHAL (1972) *Biochem. Pharmacol.* **21**, 347–353.
TRANTER, E. K. and R. B. CAIN (1967) *Biochem. J.* **103**, 22p–23p.
TREBLE, D. H. and R. A. PETERS (1962) *Biochem. Pharmacol.* **11**, 891–899.
WARD, P. F. V., R. J. HALL and R. A. PETERS (1964) *Nature (Lond.)* **201**, 611–612.
WEIMBERG, R. (1963) *Biochim. & Biophys. Acta* **67**, 349–358.
WHITE, F. H. and N. F. TAYLOR (1970) *FEBS Lett.* **4**, 268–271.
WICK, A. N. and G. S. SERIF (1959) *Ann. N.Y. Acad. Sci.* **82**, 374–377.
WILLIAMS, A. K. and R. G. EAGON (1959), *J. Bacteriol.* **77**, 167–172.
WILSON, T. H. and B. R. LANDAU (1960) *Am. J. Physiol.* **198**, 99–102.
WOOD, W. A. and R. F. SCHWERDT (1953) *J. Biol. Chem.* **201**, 501–511.
WOOD, W. A. and R. F. SCHWERDT (1954) *J. Biol. Chem.* **206**, 625–635.
WOODWARD, B., N. F. TAYLOR and R. V. BRUNT (1969) *Biochem. J.* **114**, 445–447.
WOODWARD, B., N. F. TAYLOR and R. V. BRUNT (1970) *Anal. Biochem.* **36**, 303–309.
WOODWARD, B., N. F. TAYLOR and R. V. BRUNT (1971) *Biochem. Pharmacol.* **20**, 1071–1077.
WRIGHT, J. A., D. P. WILSON and J. FOX (1970) *J. Med. Chem.* **13**, 269–272.

Discussion

Foster: Dr Taylor raised a very important point in relation to the hydrogen bonding capability of fluorine. A hydroxyl group can both donate and accept a hydrogen bond, but a fluorine substituent, if it can do anything, can only accept.

We have been particularly interested in hexokinase because there is a marked difference in the isozyme profile for normal liver and for certain hepatomas, and we asked: can this difference be utilized to devise antitumour agents? First, one needs to define the substrate specificities of the hexokinase isozymes. The information available indicates that hydroxyl groups in glucose are necessary at positions 6, 4, 3, and 1 if the compound is to serve as a substrate for hexokinase. But it is not known whether each hydroxyl group acts as a hydrogen bond donor or acceptor in the enzyme substrate complex, or whether it is possible to distinguish between these two roles. If one puts fluorine into these various positions then only hydrogen bond acceptance can occur. We find that when the fluorine is at position 3, then 3-fluoroglucose is a poor substrate for hexokinase and a poor inhibitor of glucose phosphorylation. The same applies for fluorine substituents at positions 1, 4, and 6; 6-fluoroglucose cannot be a substrate and is, in fact, a poor inhibitor. We find that 2-fluoroglucose and 2-fluoromannose are very good substrates for yeast hexokinase. Moreover, the compound with two fluorine atoms at position 2 ('2,2-difluoroglucose') is a better substrate than glucose itself. However, if a methoxy group or one or two chlorine substituents are introduced at position 2 instead of a hydroxyl group, then these substances are neither substrates nor inhibitors. 2,2-Difluoroglucose might be a better substrate than glucose because of the electronic effect of the 2-fluorine substituents on the acidity of the hydroxyl groups at positions 1 and 3. If these are more acid and hydrogen bond donation is the key role of these hydroxyl groups in the formation of the enzyme-substrate complex, then one might expect this sort of effect. This theory is consonant with the observation that when the hydroxyl groups at positions 1 and 3 are replaced by fluorine, there is a loss of substrate capability.

Taylor: Clearly a more systematic study has to be undertaken with both hydrogen analogues and fluorine analogues to sort out this problem of whether fluorine is hydrogen bonding like a hydroxyl group. In biological *in vivo* systems there are very few examples of the fluoro analogue behaving like the hydroxyl analogue. Kay and Kornberg (1969, 1971) obtained a fluoromalate-resistant mutant of *E. coli* which could also utilize the stereochemical analogue, L-tartrate. In our own enzyme and metabolic studies with 3-deoxy-3-fluoroglucose and the monodeoxyfluoroglycerols it might be argued that they also are behaving like hydroxyl analogues.

Heidelberger: Langen, in East Berlin, has made an analogue with the structure 5'-fluoro-5'-deoxythymidine. This compound is a very efficient inhibitor of thymidylate kinase, not thymidine kinase. Thymidine kinase takes thymidine to the monophosphate and thymidylate kinase takes this monophosphate to the diphosphate and eventually to the triphosphate. Langen has clearly shown that this analogue blocks this reaction (Langen and Kowollik 1968). So it looks as if the fluorine is acting here as an analogue of a phosphate group.

Taylor: Does it replace the fluorine?

Heidelberger: No, it is a stable compound. The fluoro compound acts as an inhibitor of the kinase but it is not replaced.

Kent: Could the catalytic site primarily involve the pyrimidine ring and the deoxyribofuranose ring, the C-5 position not being deeply embedded in the receptor site?

Heidelberger: Yes, but all the other compounds which have the thymidine and the sugar do not inhibit this particular enzyme.

Kent: In principle, if the enzyme can be isolated, one should be able to find out whether that fluorine is embedded in the protein or is projected away from the binding site.

Barnett: Dr Taylor, your reconstituted glycolytic pathway seems to undergo inhibition at the aldolase step. Could you elaborate on this? It seems to me that a good alkylating inhibitor of aldolase should be 3-deoxy-3-fluorofructose-

$$CH_2O \ \text{\textcircled{P}}$$
$$|$$
$$C=O$$
$$|$$
$$F-C-H \qquad + \ H_2N- \qquad \longrightarrow$$
$$|$$
$$H-C-OH$$
$$|$$
$$H-C-OH$$
$$|$$
$$CH_2O \ \text{\textcircled{P}}$$

$$CH_2O \ \text{\textcircled{P}}$$
$$|$$
$$C=N-$$
$$|$$
$$F-C-H$$
$$|$$
$$H-C-OH$$
$$|$$
$$H-C-OH$$
$$|$$
$$CH_2-O \ \text{\textcircled{P}}$$

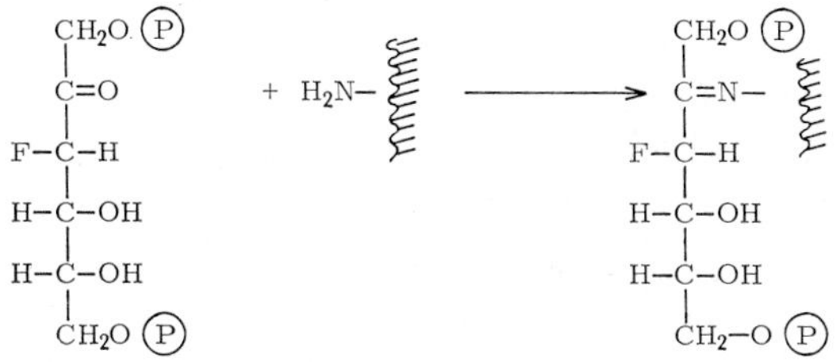

FIG. 1. (Barnett).

1,6-diphosphate (Fig. 1) which, in the straight chain form, would form a Schiff's base, with the enzyme adjacent to the fluorine, which would then become a good leaving group and might be attacked by a group on the enzyme, so stopping the enzyme working.

Taylor: If we start with 3-deoxy-3-fluoroglucose as the substrate and use a reconstituted glycolytic enzyme system, then with a certain concentration of enzymes and substrate we can demonstrate that an aldolase split of what is presumably 3-deoxy-3-fluorofructose-1,6-diphosphate occurs. However, optimum concentrations are yet to be described for the recently synthesized 3-deoxy-3-fluoroglucose-6-phosphate as the initial substrate. With this substrate the glycolytic sequence operates until the aldolase step and we appear to obtain a build-up of what is presumably 3-deoxy-3-fluorofructose-1,6-diphosphate in a similar manner to the build-up of 3-deoxy-3-fluoroglucose-1,6-diphosphate, which I have already mentioned in connection with our studies of 3-deoxy-3-fluoroglucose-1-phosphate and the enzyme glucomutase. The accumulation of fluorinated sugar phosphates may also occur intracellularly and this might become a significant phenomenon in relation to the regulation and control of metabolism. We may find that a number of fluorinated sugars act as inert substrates but exert allosteric effects on glycolytic enzymes. This could produce some interesting results with possible chemotherapeutic applications.

Peters: Dr R. D. O'Brien and I examined both the 1-deoxyfluoroglycerols and the 2-deoxyfluoroglycerols (O'Brien and Peters 1958). With the 1-deoxy-fluoroglycerol the animal went into a kind of stupor. Two lethal syntheses are occurring, and with the 1-deoxy-1-fluoroglycerol the animal goes to enormous lengths to kill itself. The 1-deoxyfluoroglycerol first has to be acted upon by the liver, after which it goes to the other parts of the body and is converted to fluoroacetate and eventually to fluorocitrate. However, we had to perfuse the whole liver. This is about the only case I know in which the whole organ has to be used. I have never understood it.

Taylor: The 3-phosphate may be one of the first steps in the metabolism of this compound. Fondy and co-workers (1970) have recently synthesized the 1-deoxy-1-fluoroglycerol-3-phosphate and they find that this phosphate can act as a substrate for L-α-glycerolphosphate dehydrogenase. We have also shown that 1-deoxy-1-fluoroglycerol acts as a substrate for glycerol kinase.

Peters: But why should it only happen in the whole liver?

Kun: Liver slices have a tendency to degrade nucleotides very rapidly, so they are poor models for most biochemical studies. Berry and Friend (1969) have devised a method for isolating liver cells. It seems that isolated liver cells behave in the same manner metabolically as a perfused organ.

Gál: Sir Rudolph, we have an example which is just the reverse situation. We have a natural metabolite, 5-hydroxytryptophan, which we cannot demonstrate *in vivo*. And yet we know it is formed, because there is 5-hydroxytryptamine. We can, however, demonstrate it in the excised organ or in slices.

References

BERRY, M. N. and D. S. FRIEND (1969) *J. Cell. Biol.* **43**, 506.
FONDY, T. P., J. S. GHANGAS and M. J. REZA (1970) *Biochemistry* **9**, 3272–3280.
KAY, W. W. and H. L. KORNBERG (1969) *FEBS Lett.* **3**, 93–96.
KAY, W. W. and H. L. KORNBERG (1971) *Eur. J. Biochem.* **18**, 274–281.
LANGEN, P. and G. KOWOLLIK (1968) *Eur. J. Biochem.* **6**, 344.
O'BRIEN, R. D. and R. A. PETERS (1958) *Biochem. J.* **70**, 188–195.

Nuclear magnetic resonance studies of macromolecules with fluorine nuclei as probes

RAYMOND A. DWEK

Department of Biochemistry, University of Oxford

Nuclear magnetic resonance studies have established that, in general, fluorine coupling constants and chemical shifts are at least an order of magnitude larger than those of the corresponding proton analogues (see Emsley and Phillips 1971). At a given magnetic field, the sensitivity of fluorine nuclei is second only to that of protons, and taking into account the small size of a fluorine nucleus (whose introduction into a molecule may cause only small perturbations) one has the ingredients for an excellent probe.

In enzymes, two approaches with fluorine probes have been used. The first involves introducing fluorine covalently on the enzyme and observing the changes in the fluorine spectrum as inhibitors are added. In principle, this is very similar to observation of the proton resonances of an enzyme in which all but a few selected protons have been deuterated. The advantage of using a fluorine probe in such cases comes from the potentially larger changes that may arise when the inhibitor binds.

The second approach is to use fluorosubstrates as probes. The chemical shift, line width and other characteristics of the substrate will alter when it binds to the enzyme since its environment is different. Unfortunately at biological concentrations the detection of these signals provides severe instrumental problems at present. There is, however, a way of obtaining information about the bound site under certain conditions. If there is fast chemical exchange between the free (bulk) substrate and the bound substrate, then the observed signal will be a weighted average of that in each environment.

Under these conditions the further addition of certain paramagnetic metal ions which bind to the enzyme results in a broadening of the n.m.r. high resolution spectrum of the substrate. This broadening is caused by paramagnetic metal ions, such as Mn^{2+} and Gd^{3+}, which have a relatively long electron spin lattice relaxation time. The broadening is a function of the distance between the metal ion and the observed nucleus and of the correlation time for the n.m.r. relaxation behaviour of the complex. If the metal binds at a conformationally strategic site one has a means of 'mapping out' the substrates around the site. This is obviously important on large enzymes where crystallographic data are unavailable, and clearly will be applicable to any nucleus, not just fluorine. The advantage of using fluorine probes in such studies rather than proton analogues is that the proton spectrum may be too complex to enable such experiments to be carried out; for instance, when the chemical shifts of all the protons occur over a small range, such broadening experiments cannot be easily done. In theory, it should be possible to substitute fluorine nuclei at strategic points on each substrate, so that their orientations relative to each other and to the different substrates can be defined relative to the metal ion.

Some examples of the uses of fluorine probes follow. They are divided into covalent probes and fluorosubstrate probes.

FLUORINE AS A COVALENT PROBE

The enzyme ribonuclease A (RNase A) can be cleaved by subtilisin at the

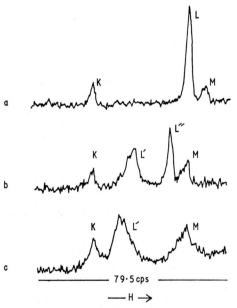

FIG. 1. ^{19}F-n.m.r. spectra of trifluoroacetylated RNase S: (a) trifluoroacetylated RNase S-peptide; (b) trifluoroacetylated RNase S-peptide associated with RNase S-protein; (c) trifluoroacetyl peptide-protein complex, after removal of Lys-1 from the peptide (from Huestis and Raftery 1971).

peptide bond between residues 20 and 21 to give the ribonuclease S-peptide, RNase S (residues 1–20) and ribonuclease S-protein (residues 21–124). The two components together have the same enzymic activity as the parent RNase A. After the specific introduction of trifluoroacetyl groups at lysine residues 1 and 7 of bovine pancreatic RNase A, the changes in the ^{19}F high-resolution spectrum (Huestis and Raftery 1971) can be used to detect environmental changes when inhibitors bind to RNase S. The labelled enzyme shows no change in its enzymic activity.

Lys-1 and 7 were acylated by means of ethyl thioltrifluoroacetate. Normally this is specific for the ε-amino group, but Huestis and Raftery used a large excess of the acylating reagent, so that some doubly acylated Lys-1 as well as the singly acylated residue resulted.

The ^{19}F spectrum of the labelled S-peptide is shown in Fig. 1(a). The assignment of the resonances was based on comparison with the model compounds ε-trifluoroacetyl-lysine and α,ε-bis(trifluoroacetyl)lysine (Fig. 2). Peak

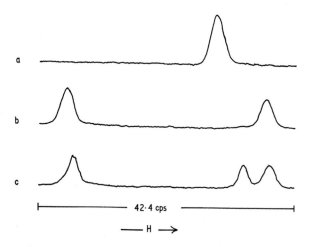

FIG. 2. ^{19}F-n.m.r. spectra of trifluoroacetylated lysine: (a) ε-trifluoroacetyl lysine; (b) α,ε-bis(trifluoroacetyl)lysine; (c) a mixture of ε-mono and α,ε-bis(trifluoroacetyl)lysine (from Huestis and Raftery 1971).

L (Fig. 1) is assigned to the ε-acylated groups of Lys-7 and the singly acylated Lys-1. Peaks K and M are attributed to the trifluoroacetyl groups on the α and ε amino groups of the double acylated Lys-1.

Addition of a molar equivalent of RNase S-protein caused changes in the spectrum (Fig. 1b). The resonance L splits into two and that of M is shifted. The shift of L″ and M was 0.07 p.p.m. downfield and that of L′ 0.22 p.p.m.

downfield. Thus trifluoroacetyl residues on the ε-amino groups experience significantly different chemical environments. This is in agreement with the X-ray data, which suggest that Lys-1 is exposed to the solvent (and consequently experiences the smaller shift) near the surface of the protein, whereas Lys-7 is inside, close to the active site.

Addition of the inhibitors PO_4^{2-} or 3'-cytidine monophosphate (3'-CMP) to the modified enzyme produced upfield shifts of resonance L' of 0.03 p.p.m. and 0.07 p.p.m. respectively. The shift induced by 3'-CMP was pH-dependent (Fig. 3). This suggests that peak L' (from the trifluoroacetyl group on Lys-7)

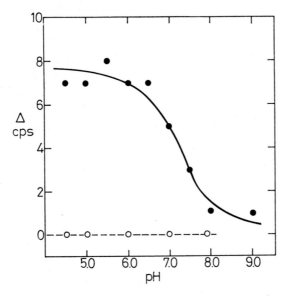

FIG. 3. Chemical shift of ε-trifluoroacetyl Lys-7 (peak L') on binding of 3'-CMP as a function of pH: (○) position of peak L' in free RNase S; (●) position of peak L' in RNase S saturated with 3'-CMP. The solid line is the theoretical titration curve of an ionizable group of pK_a 7.25 (from Huestis and Raftery 1971).

is affected by an ionizable group of pK_a 7.2–7.3 as could arise from His-119, which is known to have a pK_a of 7.4 (His-119 is known to be at the active site). His-12 is also known to be at the active site but has a pK_a of 8.0; however, no change in the chemical shift of L' is observed in the pH range 8–9 with 3'-CMP present. The charge on His-12 has no effect, therefore, on the observed resonance, and yet the X-ray data indicate that His-12 and His-119 are almost equidistant from Lys-7. Thus it seems unlikely that only His-119 would exert an electric field effect (with consequent chemical shift) on Lys-7. Hydrogen bonding of some group on the protein to the fluorine is also expected to be quite weak because the carbonyl group will tend to draw electrons from the fluorine, and this is unlikely to be the cause of the shift. One tentative explanation could be that the inhibitor binding results in a change in the conformation of His-12, bringing it closer to His-119 in the complexed form. When His-119 is deprotonated (pK_a 7.2–7.3), His-12 moves back towards its position in the native enzyme. As the distance between them increases and His-119 ceases to be strongly bound to the phosphate (as is the case with 3'-CMP above pH 5.6), the molecular model shows that Lys-7 can be as close as 6 Å to the guanidino group of Arg-39, thus changing its environment and producing a further chemical shift.

CONFORMATIONAL EFFECTS IN SELECTIVELY TRIFLUOROACETONYLATED
HAEMOGLOBIN A

The conformational transitions in haemoglobin have been the subject of much attention in terms of the ligand-dependent induced fit (Koshland 1953) and allosteric effects (Koshland, Nemethy and Filmor 1966). Nuclear magnetic resonance affords a way of studying the mechanism of such processes. The majority of the proton resonances in haemoglobin occur in a small spectral range, but several of the protons of the haem group are shifted out of this region by interaction with the unpaired electrons of the iron (Wurtrich 1971). From a study of these some progress has been made towards choosing a model for the conformational transitions resulting from the binding of oxygen.

The use of fluorine as a probe at a specific site in haemoglobin (Hb) (Raftery, Huestis and Millet 1971) affords in principle an experimentally easier method of monitoring conformational changes, since any changes in chemical shift will be larger than those of the protons. By reaction with 1-bromo-3-trifluoro-acetone, a 3,3,3-trifluoroacetonyl group ($CF_3COCH_2^-$, here designated TFA) was covalently and specifically attached to the -SH group of cysteine β-93, thus placing the probe close to the $\alpha_1\beta_2$ region of intersubunit contact. The binding of oxygen to the modified haemoglobin was essentially the same as that to native haemoglobin.

In Fig. 4, the chemical shifts of oxy and deoxy TFA-Hb are shown, together with those of various liganded forms. The small differences in chemical shifts caused by the ligands suggest similar but not identical environments for the ^{19}F probe in each case. Apart from the chemical shift of deoxy TFA-Hb, the chemical shifts were independent of pH. The pH dependence of the shift of deoxy TFA-Hb implies a group with a pK_a value of 7.4. Using the crystallo-

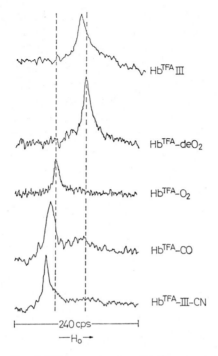

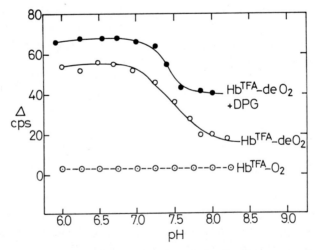

FIG. 4. ¹⁹F-n.m.r. spectra of trifluoroacetonylated haemoglobin with and without various ligands (from Raftery, Huestis and Millet 1971).

FIG. 5. Chemical shift, as a function of pH, of trifluoroacetonylated haemoglobin (Hb^{TFA}): oxygenated (Hb^{TFA}-O$_2$), deoxygenated (Hb^{TFA}-deO$_2$), and deoxygenated in the presence of 2,3-diphosphoglycerate (DPG) (Hb^{TFA}-deO$_2$) (from Raftery, Huestis and Millet 1971).

graphic results of Perutz (1970), we see that possible groups with this pK_a are the β-histidine residues 146, 143 or 92. His-β-92 is the proximal haem-linked histidine, and the bond to the iron would be expected to lower its pK_a. 2,3-Diphosphoglycerate (DPG) is thought to bind to His-β-143 and does not change the pK_a value (7.4) of the deoxy TFA-Hb complex (Fig. 5). Thus this pK_a value is assigned to His-β-146. This argument also implies that the probe will be sensitive primarily to conformational changes within the β-chain.

Raftery, Huestis and Millet (1971) used this probe to show that, contrary to previous evidence (dye binding, crystal form), methyl Fe^{3+}-Hb is different from oxy-Fe^{2+}-Hb, since the curve for the pH dependence of the ^{19}F probe chemical shift was different (Fig. 6). The X-ray structure does not yield precise informa-

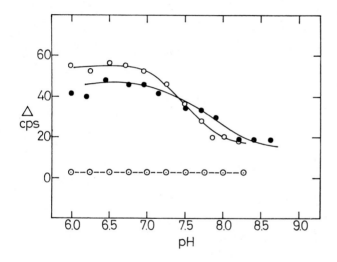

FIG. 6. As for Fig. 5 except that ● represents Met-Fe^{3+}-TFA-Hb (demonstrates that latter differs from oxy-Fe^{2+}-TFA-Hb) (from Raftery, Huestis and Millet 1971).

tion, since the oxy-Hb crystals used contained some Met-Hb (Blake *et al.* 1967).

In the above examples, precise interpretation of the chemical shift changes depends substantially on knowledge of the crystal structure. The small enzymes for which the structure is already known thus afford a means of trying out the technique and making progress in understanding the factors causing the chemical shifts. In larger enzymes, the much slower tumbling times mean that covalent probe resonances will be much broader because the dipolar interactions with other nuclei will no longer be averaged out. In such cases, the use of substrate probes has to be considered.

FLUORO SUBSTRATES AS PROBES

When the fluorine probe is on a substrate or inhibitor analogue, it may be possible to obtain information on the fluorine resonance of the substrate (or inhibitor) when it is bound to the enzyme. This may result from either the observation of an 'average' resonance signal because of rapid chemical exchange of the substrate between the enzyme and bulk sites, or, under conditions of 'slow chemical exchange', when two signals corresponding to bound and free substrate analogue can be observed.

An example of the 'slow exchange conditions' is provided by the *binding of the trifluoroacetylated analogue of chitotriose to lysozyme* (Raftery, Huestis and Millet 1971). The signal of a one-to-one mole ratio of enzyme to trisaccharide, each at 3 mM [at these concentrations, it can be calculated from the binding constants (Raftery, Huestis and Millet 1971) that the concentration of the complex is close to 100%], is shown in Fig. 7 at three pH values. The sharp

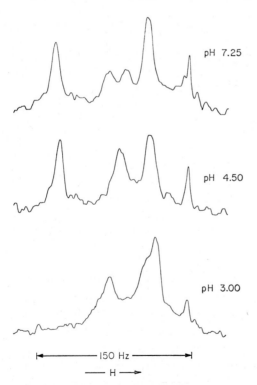

FIG 7. ^{19}F-n.m.r. spectra of solutions of 3 mM-trifluoroacetylated chitotriose and 3 mM lysozyme at 10°C at three pH values. The resonance to highest field is due to the trifluoroace-tate anion (from Raftery, Huestis and Millet 1971).

peak at highest field represents the trifluoroacetate anion, which was used as an internal reference. From the spectra at pH 4.50, the bound inhibitor resonances are identified as follows: the resonance at lowest field corresponds to the trifluoroacetyl (F_3Ac) group at the non-reducing end of the trisaccharide; the next to high field corresponds to the F_3Ac group at the reducing end of the trisaccharide (this resonance separating into the α and β anomeric forms at pH 7.25); the other resonance corresponds to the middle F_3Ac group of the trisaccharide. These assignments were made on the basis of experiments at higher temperatures, at pH 4 under conditions of fast exchange with excess oligosaccharide present, and by comparison with the spectra of the pure trisaccharide.

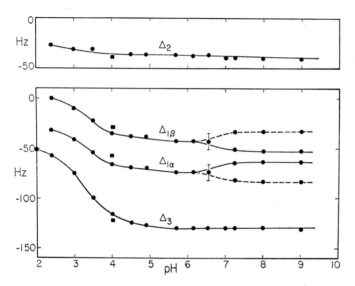

FIG. 8. Chemical shift data for the trifluoroacetyl group ^{19}F resonance for the pertrifluoro-acetylchitotriose lysozyme complex (3 mM) over a range of pH at 10 °C. The squares represent the chemical shifts measured under conditions of fast exchange at pH 4.0 and 65 °C (from Raftery, Huestis and Millet 1971).

The effects of pH on the position of the resonance is illustrated in Fig. 8, from which it can be observed that the shift of the central F_3Ac group of the trisaccharide (Δ_2) is almost independent of pH. The pH dependence of the shifts is shown in detail in Fig. 8, in which Δ_3 represents the shift of the non-reducing end F_3Ac group and $\Delta_{1\alpha}$ that of the reducing end F_3Ac group of the α-anomeric form of the trisaccharide. The pH dependence of Δ_3 implicates a group with a pK_a of 3.2. This is identified as Asp-101 in the top of the lysozyme-

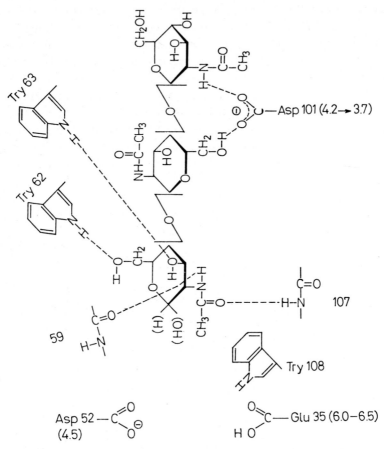

FIG. 9. Representation of hydrogen-bonded interactions of chitotriose and lysozyme as determined by crystallographic methods (from Raftery, Huestis and Millet 1971).

binding cleft, since the pK_a value of this group of 4.2 in the enzyme changes to 3.6 upon binding of chitotriose and, further, a group of pK_a 4.2 in the enzyme changes to 3.2 on binding of F_3Ac-chitotriose.

The interactions of chitotriose with specific groups on the enzyme, as determined by X-ray analysis (Perutz 1970) are shown in Fig. 9. If the fluorinated analogue binds similarly, the formation of a hydrogen bond between Asp-101 and the F_3Ac group would result in a change in the ^{19}F chemical shift of this group. From the model it is apparent that the F_3Ac group at the reducing end of the trisaccharide is the furthest F_3Ac group from Asp-101, and yet, unlike the central F_3Ac group, it experiences an obvious chemical shift. The effect is unlikely to be electronic, as is that for the F_3Ac group at the non-reducing end of the sugar. An explanation which agrees with kinetic observations using other n.m.r. methods (Dahlquist and Raftery 1972) is that the formation of the Asp-101 hydrogen bond at the reducing end of the trisaccharide causes an isomerization of the complex which alters the position of the end F_3Ac group with respect to the aromatic ring of Tyr-108 and therefore the effect of the ring current on the fluorine nuclei.

For most inhibitors or substrates binding to enzymes, conditions of slow exchange will not obtain. As mentioned earlier, even if this were the case, in larger enzymes the resonances may be too broad to detect. Under conditions of rapid chemical exchange this problem need not arise. An example where fast chemical exchange is valid is shown by the binding of N-acetyl DL-p-fluorophenylalanine to chymotrypsin. The ^{19}F spectrum of the inhibitor is complex, corresponding to an AA′BB′X multiplet (Spotswood, Evans and Richards 1967). No difference in ^{19}F chemical shift of the optical isomers was observable, but addition of chymotrypsin resulted in two sets of resonances, corresponding to the D and L isomers. Addition of the pure resolved isomers indicated that the D isomer is shifted to low field, while the L isomer is not significantly altered.

If the ^{19}F spectrum of an inhibitor (or substrate) changes as a result of binding to an enzyme, it is possible under conditions of rapid chemical exchange to obtain the binding parameters. Thus the binding of N-acetyl D-p-fluorophenylalanine (S) to chymotrypsin (E) may be determined as follows.

Consider the equilibrium:

$$E + S \rightleftharpoons ES, \tag{1}$$

where
$$K_S = [E][S]/[ES].$$

Let the chemical shift of the fully bound species be Δ, then if the fraction of S bound to the enzyme p_M, the observed shift, δ, will be given by:

$$\delta = p_M \cdot \Delta \tag{2}$$

Since $p_M = [ES]/[S_o]$, where S_o is the total concentration of S, combining (1) and (2) and assuming that $[S_o] \gg (ES)$, we obtain

$$S_o = E_o \Delta / \delta - K_S \tag{3}$$

Thus a plot of S_o against $1/\delta$ allows Δ to be obtained from the slope and K_S from the intercept on the x-axis. In the above example $\Delta = 83$ Hz (at 56.4 MHz) and $K_S \simeq 6$ mM.

Equation (3) can be simply modified to take into account the effects of competition from a second substrate. In such a case it can easily be shown that

$$\frac{1}{\delta} = \frac{1}{\Delta E_o}\left(K_S + S_o + \frac{K_S}{K_I} \cdot S_I\right) \tag{4}$$

where K_I is the binding constant and S_I total concentration of the competing substrate. In this example, *N*-acetyl D-tryptophan, allowed to compete with *N*-acetyl D-fluorophenylalanine, gave a value of 4 mM for K_I.

Returning to the low-field shift observed when *N*-acetyl D-fluorophenylalanine binds to chymotrypsin, it has been suggested that this is evidence for the binding of aromatic ring and acetyl group in a hydrophobic 'pocket' at the active site. While indeed this might be the case, there are many factors, including pH, temperature, and 'through space' interactions, that can contribute to fluorine chemical shifts, and which cannot be elucidated from a single determination of a chemical shift. The effect of pH on the bound chemical shift may lead to further information about binding sites, if one bears in mind that a detailed interpretation will depend on a knowledge of the crystal structure of the enzyme.

Compared with the magnitudes of ^{19}F shift frequently observed, the changes in chemical shift in the enzyme complexes discussed so far are small. In general, fluorine chemical shifts are large when the effects giving rise to them are transmitted through chemical bonds (for instance, in fluoromonosaccharides the shift of the equatorial fluorine at C-3 differs by about 5 p.p.m. for the α and β forms). When the shifts arise from 'through space' interactions such as ring currents and electric field effects, they are much smaller.

In the foregoing example, addition of the inhibitor or ligands may result in a change in the enzyme structure and the small shifts observed are a consequence of 'through space' interactions. It is possible that several of these interactions produce shifts in different directions, giving a small net shift. In most experiments on enzymes, shifts will result from changes in 'through space' interactions. Thus although fluorine appears to be a highly attractive enzyme probe because of the large expected shifts, the small shifts which may result in practice are relevant when designing this type of experiment.

FLUORO SUBSTRATES AS PROBES IN THE PRESENCE OF A PARAMAGNETIC ION AS A SECOND PROBE: SOME THEORETICAL CONSIDERATIONS

In larger enzymes in which the crystal structure is unknown it seems that the future of magnetic resonance in the study of these may well be in the 'mapping out' of substrates around a paramagnetic probe—preferably one at the active site. The experiments to do this involve two techniques: observation of the high-resolution spectrum of the substrate, and determination of the proton relaxation enhancement behaviour of the bulk water protons. We shall discuss the theory to this approach and then give some of the examples studied, to indicate the potential of the method.

By a combination of proton relaxation enhancement (P.R.E.) measurements and relaxation time measurements on the high-resolution spectrum of the substrate, it is possible in principle to calculate the distances of individual nuclei from the paramagnetic metal probe. The relaxation times are related to this distance by the Solomon-Bloembergen equations below. The calculation of the distances from these equations requires a knowledge of the correlation time—a parameter difficult to obtain without detailed relaxation time measurements as a function of magnetic field and temperature. However, the use of P.R.E. suggests a fairly straightforward way of obtaining reasonable estimates of the correlation time. We give therefore a brief summary of the elementary theory. The detailed theory has been presented elsewhere (Dwek 1972).

One may start by noting the equations (5, 6) which have been used to predict the nuclear relaxation behaviour of dipolar nuclei in the vicinity of the paramagnetic ions. We shall apply these first to aqueous solutions of the paramagnetic ions containing a solute and then consider what changes are necessary when a macromolecule which binds to the paramagnetic ions is added. In aqueous solutions the solute may exist in either the bulk solvent or bound to the paramagnetic ion. These sites are denoted A and M respectively, and we consider the relaxation rates initially in each site and then how the observed relaxation rate of the solution alters when chemical exchange, with the appropriate time scale, occurs between these two sites.

Bound site

The relaxation times of the bound nuclei, T_1 and T_2, are usually well represented by the Solomon-Bloembergen equations (Solomon 1955; Bloembergen 1957). Assuming $\omega_S \gg \omega_I$, these are

$$\frac{1}{T_{2M}} = \frac{1}{15} \frac{\gamma_I^2 g^2 S(S+1)\beta^2}{r^6} \left(4\tau_c + \frac{3\tau_c}{1+\omega_I^2\tau_c^2} + \frac{13\tau_c}{1+\omega_S^2\tau_c^2} \right)$$
$$+ \frac{1}{3} S(S+1) \left(\frac{A}{\hbar} \right)^2 \left(\frac{\tau_e}{1+\omega_S^2\tau_e^2} + \tau_e \right) \tag{5}$$

and

$$\frac{1}{T_{1M}} = \frac{2}{15} \frac{\gamma_I^2 g^2 S(S+1)\beta^2}{r^6} \left(\frac{3\tau_c}{1+\omega_I^2\tau_c^2} + \frac{7\tau_c}{1+\omega_S^2\tau_c^2} \right)$$
$$+ \frac{2}{3} S(S+1) \left(\frac{A}{\hbar} \right)^2 \left(\frac{\tau_e}{1+\omega_S^2\tau_e^2} \right) \tag{6}$$

The first terms arise from dipole–dipole interaction between the electron spin S and the nuclear spin I, which is characterized by a correlation time τ_c which modulates this interaction. The second terms arise from modulation of the scalar interaction $AI \cdot S$ (often called isotropic nuclear–electron spin exchange interaction) which is characterized by a correlation time τ_e. The quantities ω_S and ω_I are the electronic and nuclear Larmor precession frequencies, γ_I is the magnetogyric ratio, β is the Bohr magneton, S is the total electronic spin, r is the distance between the nucleus (fluorine or protons in these examples) and the paramagnetic ion, and $A/\hbar$ is the electron–nuclear hyperfine coupling constant in Hz.

Unbound site, no exchange

The relaxation times of the unbound nuclei, T_{1A} and T_{2A}, can, in the absence of chemical exchange, be considered to arise from two contributions: (*a*) the relaxation in the absence of the paramagnetic ion, $T_{1A(0)}$ and $T_{2A(0)}$, and (*b*) the relaxation arising from dipolar interaction with the solvated paramagnetic ion, T''_{1A} and T''_{2A}, i.e.

$$\frac{1}{T_{1A}} = \frac{1}{T_{1A(0)}} + \frac{1}{T''_{1A}} \tag{7}$$

The contribution of T''_{1A} may be estimated by averaging the dipolar contribution over the volume between the sphere of closest approach, radius d (assumed to be negligible here).

Correlation times

The correlation times in equations (5) and (6) are defined as follows:

$$\frac{1}{\tau_c} = \frac{1}{\tau_S} + \frac{1}{\tau_M} + \frac{1}{\tau_R} \tag{8}$$

and

$$\frac{1}{\tau_e} = \frac{1}{\tau_S} + \frac{1}{\tau_M} \tag{9}$$

where τ_M is the lifetime of a nucleus in the bound site, τ_R is the rotational correlation time of the solvated paramagnetic ion, and τ_S is the electron spin relaxation time.

Unbound site with chemical exchange

If the observed relaxation rates in the presence of chemical exchange are $1/T_{1 \text{ obs}}$ and $1/T_{2 \text{ obs}}$, the contributions of the paramagnetic ion to these rates, $1/T_{1P}$ and $1/T_{2P}$, may be shown to be (Luz and Meiboom 1964; Swift and Connick 1962)

$$\frac{1}{p_M q} \cdot \frac{1}{T_{1p}} = \frac{1}{p_M q} \left(\frac{1}{T_{1 \text{ obs}}} - \frac{1}{T_{1A(0)}} \right) = \frac{1}{T_{1M} + \tau_M} \tag{10}$$

and for the cases of Mn^{2+} and Gd^{3+} a similar equation holds for $1/T_{2p}$, namely

$$\frac{1}{p_M q} \cdot \frac{1}{T_{2p}} = \frac{1}{p_M q} \left(\frac{1}{T_{2 \text{ obs}}} - \frac{1}{T_{2A(0)}} \right) = \frac{1}{T_{2M} + \tau_M} \tag{11}$$

where $p_M q$ is the fraction bound. The equations assume that $p_M q \ll 1$. Actually p_M is the mole fraction of paramagnetic ion in the solution and q is the coordination number of the ligands, e.g. $Mn(H_2O)_6^{2+}$ has $q = 6$, but $Mn(H_2O)_5 AMP$ has $q = 5$, if we consider the water ligands, but $q = 1$ if we consider the AMP as the ligand.

Definition of slow, intermediate and fast exchange

In the above equations (10) and (11) we see that two limiting cases can occur, depending on the magnitudes of T_{1M} and T_{2M} relative to τ_M. We summarize thus:

$$\tau_M \gg T_{1M}, \text{ or } T_{2M} \quad - \quad \text{slow exchange}$$
$$\tau_M \simeq T_{1M}, \text{ or } T_{2M} \quad - \quad \text{intermediate exchange}$$
$$\tau_M \ll T_{1M}, \text{ or } T_{2M} \quad - \quad \text{fast exchange}$$

[We also note from equations (5) and (6) that since $T_{1M} \geqslant T_{2M}$, it is possible that fast exchange conditions could apply in equation (10) and not equation (11)].

Temperature dependence of equations (10) and (11)

Equations (10) and (11) are used to describe the observed temperature behaviour of the relaxation times. The temperature dependence of these equations arises from the temperature dependence of τ_M, τ_M, τ_S and τ_R. τ_M is given by the usual Eyring relationship.

$$\frac{1}{\tau_M} = \frac{kT}{h} \exp \left[-\frac{\Delta H^{\ddagger}}{RT} + \frac{\Delta S^{\ddagger}}{R} \right] \tag{12}$$

where k is the Boltzmann constant, T the temperature in degrees Kelvin, and $\Delta H^{\ddagger}$ and $\Delta S^{\ddagger}$ are the enthalpy and entropy of activation, respectively, for the first-order reaction of chemical exchange.

The temperature dependence of τ_S may be complex, and even appear to vary from a positive to a negative dependence with magnetic field. For simplification we assume that it is given by the equation

$$\tau_S = \tau_S^{\circ} \exp\left(\pm E_S/RT\right), \tag{13}$$

where the $\pm$ sign allows for a positive or negative dependence. τ_R is given by

$$\tau_R = \tau_R^{\circ} \exp\left(E_R/RT\right). \tag{14}$$

Consideration of the temperature dependences of τ_R, τ_M, and τ_S indicates that if $\tau_M \gg T_{1M}$ (or T_{2M}) i.e. there is slow exchange, the value of $1/T_{1p}$ or $1/T_{2p}$ given by equation (10) or (11) decreases with decreasing temperature. Conversely, if the value of $1/T_{1p}$ or $1/T_{2p}$ decreases with increasing temperature, this must mean that fast exchange is occurring ($\tau_M \ll T_{1M}$), and in such a case

$$\frac{1}{T_{1p}} = 1/T_{1\,\text{obs}} - 1/T_{1(0)} = p_M q/T_{1M} \tag{15}$$

and similarly for $1/T_{2p}$.

Relaxation rates in the presence of a macromolecule

The presence of a macromolecule (E) to which the paramagnetic ion (M) can bind introduces the equilibrium

$$M + E = ME \tag{16}$$

Thus we now have to consider nuclei in three environments: (a) those in the bulk, (b) those bound to the 'free' paramagnetic ion which has a concentration M_f, and (c) those bound to the paramagnetic ion on the macromolecule, which has a concentration M_b. If we assume that M_b and M_f are small, the concentration of nuclei at each of these two sites is small compared with the concentrations of nuclei in the bulk solvent (as is usually the case). In such a case the paramagnetic contribution to the observed relaxation rates is simply given by

$$\frac{1}{T_{iP}^*} = \left(\frac{1}{T_{iP}^*}\right)_f + \left(\frac{1}{T_{iP}^*}\right)_b \dots i = 1,2. \tag{17}$$

The explicit form of the paramagnetic ion contribution to the spin lattice relaxation rate is, from equations (17), (10) and (11),

$$\frac{1}{T^*_{1P}} = \left(\frac{p_M q}{T^*_{1M} + \tau^*_M}\right)_f + \left(\frac{p^*_M q^*}{T^*_{1M} + \tau^*_M}\right)_b. \tag{18}$$

Equations for the explicit form of $1/T^*_{2p}$ can also be written. For binding to occur, the number of water molecules coordinated to the paramagnetic ion when bound to the macromolecule (q^*) must be less than the number coordinated to the ion in the free aqueous solution (q).

Definition of proton relaxation enhancement (P.R.E.)

It is usual to define the P.R.E. factor ε^* for a given solution as:

$$\varepsilon^* = 1/T^*_{1\,obs} - 1/T^*_{1(0)} \big/ 1/T_{1\,obs} - 1/T_{1(0)} = (1/T^*_{1p})(1/T_{1p}), \tag{19}$$

where the asterisk indicates the presence of a macromolecule. $1/T^*_{1\,obs}$ and $1/T_{1\,obs}$ are the observed spin lattice relaxation rates of the solution in the presence of the paramagnetic ion, and $1/T^*_{1(0)}$ and $1/T_{1(0)}$ those of the same solutions in the absence of the paramagnetic ion. A similar enhancement factor can be defined for the spin–spin relaxation rates.

Uses of proton relaxation enhancements: binding constants and enhancement parameters

The study of the proton relaxation of the water molecules around the paramagnetic ion in the presence of different amounts of macromolecules or ligands allows the determination of the stability constant for the paramagnetic ion–macromolecule or paramagnetic ion–ligand complex. These complexes are termed *binary complexes*. The addition of a substrate to form the metal–macromolecule–substrate complex gives *ternary complexes*, whereas the further addition of a substrate will form a *quaternary complex*.

In addition to allowing the determination of binding constants, the binary, ternary and quaternary enhancement parameters of the fully-formed complexes ε_b, ε_t etc. may be obtained, which are believed to be characteristic for a given state or conformation of an enzyme.

In the next section I shall deal with the binary complexes and the determination of their binding parameters, and then indicate some of the problems that are encountered in corresponding studies with ternary complexes.

Characteristic enhancement

If X_b and X_f represent the mole fractions of the bound and free paramagnetic ion, ε_b is the enhancement of the fully-formed macromolecular–paramagnetic ion complex and ε_f the enhancement of the water protons on the unbound paramagnetic ion in solutions containing the macromolecule (assumed to be unity, Dwek 1972), we may write

$$\varepsilon^* = X_b\varepsilon_b + X_f\varepsilon_f \tag{20}$$

Determinations of ε_b and binding constant of paramagnetic ion to a macromolecule

Consider the equilibrium

$$M + E \rightleftharpoons ME, \tag{21}$$

where M and E represent the paramagnetic ion and the macromolecule concentrations, and the number (n) of M binding sites on the enzyme is

$$K_D = [M_f] \, [nE_f]/[ME], \tag{22}$$

where, as usual, the subscript f indicates the free concentrations. By the use of equations (20) and (21) it can be readily shown that

$$1/\varepsilon^* = 1/\varepsilon_b \, [K_D/[(nE_f) + K_D/\varepsilon_b] + (nE_f)/[(nE_f) + K_D/\varepsilon_b)].$$

The value of K_D and ε_b can then be obtained from a titration of the enhancement, with enzyme concentrations at fixed metal concentrations.

Nature of ε_b

If we assume that chemical exchange of the water protons is rapid it follows from equations (6), (15), (19) that

$$\varepsilon_b = q^*T_{1M}/qT_{1M}^* \simeq q^*\tau_c^*/q\tau_c \tag{23}$$

where q^* and q are the number of water molecules in the first hydration sphere of the metal ion when it is bound to the macromolecule and when it is free in the solution respectively; τ_c^* and τ_c are the corresponding correlation times. Thus τ_c^* can be evaluated. If conditions of slow chemical exchange apply, an upper limit for $T_{1M}^* \, (<\tau_M^*)$ and thus τ_c^* can be obtained.

Nature of paramagnetic probes

If conditions of fast chemical exchange obtain, then from equation (23) an enhancement will be obtained if there is an enhancement in τ_c^*. From equation (8) τ_c is made up of three contributions: τ_R, τ_M and τ_S. In general τ_S and τ_M do not vary much from complex to complex for a given metal ion, so that if either of these correlation times dominates τ_c initially, this will still be so for τ_c^* in the complex. On the other hand, the value of τ_R will change according to the size of the complex. For instance in the Mn^{2+} aquo ion the value of τ_R is 3×10^{-11} sec at 25 °C and those of τ_S and τ_M about 10^{-8} sec. Thus in the aquo ion τ_R will dominate τ_c. In a macromolecular complex, the tumbling time of the complex, which may be loosely identified as τ_R^*, will become longer, perhaps even less than 10^{-8} sec, so that τ_R^* may no longer dominate τ_c^*. In any event $\tau_c^* > \tau_c$ and an enhancement is obtained.

Clearly, the most suitable metal probes are going to be those for which τ_R dominates τ_c in the aquo ion. At present the most suitable enhancement probes are Mn^{2+} and Gd^{3+}, which fulfil this condition. Probes like Co^{2+} and Ni^{2+}, for which τ_S is about 10^{-12} sec, will clearly be unsuitable, since in this case τ_S will dominate τ_c.

Ternary complexes

In a solution that contains enzyme and ligand metal ion, the following equilibria must be considered:

Equilibria	*Enhancement parameter*	
$K_D = \dfrac{[E_f][M_f]}{[ME]}$	ε_b	(24)
$K_1 = \dfrac{[M_f][S_f]}{[MS]}$	ε_s	(25)
$K_2 = \dfrac{[MS][E_f]}{[MES]}$	ε_t	(26)
$K_3 = \dfrac{[ME][S_f]}{[MES]}$	ε_t	(27)
$K_{A'} = \dfrac{[ES][M_f]}{[MES]}$	ε_t	(28)
$K_S = \dfrac{[E_f][S_f]}{[ES]}$		(29)

where ε_b, ε_s and ε_t are the enhancements of the complexes ME, MS and MES, respectively, and S represents ligand (substrate or modifier).

The metal ion may be considered as having three sites for binding—E, S and ES. The enhancement of such a solution is given by

$$\varepsilon^* = \frac{[M_f]}{[M_f]} \cdot \varepsilon_f + \frac{[ME]}{[M_t]} \cdot \varepsilon_b + \frac{[MS]}{[M_t]} \cdot \varepsilon_S + \frac{[MES]}{[M_t]} \cdot \varepsilon_t. \tag{30}$$

Noting that

$$[M_t] = [M_f] + [ME] + [MS] + [MES], \tag{31}$$

and substituting from equations (24) to (29) in equation (30), one can obtain expressions like that in equation (32) (Dwek 1972):

$$\varepsilon^* = \frac{1 + \dfrac{[E_f]}{K_D} \cdot \varepsilon_b + \dfrac{[ES]}{K_{A'}} \cdot \varepsilon_t + \dfrac{[S_f]}{K_1} \cdot \varepsilon_S}{1 + \dfrac{[E_f]}{K_D} + \dfrac{[ES]}{K_{A'}} + \dfrac{[S_f]}{K_1}}. \tag{32}$$

For ternary complexes the evaluation of the parameters ε_t etc. is slightly more complex than in the binary case because of the several coupled equilibria that exist. If the values of K_1, K_D and K_S are known, together with the corresponding enhancement factors, it is possible, in principle, for a trial value of K_2 or K_3 to calculate the equilibrium composition of a solution containing enzyme, metal ion and substrate. However, such a calculation involves simultaneous solutions of equations (24) to (29) (Reed, Cohn and O'Sullivan 1970). The non-linear form of these equations dictates a numerical solution and one method that has been used is the computer adaptation of the standard successive approximation method for complex equilibria. In some cases, however, graphical solutions may be used.

As in the binary case, the value of ε_t allows one to calculate the correlation time in the complex from an equation similar to (23).

I shall now consider some particular examples.

BINDING OF METHYL *N*-FLUOROACETYL-β-D-GLUCOSAMINIDE AND OF *N*-FLUOROACETYLGLUCOSAMINE TO LYSOZYME

Methyl *N*-fluoroacetyl-β-D-glucosaminide (β-Me GlcNAcF) has a ^{19}F spectrum at 225 p.p.m. upfield from $CFCl_3$ which consists of a triplet (J=47 Hz) (Kent and Dwek 1970). The addition of lysozyme (Butchard *et al.* 1972) results in a downfield chemical shift. The variation of chemical shift can be used to determine the binding constant K_S, as in the example with chymotrypsin, by means of the relationship:

$$S_o = E_o\Delta/\delta - K_S,$$

where S_o is the total concentration of substrate, E_o the enzyme concentration, δ the observed chemical shift and Δ that of the fully formed complex. It must be stressed that this equation assumes that fast chemical exchange occurs between the bulk and bound substrate/inhibitor. The results of such a plot for β-Me GlcNAcF are shown in Fig. 10; $K_S = 50$ ($\pm$ about 15) mM and Δ (at 84 MHz) $= 140$ Hz (Dwek, Kent and Xavier 1971).

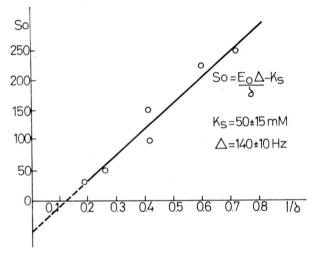

FIG. 10. Plot of $S_o = E_o\Delta/\delta - K_s$ for the binding of methyl N-fluoroacetyl-β-D-glucosaminide to lysozyme at pD $= 5.0$ (imidazole buffer, 100% D_2O).

The paramagnetic probe chosen was Gd^{3+}, since X-ray crystallography had shown this ion to bind at the active site. Also the chemistry of the lanthanides parallels that of the actinides, so that if the position of a heavy atom is known from X-ray work, it is very likely that the corresponding lanthanide ion (Gd^{3+} in this case) binds at the same site.

The addition of Gd^{3+} results, as expected, in a broadening of the pure β-Me GlcNAcF resonance and a much greater broadening of the resonance when the enzyme is present (Fig. 11, C and D). The broadening for the pure substrates probably arises from dipolar interactions, as mentioned above, or from the formation of a very weak complex, and must be subtracted from the

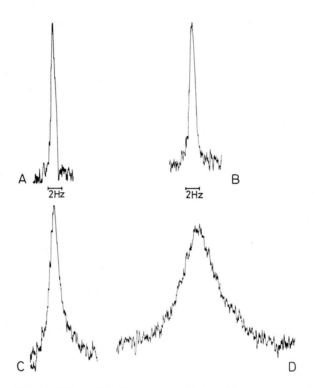

FIG. 11. Effect of Gd^{3+} on one of the ^{19}F resonances of methyl N fluoroacetyl-β-D-glucosaminide (200 mM) in the presence and absence of lysozyme. (A): β-Me GlcNAcF alone; (B): + lysozyme (3 mM); (C): + Gd^{3+} (1 mM); (D): + lysozyme + Gd^{3+} (1 mM).

line width of the ^{19}F in the enzyme solution to obtain the true broadening arising from the paramagnetic ion.

The proton relaxation enhancement work indicates that the Gd^{3+}–lysozyme complex has a dissociation constant of 11 mM and a characteristic enhancement of 3.8 at 23 °C (Fig. 12). This enhancement does not change, within the experimental error, on addition of the substrates and so, using the known value of τ_c for aqueous Gd^{3+} solutions of 4.5×10^{-11} sec, the value of q as the number of hydrated water molecules, 9, and $q^* = 7$ (on binding), we obtain $\tau_c^* = 2.2 \times 10^{-10}$ sec (by equation 23). Previous experiments have shown that only the dipolar term in equation (6) is important for proton with Gd^{3+}. If this is true here, from the value of the broadening, r can be calculated as 5.1 Å. Preliminary experiments on a solution containing α- and β-Me GlcNAcF give a value of 5.2 Å for the α-anomer, and a value of Δ of about 120 Hz.

N-fluoroacetylglucosamine (GlcNAcF) has resonances centred about 225

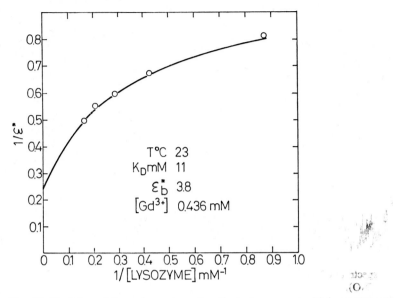

FIG. 12. Variation of the water proton relaxation enhancement with lysozyme concentration.

p.p.m. upfield from $CFCl_3$. The ^{19}F spectrum appears as the expected triplet of an A_2X pattern ($J = 46$ Hz), as for the β-methyl derivative. However, two sets of resonances are observed, corresponding to two X patterns separated by 0.4 p.p.m. These arise from the α and β anomers and are assigned as shown in Fig. 13. The assignments are based on comparison with the spectra for α- and β-Me GlcNAcF and the chemical shifts of 6-deoxy-6-fluoro-D-glucose, which are at 230.6 p.p.m. for the α-anomer and 229.9 p.p.m. for the β-anomer (Bessell *et al.* 1971).

The addition of lysozyme to a solution of GlcNAcF results in a broadening and a shift to low field of all resonances. However, those of the β-anomer shift about twice as much as those of the α-anomer. The different shifts could be a consequence of different binding constants of the anomers to the enzyme or of different chemical shifts of the anomers in the fully formed GlcNAcF–lysozyme complexes. If we assume that conditions of fast chemical exchange obtain, then the chemical shifts of the fully formed complexes, Δ, at 84 MHz (assuming that all the enzyme is equally saturated with each anomer) are $\Delta_\alpha = 320$ and $\Delta_\beta = 160$ Hz.

The addition of Gd^{3+} causes all the resonances to broaden and, as in the case of β-Me GlcNAcF, the distances from Gd^{3+} can be calculated. These are $r_\alpha = 5.6$ Å and $r_\beta = 5.0$ Å; thus the orientations of the two anomers when bound to lysozyme are probably different. This may well explain the difference

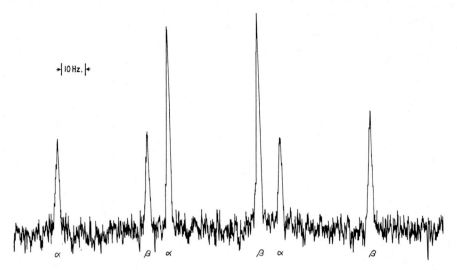

FIG. 13. ^{19}F spectrum of N-fluoroacetylglucosamine at 84 MHz (25 °C, pD = 5.0, imidazole buffer, 100% D_2O).

in the chemical shifts of the fully bound forms. Further, the values of r for α- and β-Me GlcNAcF also indicate that the orientation of these is similar to that of β-GlcNAcF while that of α-GlcNAcF appears to indicate that this binds in a slightly different manner. This is consistent with the results on the non-fluorinated monosaccharides, for which α-GlcNAc has been shown previously to bind with a different orientation from the others in subsite C (Dahlquist and Raftery 1968). Low-resolution (6 Å) X-ray studies have shown that β-Me GlcNAc binds to lysozyme in the same way as β-GlcNAc. Using a model of lysozyme with a β-Me GlcNAc molecule positioned according to the high-resolution coordinates of β-GlcNAc, the distances between the acetamido and glycosidic methyl groups from the binding site of Gd^{3+} were measured, r acetamido was 7.3 ± 0.8 Å and r glycosidic 4.4 ± 0.4 Å. If β-GlcNAcF (free sugar or methyl glycoside) binds similarly then we would expect this distance to be about the same here. That all these distances are shorter suggests that there is hindered rotation of the $-CH_2F$ group in the enzyme site such that the fluorine is orientated towards the Gd^{3+}, which would reduce the expected distance by about 1.4 Å.

The $Gd^{3+}-F$ distances in α- and β-GlcNAc F_3 support this explanation. Here the values obtained will be an average of the distances of the three fluorines— and these average values (Table 1) are in good agreement with those obtained from the X-ray data and from the n.m.r. data of the protons of the acetamido methyl group of β-Me GlcNAc. The values of the chemical shifts of the fully

TABLE 1

Values of the chemical shift (Δ) of some sugar lysozyme complexes of the Gd^{3+} ... F distance

	ΔHz†*	r ($\mathring{A}$)
α-GlcNAcF	−326	5.5 ± 0.2
β-GlcNAcF	−160	5.0 ± 0.2
α-MeGlcNAcF	−140	5.6 ± 0.2
β-MeGlcNAcF	−140	5.1 ± 0.2
α-GlcNAcF$_3$	+ 90	7.2 ± 0.2
β-GlcNAcF$_3$	− 30	7.2 ± 0.2

† Negative values indicate a shift to low field.
* Measured at 84.6 MHz at 22 °C.

formed GlcNAcF$_3$–lysozyme complexes of the Gd^{3+}–F distances are given in Table 1. Reference to the lysozyme model suggests that the shifts can be explained on the basis of ring current effects. The downfield chemical shift of the β-anomer is a result of the influence of tryptophan-108 on the CF$_3$ group (Try-108 is known to tilt and move away from the CF$_3$ group when the inhibitor binds). The upfield chemical shift of the α-anomer arises from interaction of the CF$_3$ group and Try-63.

This example illustrates that with the use of distance information from broadening studies, complemented by the geometric information obtained from chemical shifts from 'through space' (dipolar) interactions, it is possible in principle to locate accurately the positions of the nuclei studied. Certainly in this example the position of the inhibitor CF$_3$ groups in the inhibitor–enzyme complexes

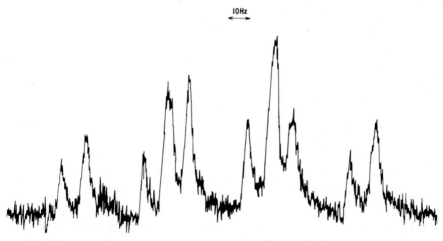

10Hz

Fig. 14. ^{19}F spectrum of N-fluoroacetylglucosamine at 84 MHz (25 °C, pD = 5.0, imidazole buffer, 25% D$_2$O).

must be almost the same in solution as expected in the crystal from consideration of molecular models of α- and β-Me GlcNAc–lysozyme complexes.

We conclude this example with a warning against attempting to derive too
much information from fluorine chemical shifts. Fig. 14 shows a spectrum of
GlcNAcF in 25% D_2O. Analysis of the spectrum indicates that there are now
four sets of triplets, corresponding to a proton or deuteron on the nitrogen
atom (this can be confirmed by observing the N—H resonance in the proton
spectrum). The shift between the NH or ND forms is 11 Hz. The exchange rate
between the two forms can be increased on heating, until at 80 °C only two sets
of resonances are observed, their chemical shifts being at the mean of the two
positions of those for the NH or ND compound. This isotope effect is one of
the largest reported and indicates the extreme sensitivity of fluorine to any
changes in environment, when the effect can be transmitted through chemical
bonds.

BINDING OF FLUORIDE IONS TO CARBOXYPEPTIDASE A

The second example is the binding of fluoride ions to carboxypeptidase A
(Navon et al. 1970) which hydrolyses C-terminal residues if they contain
branched aliphatic or aromatic side chains. Unlike the other examples, this
enzyme is a metalloenzyme, and the naturally occurring zinc atom can be
replaced by Mn^{2+}, which provides a paramagnetic probe at the active site.
The activity of the manganese enzyme is 35% that of the zinc enzyme. Fluoride
ions do not affect the activity of either enzyme and the effects on the fluorine
nuclear relaxation times resulting from interaction with the paramagnetic
metal ion provide information on the mode and strength of fluoride binding.
These results, in conjunction with proton relaxation enhancement measurements, give an indication of the total number of metal ligands.

The binding of fluoride to the enzyme is obtained by a titration, at constant
concentration of the manganese enzyme with fluoride ion. Since carboxypeptidase A is insoluble in solutions containing low salt concentration, the
titration was carried out by keeping the sum of NaCl and KF concentrations
constant and varying only their proportions. A plot of T_{2P} of the fluoride ion
against its concentration gave a straight line (Fig. 15), whose intercept on the
x-axis corresponds to the dissociation constant for the enzyme–fluoride complex; it had the value 0.1 ± 0.05 M. Interestingly there appears to be no competitive effect from Cl^- ions over the entire range of salt concentrations.

The effect of the strong inhibitor β-phenylpropionate upon the ^{19}F line widths

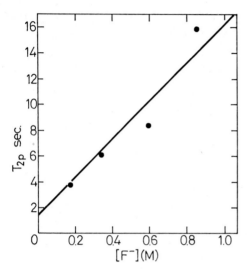

FIG. 15. T_{2p} of ^{19}F as a function of fluoride ion concentration. The total salt concentration was constant since $[NaCl]+[KF] = 1.0$ M (0.05 M-tris buffer, pH 7.5, 3 °C) (from Navon et al. 1970).

is shown in Fig. 16. The decrease in line width parallels the decrease in specific activity. From this titration a value of the inhibitor dissociation constant of 0.36 mM can be obtained. This value is identical with that obtained from titra-

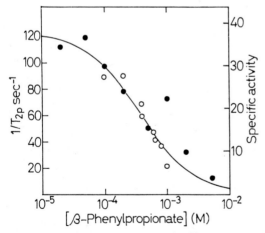

FIG. 16. Measured values of $1/T_{2p}$ for ^{19}F, and enzymic activity of manganese carboxypeptidase A with respect to 10^{-3} M-hippuryl-L-phenylalanine, as a function of the β-phenylpropionate concentration in manganese carboxypeptidase A. The curve is a calculated one for $K_1 = 2800$ M^{-1}. (○) specific activity (in micromoles of product produced per minute per milligramme of protein); (●) $1/T_{2p}$ (from Navon et al. 1970).

tion of the water proton relaxation rates with β-phenylpropionate in the absence of fluoride ion. In fact the water relaxation rates were found to be independent of the fluoride ion concentration. Thus F^- binding does not affect the inhibitor binding. On the other hand the changes in line width when β-phenylpropionate binds to the enzyme could arise from (a) displacement of the F^- by the inhibitor or (b) an overall conformational change by the inhibitor in which the F^- moves further away from the metal ion or (c) a slowing down of the F^- exchange rate.

A temperature variation of the ^{19}F relaxation times in the enzyme–fluoride complex appeared to indicate that conditions of slow exchange obtained. Thus only an upper limit for T_{2M}, and hence r, is obtained in these experiments. Use of a correlation time of 1.7×10^{-9} sec obtained from P.R.E. measurements gives a value of $r = 2.1$ Å. This, compared with the crystallographically determined Mn–F distance of 2.12 Å, would again indicate that under the conditions of these experiments the fluoride ion is directly coordinated to the manganese ion. The very strong interaction between Mn and F suggests that the explanation for the changes in the ^{19}F line width on addition of the inhibitor is either a or c in the previous paragraph. It does, however, pose a problem, for the binding of F^- does not affect the activity of either the manganese or the zinc form of carboxypeptidase A, even though the fluoride binds at the active site of the enzyme. Perhaps it displaces a chloride ligand, for X-ray data on the zinc enzyme indicate that the zinc is bound to three ligands on the protein and possibly to one (or more) other ligands which are 'not part of the protein', e.g. Cl^- or H_2O.

Relaxation measurements on the solvent water protons in the manganese enzyme indicated that there was at least one water molecule liganded to the manganese. This, with the present ^{19}F results (in which the P.R.E. is unaffected), indicates that if the manganese enzyme is similar to the zinc one, there must be at least five metal ligands.

One final caution, however, is that the probe manganese is not an ideal replacement for zinc, particularly since the former has a strong tendency to 6-coordination and the latter to 4-coordination, and the two enzymes may well have slightly different structures. Nevertheless, this example does illustrate the type of experiments that can be done.

THE FLUOROKINASE REACTION

Fluoride and fluorophosphate are the substrate and product of the fluorokinase reaction (Mildvan, Leigh and Cohn 1967), a side reaction catalysed by pyruvate kinase

$$F^- + ATP^{4-} \xrightarrow[HCO_3]{Mg^{2+}K^+} FPO_3^{2-} + ADP^{3-}$$

The chemistry of the paramagnetic metal ion Mn^{2+} closely resembles that of Mg^{2+}, so that it is an obvious choice for a paramagnetic probe in this enzyme.

The interactions of F^- and FPO_3^{2-} with manganese and with the manganese–pyruvate kinase complex were examined by fluorine n.m.r. and by proton relaxation enhancement in an attempt to detect an enzyme-metal–fluorophosphate bridge.

The values of the characteristic enhancement parameters for the various manganese complexes discussed here are listed in Table 2, together with the appropriate binding constants and correlation times for distance calculations.

TABLE 2

Dissociation constants (K), enhancement parameters (ε) and values of the coordination number (q^*) and correlation time (τ_c^*) used in distance calculations for the action of fluorokinase

Complex	$K(mM)$	ε	q^*	$\tau_c^* \times 10^{-11}$ sec‡
F–Mn	400	0.9	5	3
FPO$_3$–Mn	1.8	1.0	5	3.6
FPO$_3$–Mn–pyruvate kinase	28	1.5	2 or 1	13.5 or 27.0

‡   calculated from equation (22) with $\tau_c = 3 \times 10^{-11}$ sec.

The value of ε_b for fluoride is about $^5/_6$ and is consistent with the replacement of one water molecule in the hydration sphere of the Mn^{2+}—i.e. with formation of the complex $Mn(H_2O)_5F$. The value of τ_c^* for the complex is, in this case, identical with τ_c for the Mn^{2+} aquo ion (i.e. 3×10^{-11} sec, which is known from previous measurements). In the Mn–FPO_3 complex the value of ε_b is consistent with a value of $\tau_c^* = 3.6 \times 10^{-11}$ sec, with $q^* = 5$. For the enzyme–Mn–FPO_3 complex, the value of q is, a priori, unknown. Independent studies have shown that in the Mn–enzyme complex, the Mn^{2+} is attached through three ligands to the enzyme (Reuben and Cohn 1970) and possible values are then $q^* = 2$ (if coordination is through either the fluorine or one oxygen atom) or $q^* = 1$ (if coordination is through two oxygens).

The addition of Mn^{2+} to a solution of fluoride ion results in a marked increase in the observed fluorine relaxation rate (Fig. 17). The addition of pyruvate kinase, which binds about one half of the manganese in the solution,

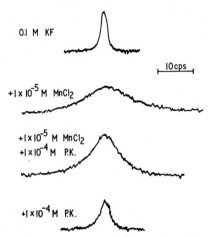

FIG. 17. Effect of manganese and pyruvate kinase (P.K.) on the line width and radiofrequency power at maximal signal amplitude of the n.m.r. spectrum of fluoride ion (0.03 M-tris-HCl, pH 7.5, 27 °C) (from Reuben and Cohn 1970).

decreases the effect of the paramagnetic ion. Under these conditions it seems that the effect of the enzyme is consistent with competition for the manganese ions between the enzyme and the fluoride complexes, with no indication of a ternary enzyme–Mn–F bridge structure, though more detailed measurements would be required to check this.

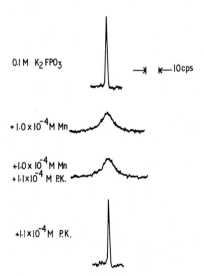

FIG. 18. Effect of manganese and pyruvate kinase (P.K.) on the line width and power at maximal signal amplitude of the n.m.r. spectrum of the fluorine nucleus of K_2FPO_3. Only the high field component of the doublet is shown (from Reuben and Cohn 1970).

In contrast to the above findings, the addition of pyruvate kinase to a solution of Mn^{2+} and FPO_3^{2-} results in an enhancement of the effect of the manganese ions on the ^{19}F relaxation time (Fig. 18). The control experiment with Mg^{2+} indicated that this enhancement must result from interaction with the paramagnetic ion. The calculated F–Mn distances in the various complexes from the corresponding values of $^1/(p_Mq)T_{1p}$ are given in Table 3. The value for r(F–Mn) of 2.12 Å is in excellent agreement with the distance determined crystallographically, which indicates that the method is viable. The r(FPO$_3$–Mn) value of 5.0 Å is well above the value expected if Mn^{2+} directly binds to fluorine, but is consistent with one or two Mn–O bonds in the Mn–FPO$_3$ complex.

TABLE 3

Values of $^1/(p_Mq)T_{1p}$ and r in various complexes of fluorokinase at 27 °C

Complex	$^1/(p_Mq)T_{1p}$ (sec^{-1})	r $(Å)$
F–Mn	2.7×10^5	2.12
FPO$_3$–Mn	1.8×10^3	5.0
FPO$_3$–Mn-pyruvate kinase	3.2×10^3	5.6 $(q=2)$
		6.1 $(q=1)$

Similarly in the enzyme–Mn–FPO$_3$ complex, the value of r is still consistent with this interpretation.

For completeness, we note that these distances have been calculated using the $^1/T_{1p}$ term. This is because it can be shown that the scalar term in equation (6) is negligible in comparison with the dipolar term. This is not generally the case in equation (5) for solutions of Mn^{2+}. There, the scalar term can dominate the value of $^1/T_{2M}$. So unless the value of the scalar coupling constant is known in equation (5), distances calculated using only the dipolar term (as in the first example) would be too small. However, the scalar interaction is transmitted through chemical bonds and rapidly decreases as the number of bonds increases. Thus in MnF its value would be expected to be large, but in Mn–FPO$_3$ (in which the fluorine is three chemical bonds away from the Mn^{2+}) it will be very much smaller.

SUMMARY

Fluorine nuclear magnetic resonance parameters are often one or two orders of magnitude larger than those of protons.

The sensitivity of the fluorine magnetic resonance parameters to environment suggests the use of fluorine as a probe in the study of the structure and function of enzymes. The fluorine may either be covalently bound to the enzyme at a conformationally strategic position, as illustrated for the binding of trifluoro-acetylated S-peptide ribonuclease to ribonuclease S-protein and for a trifluoro-acetonyl group at the cysteine β-93 position in haemoglobin. The fluorine may also be incorporated into the substrate as in the case of the binding of the trifluoroacetylated analogues of chitotriose to lysozyme.

Fluorine relaxation times are increased by the addition of paramagnetic metal ions, which have relatively long electron spin lattice relaxation times. The increase in the relaxation times is a function of the distance between the metal ion and the observed nucleus and of the correlation time for the n.m.r. relaxation behaviour of the complex. The introduction of a metal ion such as Mn^{2+} and Gd^{3+} into enzymes thus allows molecular conformation determinations of inhibitor enzyme complexes with respect to the metal site. Examples of this are the binding of several N-fluoroacetylglucosamines to lysozyme, the binding of fluoride ions and carboxypeptidase A and the fluorokinase reactions.

ACKNOWLEDGEMENT

I wish to thank Dr P. W. Kent for stimulating my interest in fluorinated compounds and also to thank Dr M. A. Raftery for many helpful discussions and for permission to quote his work prior to publication.

References

BESSELL, E. M., A. B. FOSTER, J. H. WESTWOOD, L. D. HALL and R. N. JOHNSON (1971) *Carbohydr. Res.* **19**, 39.
BLAKE, C. C. F., L. N. JOHNSON, G. A. MAIR, A. C. T. NORTH, D. C. PHILLIPS and V. R. SARMA (1967) *Proc. R. Soc. B* **167**, 378.
BLOEMBERGEN, N. (1957) *J. Chem. Phys.* **27**, 572.
BUTCHARD, G., R. A. DWEK, P. W. KENT, R. J. P. WILLIAMS and A. V. XAVIER (1972) *Eur. J. Biochem.* In press.
DAHLQUIST, F. W. and M. A. RAFTERY (1968) *Biochemistry* **7**, 3269.
DAHLQUIST, F. W. and M. A. RAFTERY (1972) *Biochemistry*. In press.
DWEK, R. A. (1972) *Advances in Molecular Relaxation Processes*. In press.
DWEK, R. A., P. W. KENT and A. V. XAVIER (1971) *Eur. J. Biochem.* **23**, 343–348.
EMSLEY, J. W. and L. PHILLIPS (1971) In *Progress in Nuclear Magnetic Resonance Spectroscopy*, vol. 7, ed. J. W. Emsley, J. Feeny and L. H. Sutcliffe. Oxford: Pergamon Press.
HUESTIS, W. H. and M. A. RAFTERY (1971) *Biochemistry* **10**, 1181.
KENT, P. W. and R. A. DWEK (1970) *Biochem. J.* **212**, 11p.

KOSHLAND, D. E., Jr (1953) *Biol. Rev.* **28**, 146.
KOSHLAND, D. E., G. NEMETHY and D. FILMOR (1966) *Biochemistry* **5**, 365.
LUZ, Z. and S. MEIBOOM (1964) *J. Chem. Phys.* **40**, 2686.
MILDVAN, A. S., J. S. LEIGH and M. COHN (1967) *Biochemistry* **6**, 1805.
NAVON, G., R. G. SHULMAN, B. J. WYLUDA and T. YAMANE (1970) *J. Mol. Biol.* **51**, 15.
PERUTZ, M. (1970) *Nature (Lond.)* **228**, 726.
RAFTERY, M. A., W. H. HUESTIS and F. MILLET (1971) *Cold Spring Harbor Symp. Quant. Biol.*
  36, 541-550.
REED, G. H., M. COHN and W. J. O'SULLIVAN (1970) *J. Biol. Chem.* **245**, 6547.
REUBEN, J. and M. COHN (1970) *J. Biol. Chem.* **245**, 6539.
SOLOMON, I. (1955) *Phys. Rev.* **99**, 559.
SPOTSWOOD, T. Mc. L., J. EVANS and J. H. RICHARDS (1967) *J. Am. Chem. Soc.* **89**, 5052.
SWIFT, T. J. and R. E. CONNICK (1962) *J. Chem. Phys.* **37**, 307.
WURTRICH, K. (1971) *Struct. & Bond.* **8**, 53.

Discussion

Bergmann: It would be interesting to investigate the cholinesterase–phospho-fluoridate system by your method.

Dwek: Acetylcholinesterase has been labelled by isopropyl fluorophosphate and studied by electron spin resonance (Hsia, Kosman and Piette 1969). Instead of adding a metal paramagnetic centre as I have described one can do these experiments by covalently binding a spin label to an SH group.

Kent: The specific binding of small molecules to a large molecule could be a common phenomenon in biochemistry. The only cases that can be readily examined are those in which there is some kinetic change, such as at an enzyme centre or an antibody centre. Apart from a few examples of fluorescent probes there is no other alternative general method of investigation such as ^{19}F-techniques offer.

Peters: Is this happening especially in mitochondria?

Kent: One thinks that these techniques will become applicable to mito-chondria.

Dwek: Dr Radda's group at Oxford have been working on these techniques as applied to membranes (Radda 1971).

Heidelberger: How much protein do you need in the n.m.r. tube for these studies?

Dwek: It depends on the type of experiment. The broadening studies or spin echo studies require about 0.2 ml of 50 μM-enzymes.

Heidelberger: Is it safe to extrapolate from such high concentrations to a biological situation?

Dwek: Do we really know what the biological situation is?

Hall: I agree that fluorine parameters have enhanced sensitivity compared with proton parameters. However, one has to distinguish clearly between the sensitivity of fluorine chemical shifts towards intramolecular and intermolecular influences. For the organic chemist primarily interested in intramolecular effects within one small molecule, fluorine is indeed many orders of magnitude more sensitive than protons. But for the biochemist who is mostly interested in the intermolecular type of interaction this enhancement is greatly reduced, although fluorine chemical shifts are still more sensitive than proton chemical shifts.

The sensitivity stretches into spin-coupling constants and these are very much

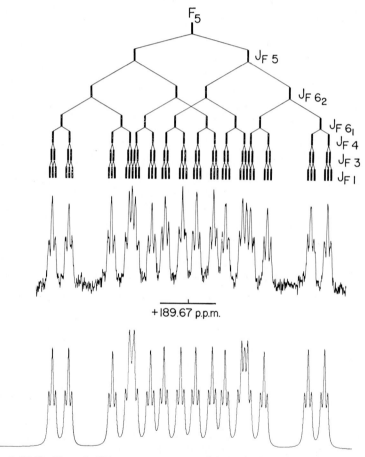

Fig. 1 (Hall). Normal ^{19}F-n.m.r. spectrum of 3,6-anhydro-5-deoxy-5-fluoro-1.2-*O*-isopropylidene-α-L-idofuranose in benzene-d_6 solution (upper trace). Simulated spectrum (lower trace). [Reproduced by permission of the National Research Council of Canada from the *Canadian Journal of Chemistry*, **48**, p. 454 (1970).]

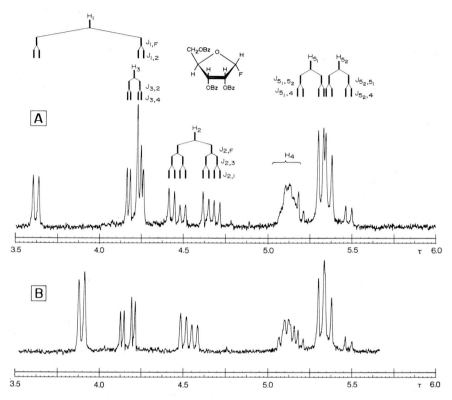

Fig. 2 (Hall). Partial ^{1}H-n.m.r. spectra (100 MHz) of 2,3,5-tri-O-benzoyl-β-D-ribofuranosyl fluoride in acetone-d_6 solution. (A) normal spectrum; (B) spectrum obtained with simultaneous irradiation at the ^{19}F frequency (94.083 045 MHz). [Reproduced by permission of the National Research Council of Canada from the *Canadian Journal of Chemistry*, **48**, p. 1159 (1970).]

larger for fluorine–proton interactions than for proton–proton interactions. One has to design the type of fluorine probe one is going to use very carefully. Fig. 1 shows the normal fluorine n.m.r. spectrum, the single fluorine from a 5-fluoro-L-idose derivative (Hall and Steiner 1970). This fluorine couples extensively with protons in the same molecule, so the sort of experiments Dr Dwek has been describing would be impossible if one were to try to use a fluorine resonance of this sort. Either one has to put a particular type of fluorine in, or indulge in more esoteric n.m.r. instrumentation. Fortunately there is now a technique known as heteronuclear spin decoupling (Hoffman and Forsen 1966). Fig. 2 shows the normal n.m.r. spectrum of a ribosyl fluoride derivative and it is quite complicated (Hall, Steiner and Pedersen 1970). Instrumentation is now commercially available, or can be constructed, which makes it possible to observe the proton spectrum of a derivative during simul-

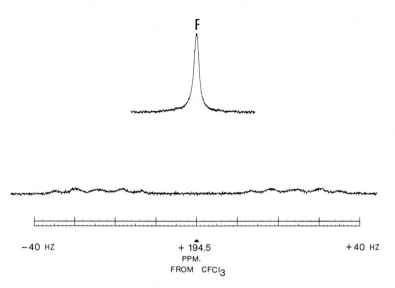

F

−40 HZ   + 194.5   +40 HZ
                               PPM.
                           FROM CFCl₃

Fig. 3 (Hall). ¹⁹F-n.m.r. spectra of a 2-fluorocholestanone derivative. Normal spectrum (lower trace); spectrum after irradiation by noise-modulated decoupling (upper trace). [Reproduced by permission of the National Research Council of Canada from the *Canadian Journal of Chemistry*, **48**, pp. 35–39 (1970).]

taneous irradiation at the resonance frequency of the fluorine (Burton and Hall 1970); this effectively decouples the fluorine from the system and the proton n.m.r. spectrum becomes a lot simpler. For example, there is a collapse of the quartet centred at τ 3.86 to a doublet and a collapse of the octet at τ 4.53 to a quartet. So now the organic chemist has an easy means of distinguishing between proton–proton couplings and proton–fluorine couplings. Fig. 1 showed that such couplings are very extensive, and one has to distinguish between them before one can make any structural assignment. We are interested in using these fluorine n.m.r. techniques to study the interactions between membranes and fluorinated substances. The trace on the lower part of Fig. 3 shows the normal fluorine resonance of a 2-fluorocholestanone derivative (Grant and Hall 1970) which by normal steroid standards gives an extremely simple fluorine n.m.r. spectrum. Even so it is multiple and one needs large molar concentrations before one can detect this resonance. One can use a variant of this heteronuclear decoupling technique, known as noise-modulated decoupling (Ernst 1966; Burton and Hall 1970), to irradiate simultaneously all the protons of this steroid system. In this method audiofrequency-noise of defined band width is mixed with the required radiofrequency, which is thereby spread over a wider band width. In the work shown in Fig. 2 we were irradiating a single fluorine substituent over a band width of about 100 Hz. In this case, to decouple this fluorine from all the protons in the molecule, one has to irradiate over about 400 Hz. We irradiated by noise-modulated decoupling by which one can collapse this fluorine multiplet down to a singlet, as shown in the upper trace of Fig. 3. Having got this single fluorine resonance, which can be detected far more readily than the fluorine multiplet, one can go on to perform some of the types of experiments which Dr Dwek discussed. So I suggest that this hetero-nuclear decoupling technique with noise modulation generalizes the use of fluorine as a structural probe. If one does not have access to this technique then one has to build into the molecule of interest a fluorine substituent which gives a singlet or other simple multiplet.

Until recently the apparatus for measuring the relaxation times, T_1 and T_2, was only available in physics and chemical physics laboratories. I would like to show how these experiments can be greatly simplified and ultimately reduced *ad absurdum*. In my laboratory we converted a standard high resolution n.m.r. machine to a pulse n.m.r. machine (Freeman and Wittekock 1969; Burton, Grant and Hall 1972) to enable us to measure T_2 and T_1. We then dispensed with all the normal monitoring devices and substituted a loudspeaker. We can then record the experiments on tape; for example we did this when measuring the two relaxation times of benzene in the presence of tetramethylsilane and in fluorine experiments similar to those Dr Dwek mentioned, where we used lysozyme as a model and looked at the N-trifluoroacetate derivative of D-glucosamine. Our results were similar to Dr Dwek's. There is a background noise on the tape due to the resonances of tetramethylsilane, and superimposed on that is the T_2 relaxation experiment involving benzene. One can hear the initial pulse which prepares the benzene sample for a second pulse, and then a spin echo. The resonance of tetramethylsilane gives a continuous low-pitched sound. Measurements of T_1 and T_2 such as these are, as Dr Dwek has implied, going to be extremely important in future work on biomolecular associations. Furthermore, they can become generally available without laboratories neces-sarily spending a great deal of money.

Peters: Did you suggest earlier (p. 164) that one would be able to detect the kinds of fluorine compounds in a plant extract?

Hall: Yes. The only prerequisite for such an assay is that one have a sufficient molar concentration of the fluorinated species. Improved technology is now becoming commercially available, but we could not detect nanogramme quan-tities. However, I think one can look towards some useful, routine assays which can be performed quickly and inexpensively, without destruction of the material. The initial detection of a fluorine resonance would immediately give some insight as to what type of fluorine was contained in the sample and this might predicate how one would go about isolating that fluorinated material.

Dwek: I agree.

Heidelberger: I have two fluorinated nucleotides. I now have the enzyme with which they interact in pure form. FdUMP has one fluorine and a hydrogen next to it to couple with. The trifluoromethyl one should be even simpler because there are no protons with which to couple. Dr Dwek, what sort of information on the interaction of my nucleotides with the enzyme can you obtain? For example, can you find out whether the analogue at the active site changes the conformation of the enzyme?

Dwek: First one would see whether a metal ion would go in. Or we might look for a reactive SH group.

Heidelberger: There is a reactive SH group but it is essential.

Dwek: In that case it might be possible to block that one and label another SH group. Then after the substrate has been added n.m.r. investigations can be performed and you will be able to obtain the conformation.

Heidelberger: What shall I learn about the conformation?

Dwek: In principle it should be possible to find the conformation of the molecule in the enzyme.

Bergmann: For this you have to know the structure of the enzyme.

Dwek: Not necessarily. We have no idea of the structure of phosphorylase and yet we were able to show that the allosteric sites and the substrate sites act close to each other and that all these occupy a small portion of any one subunit (Bennick *et al.* 1972). Therefore we can talk about allosteric interactions between site–site and between different subunits. This is quite important because in the future there might be such a thing as an allosteric drug.

Kent: One is potentially in a position to map the shape of the substrate and its relative position in the macromolecule, say as a function of the pH change of the system. But the concept of micro pH within the macromolecule has yet to be explored.

Heidelberger: People would very much like to know how a substrate or an inhibitor influences the conformation of an enzyme, if it does.

Hall: Then you must look at the enzyme rather than the 'probe' molecule with which it is interacting. Such experiments are possible in principle. One should look at the n.m.r. spectrum of the enzyme itself, preferably at the ^{13}C nuclei that actually constitute the backbone of the enzyme system so that one can see how these are influenced by the association with the substrate. The problems in doing such experiments are formidable because ^{13}C nuclei are intrinsically less sensitive towards detection than either proton or fluorine substituents. Changing the topic slightly, I should point out that the pulse experiments on the tape to which I referred can be interpreted in terms of the rate of association and dissociation of the reversible inhibitor going onto an enzyme (Sykes 1969). Are these values of any use, because they are intrinsically available from such experiments?

Heidelberger: Such numbers will be very useful and will add to human knowledge. But whether or not they will help me to devise a better drug is, I think, debatable.

Gál: We have an enzyme which is about 268×10^5 gramme molecular weight, has four different subunits and has 19 grammes of fluorine in it, incorporated in a subunit. Dr Dwek, which subunit contains the fluorine, what are the limits of sensitivity of detection, and how much protein would you require for the analysis?

Dwek: That cannot be investigated by straightforward fluorine n.m.r. spectroscopy; some further biochemical information is necessary.

Taylor: Is the difficulty with a small molecule binding to a large molecule helped by using a combination of fluorine and phosphorus resonances? I am thinking of fluorinated sugar phosphates and their binding to the enzymes.

Dwek: We have worked on phosphorus with phosphorylase, studying the binding of the substrates AMP and IMP. Phosphorus is easy to use; it has a spin $I = \frac{1}{2}$ like that of protons and has large chemical shifts like fluorine.

Hall: That statement is very misleading. It is not easy to measure phosphorus resonances as there is only 4% of the intrinsic sensitivity of the equivalent number of protons. Therefore, large quantities of material are generally required.

Dwek: We have all these drawbacks. But Professor R. E. Richards' group at Oxford have obtained results quite easily on samples of about 10 mM-phosphorus.

Hall: As professional users of n.m.r. techniques we need not worry because we have the necessary equipment such as Fourier transform spectrometers, but not everyone has these.

Dwek: But we must work on the assumption that in the future the equipment will be available.

Hall: Dr Taylor, we have been working with nucleotides and other organophosphates and we have evolved n.m.r. spectroscopy of phosphorus to the point where we can work out partial conformations of nucleotides. For example, by looking at the ^{31}P-n.m.r. spectrum of 3',5'-cyclic adenosine phosphate we can work out a conformation of that substance in aqueous solution (Hall and Malcolm 1968, 1972). This technique has reasonable generality. With phosphorus there are problems such as the signal to noise ratio, but these can be overcome.

Dwek: Dr R. J. P. Williams and his group (Barry *et al.* 1971) have worked out the conformation of AMP and metal ion in solution and this is now being extended to higher nucleotides. 'Conformations in solution' might be a good name for all this.

Heidelberger: Undoubtedly the conformations of nucleotides are very dependent on the experimental conditions under which they are measured. With such simple parameters as syn and anti conformations of nucleotides, X-ray diffraction and the n.m.r. give completely opposite answers. Neither answer is ideal in my opinion because the X-ray diffraction has to be done on a crystal, and ordinarily we do not know whether the same conformation exists in dilute solution, and the n.m.r. studies are done in extraordinarily concentrated solutions. So how relevant is the concentration in the crystal and how relevant is the concentration in a very concentrated solution to the situation in biology where the solutions are usually dilute?

Hall: Once we can get this improved sensitivity of 10^3 we will be able to use concentrations which will seem more realistic to the biochemist.

Taylor: But, Professor Heidelberger, this is the situation even with the more conventional biochemical methods we have for investigating metabolite transformations in the cell. The *in vivo* situation is so very often different from the *in vitro* situation. It is the same conflict involving the use of another technique.

Heidelberger: Here are two techniques that require highly expensive instruments, used to determine the conformation of a simple nucleotide, and they give opposite answers.

Bergmann: This phenomenon is not limited to the nucleosides and nucleotides. There are many other compounds which have different conformations in the solid state and in solution. This is because different intermolecular forces are acting; in solution there is an interaction with solvent molecules which does not exist in the solid state. Also the quantum-mechanical calculations made by the Pullmans (1963) and others have shown that it is reasonable to expect differences between the solid state and the solution. I do not think it has anything to do with the utility or the efficiency of the instruments.

Heidelberger: But the protein X-ray crystallographers assume that the structure of the protein in the crystal is the same as in the solution.

Bergmann: Yes, because the crystalline proteins do not contain large amounts of water, thus in effect they can be considered as aqueous solutions. This is not necessarily so for the nucleotides.

References

BARRY, C. D., J. A. GLASEL, A. C. T. NORTH, R. J. P. WILLIAMS and A. V. XAVIER (1971) *Nature (Lond.)* **232**, 236.

BENNICK, A., I. D. CAMPBELL, R. A. DWEK, N. C. PRICT, G. K. RADDA and A. G. SALMON (1972) *Nature (Lond.)* **234**, 140.

BURTON, R., C. W. M. GRANT and L. D. HALL (1972) *Can. J. Chem.* In press.

BURTON, R. and L. D. HALL (1970) *Can. J. Chem.* **48**, 59–66.

ERNST, R. R. (1966) *J. Chem. Phys.* **45**, 3845.

FREEMAN, R. and S. WITTEKOCK (1969) *J. Magn. Resonance*, **1**, 238–276.

GRANT, C. W. M. and L. D. HALL (1970) *Can. J. Chem.* **48**, 3537–3541.

HALL, L. D. and R. B. MALCOLM (1968) *Chem. & Ind.* 92.

HALL, L. D. and R. B. MALCOLM (1972) *Can. J. Chem.* In press.

HALL, L. D. and P. R. STEINER (1970) *Can. J. Chem.* **48**, 451–458.

HALL, L. D., P. R. STEINER and C. PEDERSEN (1970) *Can. J. Chem.* **48**, 1155–1165.

HOFFMAN, R. A. and S. FORSEN (1966) in *Progress in N.M.R. Spectroscopy*, vol. 1, p. 15–204, ed. J. W. Emsley, J. Feeney and L. H. Sutcliffe. Oxford: Pergamon.

HSIA, J. C., D. J. KOSMAN and L. H. PIETTE (1969) *Biochem. Biophys. Res. Commun.* **36**, 75.

PULLMAN, A. and B. PULLMAN (1963) *Quantum Biochemistry*. New York: Wiley.

RADDA, G. K. (1971) *Biochem. J.* **122**, 385.

SYKES, B. D. (1969) *Biochemistry* **8**, 1110.

Chemistry of fluorosteroids and their hormonal properties

A. WETTSTEIN

CIBA-GEIGY Ltd., Basle

> *The most fundamental and lasting objective of synthesis is not production of new compounds, but production of properties.*
>
> (G. S. HAMMOND, 1971)

This is a review of the fluorosteroids from the viewpoint of the pharmacologically inclined organic chemist. The definition of these compounds is obvious: they contain one or more atoms of that smallest halogen element whose covalence volume is nearest to hydrogen. As the most reactive halogen, it forms one of the firmest bonds with carbon, in this instance with a carbon atom of the steroid nucleus or side chain, a bond whose binding energy surpasses that of hydrogen to carbon despite its somewhat longer internuclear distance.

Introducing fluorine into steroid hormones has been one of the most successful ways in which the chemists have manipulated these molecules. The main result of these efforts is that the quantitative efficacy of the natural hormones can be greatly enhanced—a finding which we owe to the biologists and clinicians. This finally killed the old myth of the natural products. On the other hand, the effects of fluorinated and natural hormones could be qualitatively differentiated: it was possible to enhance or reduce in the synthetic analogue one or the other of the manifold effects composing the biological spectrum. For example, it was possible in the adrenocorticoids to change the equilibrium of the glucocorticoid, anti-inflammatory and mineralocorticoid activity, and in the androgens that of their sex-specific and anabolic effects; in the oestrogens, to enhance the blood lipid-reducing effects while diminishing the uterotrophic properties; and in the gestagens, to enhance their central anti-ovulation activity.

The outstanding therapeutic properties of some of the early fluorinated steroid hormones have led to the development of an immense number of new derivatives in the last 20 years (see review by Chen and Borrevang 1970).

METHODS FOR INTRODUCING FLUORINE INTO STEROIDS*

   The first fluorination of a steroid hormone, with until now unsurpassed effect on the biological activity, led to 9α-fluoro-11β-hydroxy corticoids (Fig. 1, partial formula 3, X = F). As so often happens, this most important discovery was made by Fried and Sabo (1953, 1954) at the Squibb Company in connection with another project, that of developing a new synthesis for cortisol by reduction of the iodohydrin (3; X = I). By biologically testing all the corresponding halohydrins (3) Fried and Sabo discovered that the compounds with the smaller halogen atoms, that is the chloro and the fluoro derivatives, have higher glucocorticoid activity than the non-halogenated compounds, the fluoro compounds being about 10 times as active as the non-halogenated ones. The antiphlogistic activity is enhanced in the same proportion, but unfortunately the mineralocorticoid activity is increased even more. This stimulated a world-wide search for more suitable fluoro corticoids.

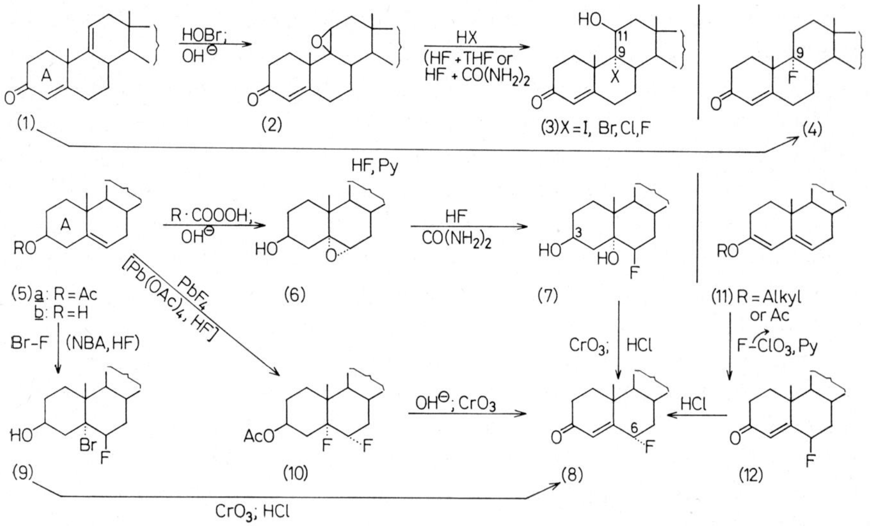

FIG. I

   Fluorination by a trans-diaxial opening of an epoxide to yield the vicinal fluorohydrin is still one of the most important methods. Such fluorination can lead not only to the 9α-fluoro-derivatives but also from the epoxide (6) to the highly active 6α-fluoro-derivatives (8). In this case the tertiary hydroxyl of the

---

* Only the most important and interesting methods and the newer modifications are dealt with (see also Chamberlin 1963; Schiemann and Cornils 1969).

fluorohydrin (7) is readily split off after oxidation of the 3-hydroxyl to yield a conjugated double bond, whereupon the 6β-fluorine atom easily undergoes rearrangement to the 6α-configuration. The opening of the epoxides originally gave only moderate yields, because hydrogen fluoride is only very slightly dissociated in the usual solvents. This could be overcome by the use, for example, of the Lewis-base tetrahydrofurane or the liquid originating from HF and urea (CIBA 1963), as indicated by the transformations of compound (2) into (3) and (6) into (7).

Other fluorination methods are important too. Among them, the direct addition of fluorine-containing reagents to olefins avoids the necessity of converting the latter first to epoxides, a reaction which if carried out with the aid of peracids does not always proceed with the requisite high stereospecificity. For example, the mixed halogen bromine-fluorine originating from *N*-bromoacetamide and HF (in CH_2Cl_2 and THF) added to (5b) gives compound (9), which is easily transformed to (8). This sequence was used especially by the Syntex groups (e.g. Crabbé *et al.* 1961) for the production of a number of more advanced corticoids such as flurandrenolone, fluocinolone and flucortolone. The reaction of lead tetrafluoride, obtained from lead tetra-acetate and HF (in CH_2Cl_2), with compound (5a) was also found in the Syntex laboratories (e.g. Bowers *et al.* 1960, 1962). Although it gives low yields, it is notable as one of the rare examples of *cis*-addition, probably through a transition state similar to the osmate esters obtained with OsO_4. The product (10) again easily yields (8). It is also possible to add hydrogen fluoride (in pyridine) to an olefin and thus to transform e.g. (1) into (4). I shall come back to the biological properties of such 9α-fluorosteroids lacking the 11β-hydroxyl group.

Reagents which yield a positively-charged fluorine atom are very scarce. For example, hypofluorous acid, HOF, the analogue of HOBr, is not known; thus the direct transformation of (1) into (3, X=F) is not possible. Perchloryl fluoride, however, is such a compound, being polarized as indicated in Fig. 1. It has been widely used by Nakanishi and co-workers (1959) for the fluorination of steroid enolethers and enolesters (11) to give (12) and hence (8). Other activated steroid olefins, like enamines and α-formylated or α-oxalylated ketones, also react quickly as well as phenols. In this way steroids fluorinated at the 2-, 4-, 6-, 10-, 16- and 21-positions have been obtained. Perchloryl fluoride has the drawback that it often leads to mixtures of isomers and can cause violent explosions.

For this reason Barton (Barton *et al.* 1968*a, b*, 1969*a*) now prefers fluoroxytrifluoromethane, i.e. trifluoromethyl-hypofluorite (also commercially available), for these electrophilic fluorinations (Fig. 2). As Tanabe and Crowe (1969, see also Barton *et al.* 1969*b*) had meanwhile developed methods for the

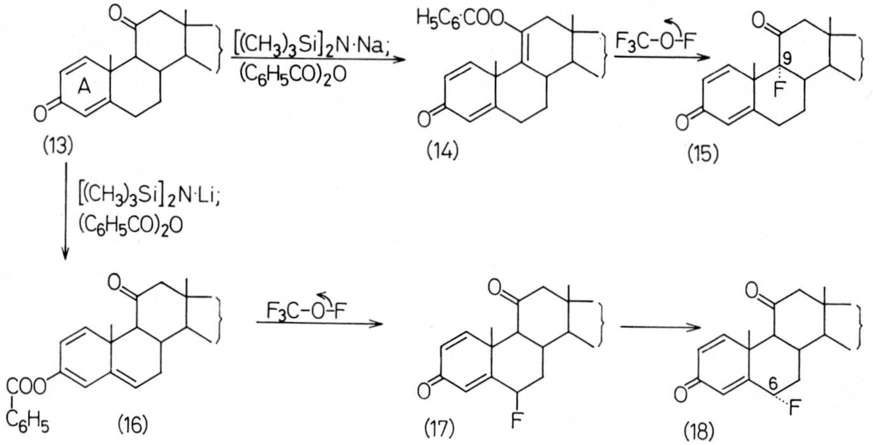

FIG. 2

selective enolization and then acylation of a prednisone derivative (13) with different alkali salts of hexamethyl-disilizane into (14) and (16) respectively, these compounds could thus be converted into the 9α-fluoro- and the 6α-fluoro-derivatives (15) and (18). The advantage of F_3COF is that one can work in glass, and according to Barton (1969c), no explosions have taken place in 15 man-years of working with the reagent. The method leads, however, to the less desirable 11-oxo- instead of the 11-hydroxy-derivatives and, for the synthesis of the latter, is no simpler than the classical Fried-Sabo procedure. Recently, Kollonitsch, Barash and Doldouras (1970) at Merck have found that light-induced, and therefore presumably free-radical, interaction of fluoroxy-tri-fluoromethane represents a useful method for the substitution of hydrogen

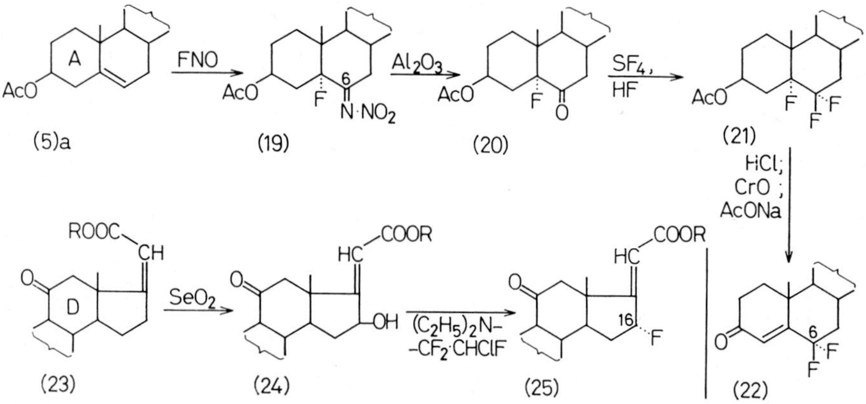

FIG. 3

atoms by fluorine. This reaction has probably not yet been applied to steroids.

Some other methods which have been used in the preparation of fluoro-steroids are included in Fig. 3. Addition to the 5,6-double bond in (5a) with the highly reactive gas nitrosyl fluoride (FNO)* has been performed by Boswell and co-workers (Boswell 1965, 1966, 1968; Boswell, Johnson and McDevitt 1971) at du Pont. Further reaction of the α-fluoroketone (20), obtained from the fluoro-nitrimine (19), with the gaseous and highly toxic sulphur tetrafluoride (SF$_4$) gave compound (21). The latter reaction is an example of the replacement of an oxo group by two fluorine atoms by this unique reagent. Further transformation of (21) including dehydrofluorination yielded the 6-difluoro structure (22).

Of greater practical importance, however, is the replacement of a primary or secondary hydroxyl group, or a heavier halogen atom, by fluorine. A new reagent for the first reaction is 2-chloro-1,1,2-trifluoroethyl-diethylamine, (C$_2$H$_5$)$_2$N·CF$_2$·CHClF. Thus, for example, the 16β-hydroxyl of (24), introduced into the intermediate (23) of corticoid syntheses performed at the Upjohn Company, has been converted with steric inversion into the 16α-fluorine group of (25) by means of this reagent (Spero *et al.* 1968; Mackeller and Slomp 1968). Compound (25) is the basis for the synthesis of the biologically interesting 16α-fluorinated corticoids.

Pregnane derivatives fluorinated in the 21-position are also biologically important, and I shall conclude this short review of fluorination methods with examples of the synthesis of such 21-fluorides (31) (Fig. 4). The standard procedure for synthesizing (31) from a ketol (27) is the reaction of its sulphonate (28), either directly or after transformation to the iodide (30), with an inorganic fluoride. In many instances, however, it is more advantageous to introduce the fluorine at the methylketone stage (33). For this purpose three classical methods, again leading through the iodide (30), are available: (*a*) the elegant direct iodination with iodine and CaO according to Ringold and Stork (1958); (*b*) the iodination of the oxalylated or formylated derivative (32) (Ruschig); and (*c*) the treatment of the Δ^{20}-enolester (34) with *N*-iodosuccinimide (Gallagher, Moffet, Djerassi). Recently Regitz (Regitz 1967; Regitz and Menz 1968; Regitz and Rüter 1968) has developed a method for the transfer of a diazo group to activated ketones. With tosylazide and trimethylamine, for example, compound (32) is thus smoothly converted into the diazoketone (29); formerly this could be obtained only from the corresponding etio-acid. At CIBA-GEIGY (1971) it has now been found that this step, followed by treatment of (29) with

---

* Addition of other fluorine-containing groups, for example difluorocarbene, is not discussed here.

HF, is often an advantageous way of transforming (33) and similar types of compounds into (31).

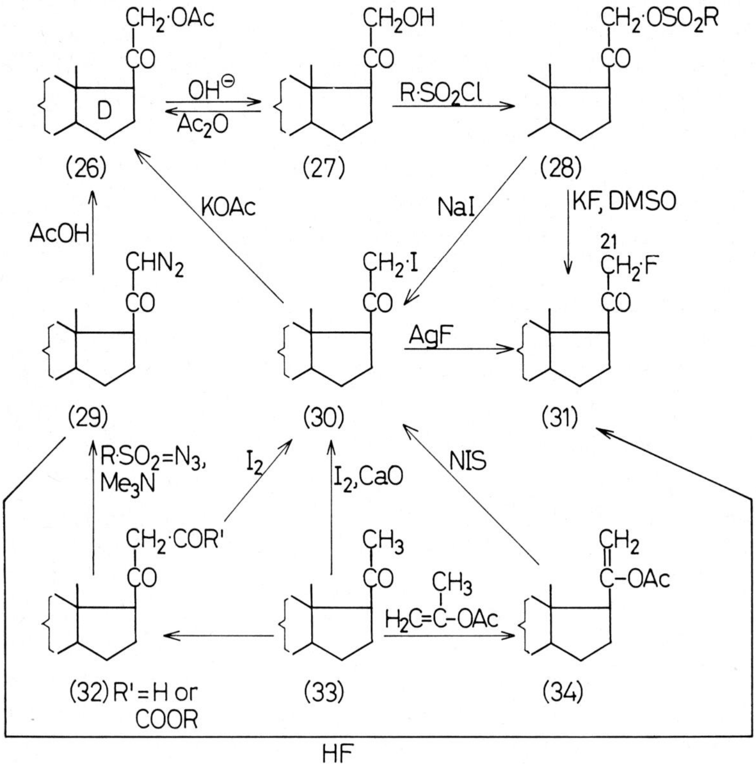

FIG. 4

BIOLOGICAL ACTIVITY OF FLUORINATED STEROID HORMONES*

It can be fairly stated that the objectives of synthesis as defined in the quotation from Hammond's paper (1971) (p. 281) have been attained in high degree. As stated before, the activities of the hormones were in many cases not only retained but also enhanced and qualitatively modified.

In the field of androgens (Dorfman 1966), oestrogens and gestagens (Miyake and Rooks 1966) only a few examples are mentioned here. These compounds stimulate the sexual target organs on the one hand and on the other behave as anti-androgens (Dorfman 1965), anti-oestrogens, and inhibitors of gonado-

---

* Only some reviews but no single investigator can be cited in this section. For the field in general see Chen and Borrevang (1970) and Buu-Hoï (1961).

tropin (Dorfman and Kincl 1966) and thus of ovulation and general fertility (Drill 1966; see also Rudel and Kincl 1966). Furthermore, other biological activities such as their protein anabolic (myotrophic) effects (Kincl 1965) and their blood lipid-reducing effects (Cook 1964) are of special interest. It is therefore not possible to give here the complete spectrum of activity for a single compound.

Androgens and oestrogens

The prototype of a fluorinated androgen with high oral androgenic and anabolic activity is the 9α-fluoro-11β-hydroxy-methyltestosterone (fluoxymesterone, Fig. 5, 35), designed on the model of the corresponding fluorocortisol. Introduction of either the 11β-hydroxyl or the 9α-fluorine alone into androgens somewhat lowers both these activities. The fluorohydrin moiety, however, enhances them both, the anabolic rather more than the androgenic activity, so that an anabolic-androgenic ratio of about 2 is attained. The very strong androgen (35), with relatively weak action on the hypophysis, is therefore used in the treatment of hypogonadic conditions or metastatic breast cancer, but less as an anabolic agent. For the latter purpose there are considerably more active steroids with more favourable anabolic-androgenic ratios. One attractive molecular manipulation of (35) seems to be still untried, namely synthesis of the corresponding 19-nor-compound lacking the angular methyl group. In the parent 19-nor molecule without the fluorohydrin moiety (normethandrolone), the androgenic activity is much reduced but the myotrophic activity is fully retained. The synthesis of 19-nor-fluorohydrins is difficult,

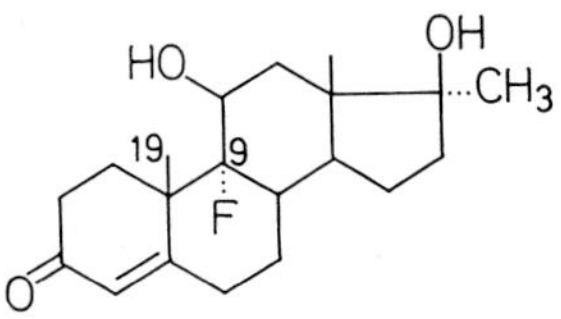

(35) Fluoxymesterone

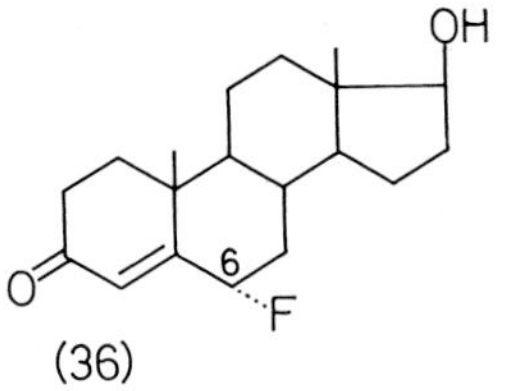

(36)

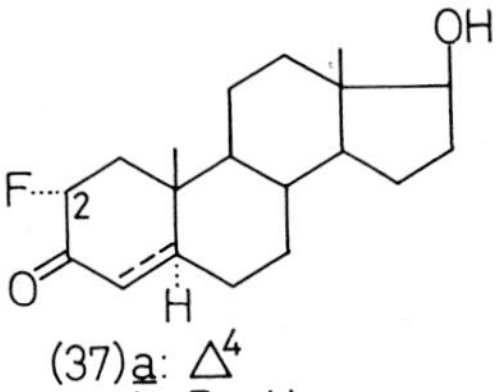

(37)a: Δ⁴
b: 5α-H

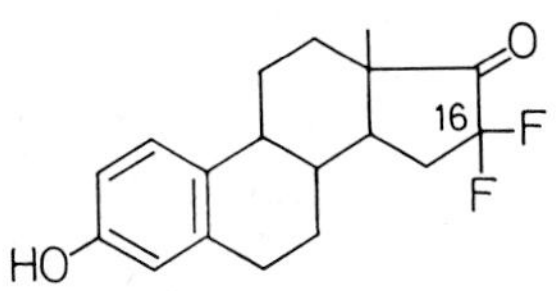

(38)

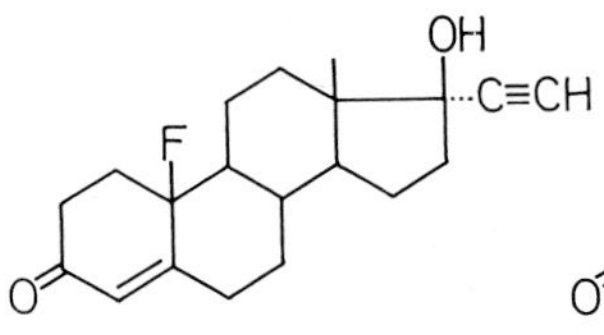

(39)

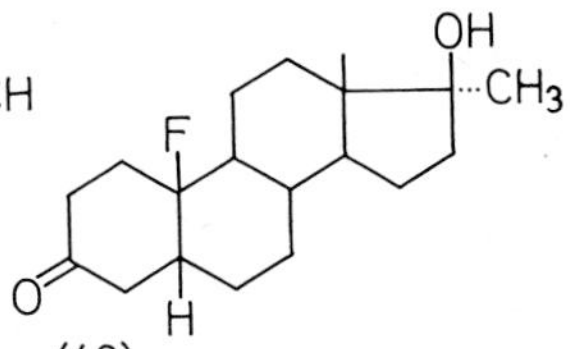

(40)

FIG. 5

however, because the normal procedure for elimination of the 11-hydroxyl does not lead to the $\Delta^{9(11)}$- but to the $\Delta^{9(10)}$-olefin, and the 9α-fluoro-19-nor-compound probably splits off HF easily.

Fluorine substitution of testosterone in other positions did not lead to clinically useful compounds, although the 6α-fluoro-derivative (36) exerts slightly enhanced myotrophic, androgenic and especially anti-gonadotropic effects. High central activity is also displayed by the 6-*gem*-difluoro-derivative. In the 2α-derivative (37a) these activities are considerably reduced; but like (37b), (37a) has experimental interest as an inhibitor of the androgen-resistant mammary fibroadenoma.

In the oestrogen series, where activity is much less closely correlated with structure, fluorination aimed mainly at reducing the uterotrophic effects in preference to the blood lipid-reducing or antitumour activity. This was achieved with a number of 16,16-difluoro compounds, of which (38) has only 0.1 % of the uterotrophic but twice the serum cholesterol-reducing activity of oestradiol. Furthermore, it is interesting that although 10β-fluoro compounds like (39) are surprisingly stable (because the already highly polarized C—F bond is not easily further polarizable), compound (39) is a relatively active oestrogen, compound (40) a very active anti-androgen.

Gestagens

There are two distinct groups of synthetic gestagens: the analogues of progesterone, and those of testosterone containing a hydrocarbon residue in the 17α-position. The first group is closely related to the corticoids, which may be defined here as 21-hydroxylated progesterones. In each of the two gestagen groups valuable fluorine derivatives are known. In the progesterone 9α,11β-fluorohydrin (Fig. 6, 41, X=F), the fluorine atom again has about a 10-fold potentiating effect, in this case on the subcutaneous gestagenic activity, and there is a certain amount of glycogenetic, antiphlogistic and sodium-retaining activity. Substitution of other halogens for fluorine in (41) has the interesting effect that the subcutaneous gestagenic activity rises with the atomic size (Br > Cl > F), that is, inversely to the electronegativity, whereas the glyco-genetic activity rises with the electronegativity (F > Cl > Br). A similar pattern of activity is also present in the 12α-halogenated 11β-hydroxy proges-terones (42). Surprisingly in (43) the 9α-fluorine atom alone causes a strong rise in the subcutaneous gestagenic action. This is in accordance with the effects in mineralocorticoids but contrary to those in androgens, and is interesting in connection with the rationale for the action of the 9α-fluorine atom.

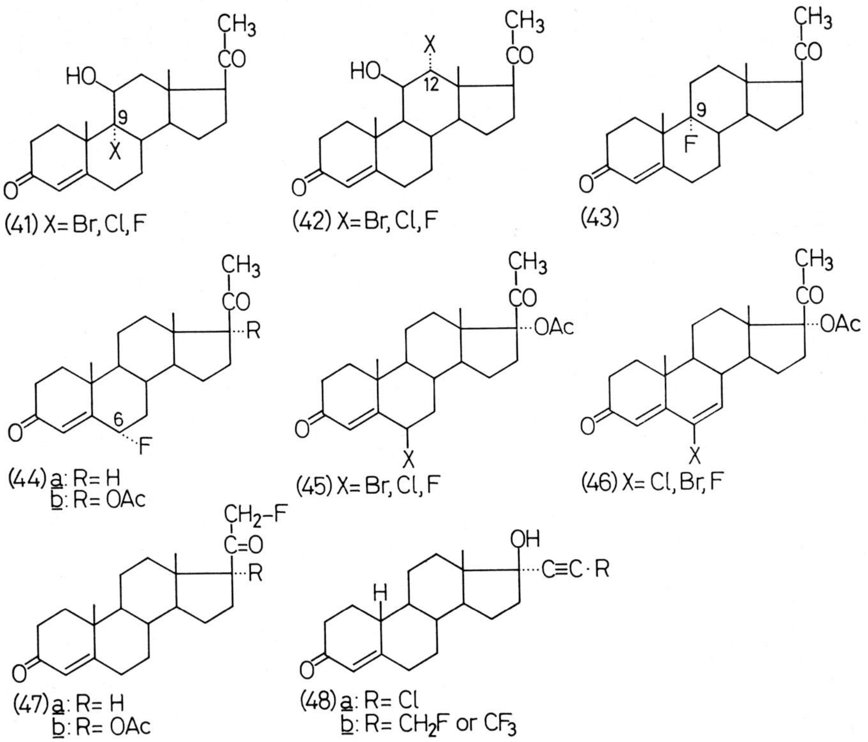

(41) X= Br, Cl, F

(42) X= Br, Cl, F

(43)

(44) a: R= H
b: R= OAc

(45) X= Br, Cl, F

(46) X= Cl, Br, F

(47) a: R= H
b: R= OAc

(48) a: R= Cl
b: R= CH₂F or CF₃

FIG. 6

Fluorination in the equatorial 6α-position is also useful for improving the progestational activity. With progesterone it leads in (44a) to a 5-fold increase in subcutaneous activity, and with the 17α-acetoxy-progesterone in (44b) to a 15-fold increase in oral activity. The 6β-halogenated (i.e. axially substituted) analogues (45)* are much less active than the 6α- substituted compounds. The loss of activity of the former becomes less as the size of the halogen increases, which argues against a primary stereochemical explanation of the biological effects of 6α-halogenation. The same is true for the compounds (46) which have an additional 6-double bond: the fluoro derivative (46) has 15 times higher oral activity** than (44b), as does the bromo derivative, but this is surpassed by an additional factor of 3 in the chloro derivative, chlormadinone acetate, which is a strong anti-ovulation agent.

---

* Introduction of two fluorine atoms into the 6-position thus does not much change the progestational activity.
** The corresponding norethisterone derivative has also a remarkable activity.

Many progesterone derivatives fluorinated in the 21-position have also been synthesized, and they too show higher progestational and corticoid-type, especially antiphlogistic, activity. Examples are the compounds (47) and the 21-fluorinated derivative of chlormadinone acetate (45, X=Cl). 21-chloro-, 21-bromo-, 21-difluoro- and 21-trifluoro-derivatives show much less activity, if any. The considerably higher activity of some norethisterone derivatives (48), halogenated or fluoromethylated in the 21-position, is also noteworthy.

Corticoids

Fluorination of the corticoids has by far the greatest practical importance. For example, the glycogenic, the anti-inflammatory but also the sodium-retaining properties are considerably enhanced, as mentioned before, in parallel with the electronegativity of the halogen introduced next to the 11β-hydroxyl. The prototype for high activity is thus 9α-fluoro-cortisol (Fig. 7, 49). The same effect is seen with 12α-fluorination.* The direct analogue of (49), the cortisol derivative (50), is, however, inactive because the axial 12α-fluorine atom forms a strong hydrogen bond with the pseudo-axial 17α-hydroxyl. If the latter is

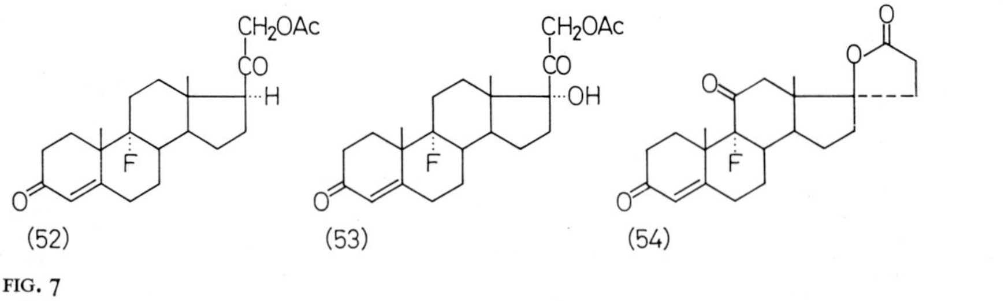

FIG. 7

blocked as in (51), the compound has the same activity as the corresponding 9α-fluoro-derivative. Of theoretical interest is that the introduction of a 9α-fluorine atom into cortexone acetate yielding (52), which has no 11-oxygen

---

* The interesting 9α,12α-difluoro-11β-hydroxy-corticoids are mentioned only in a patent (see Barton *et al.* 1969*a*) which does not give chemical and biological properties.

function, leads to 12 times higher mineralocorticoid activity, analogous to the higher progestational effect of 9α-fluoro-progesterone (43). However, for high glucocorticoid activity an isolated 9α-fluorine atom is not sufficient: compound (53) has only 5–10% of the activity of cortisone acetate. In this context, the remarkable enhancement of the mineralocorticoid-blocking property by the introduction of the 9α-fluoro-11-oxo-moiety into spirolactones such as (54) should be mentioned.

The effects of the numerous fluorinations in other positions of the cortisol molecule are summarized in Table 1. 6α-Fluorination has a very beneficial effect. It improves the glucocorticoid and anti-inflammatory activity and also reduces the mineralocorticoid activity, though not as much as 9α-fluorination increases it. The mineralocorticoid effect of the 9α-fluorine cannot,

TABLE 1

Effects of fluorinations in different positions of the cortisol molecule

Position of fluoro-substituent	Change in activity		
	glucocorticoid	*anti-inflammatory*	*mineralo-corticoid*
9α	++	++	+++
9α (without 11-OH)	−−		+++
12α (17-OH blocked)	++	++	+++
6α	++	++	−−
6α + 9α	+++	+++	+
2α	−	−	
4		−−	
15β		+	−−
15β + 9α		++	−−
16α	++	++	
16α + 6α	+++	+++	
16α + 6α + 9α		++++	−−−
F in place of OH			
11β		−−	
17α	−−		
21	−−	−−	−−−

therefore, be overcome by adding a 6α-fluorine atom. This aim can be achieved with a 15β-fluorine group; however, this has a lower favourable effect on the anti-inflammatory activity. In this respect the 16α-fluorine atom is better. In combination with 6α- and 9α-fluorine it has led to extremely high effects: positive on the antiphlogistic and negative on the sodium-retaining properties.

Replacement of one of the naturally occurring hydroxyls of cortisol by fluorine usually lowers the desirable activities. However, since the 21-fluorination reduces the sodium retention even more, and as 21-fluoro-derivatives often have a higher anti-inflammatory activity than the 21-unsubstituted pregnenes, such replacement can nevertheless be of practical importance.

To get rid of the undesired mineralocorticoid activity, fluorination has usually been combined with further molecular manipulations. These manipulations should also retain, enhance, or not reduce the desired activities too much. Introduction of a 1,2-double bond effects some improvement. Mineralocorticoid activity is completely abolished, with some loss of the glucocorticoid activity, by the introduction of a 16α-hydroxyl group; the acetals and ketals of these 16α, 17α-dihydroxylated corticoids have high topical antiphlogistic activity. But the most successful attempt to eliminate undesired activity while retaining or enhancing the desired activities has been the introduction of methyl groups, especially in the 6α-, 16α-, or, surprisingly, the 16β- and 17α-positions. Such methyl groups may enhance the glucocorticoid and anti-inflammatory activities by factors of up to 10. A 16-methylene group is also an advantageous substituent. These further substitutions of fluoro corticoids have led to a plethora of clinically very valuable preparations which cannot be dealt with in detail here. It is biochemically interesting that the introduction of a 2α-methyl group increases the glucocorticoid activity only slightly, but increases the mineralocorticoid activity 20–25 times. Thus, from 9α-fluoro-cortisol a compound with 100 times higher mineralocorticoid activity, surpassing even that of aldosterone, is obtained.

MOLECULAR PROPERTIES EXPLAINING BIOLOGICAL ACTIVITIES

Knowing the principal facts about structure-activity relations, one can make proposals to rationalize the biological effects of different fluorinations on a molecular basis. One should start from the chemical and physicochemical properties of such derivatives and investigate how they affect the biochemical processes. These processes consist of enzymic metabolic transformations of the compounds themselves, and also of their interactions in the biochemical mechanisms connected with the action of the hormones in the target cells. In the latter connection there are, unfortunately, very few direct experimental comparisons between unsubstituted and fluorinated hormones.

With regard to the chemical and physicochemical properties of fluorinated compounds, the only admissible general statement is that the quantitatively and qualitatively modified activities resulting from the fluorination in different positions of different hormone types cannot be attributed to a single common property. Stereochemical and electronic alterations have been considered especially (Ringold 1961). In most discussions of the problem it has been tacitly assumed that the small fluorine atom does not change the *stereochemical pattern* of the rigid steroid molecule, although 1,3-diaxial interactions can distort the valence angles of simple cyclohexanes and thus lead to modified conformations. With steroids it is indeed very doubtful whether an axial fluorine atom can have steric effects which would suffice to change the attachment to the complementary receptor, approaching the androgens probably from the α-side, and gestagens and corticoids (Sarett, 1959, 1965) from the β-side. For an equatorial fluorine atom this is out of the question, anyhow, but it can still have a beneficial effect on the activity, as does the 6α-fluorine atom. The overall stereochemistry of the basic molecule is, however, important for the effect of an additional fluorine atom: in the *abeo*-corticosteroids (Fig. 8, 55), for example, containing a 5-membered ring *A*, and a distorted 7-membered ring *B*, fluorination in the 6- or 9-position does not enhance the activity (Anner *et al.* 1970). However, the principle of 6-halogenation seems to act normally in the *retro*-progesterone compounds (56) which have the stereochemically more different 9β,10α-configuration but a ring *B* with normal chair conformation (Westerhof *et al.* 1965)*.

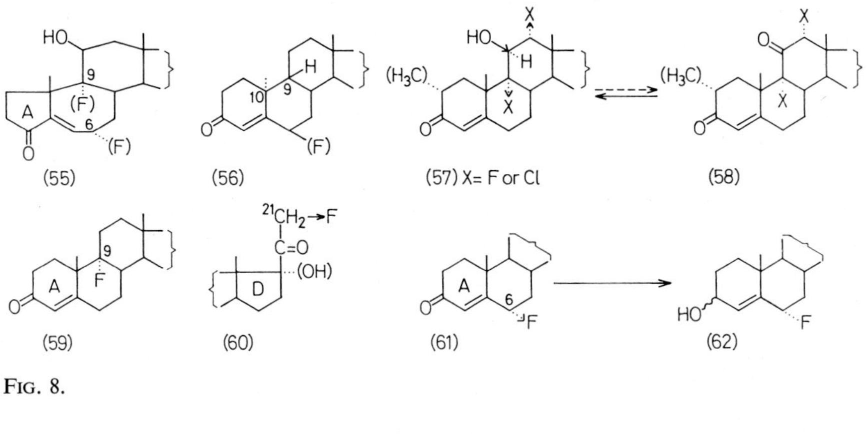

FIG. 8.

The extreme electronegativity of the fluorine atom which tends to abstract electrons from neighbouring groups, the latter becoming positively and the fluorine negatively charged, causes *electronic alterations* which can explain altered biological activity in certain cases. This has been shown especially for the 9α-fluoro- and the 12α-fluoro-11β-hydroxy-corticoids (57) (Fried and Borman 1957). In these compounds the 11β-hydroxyl (which is essential for the

---

* Here too, the more stable equatorial fluorine atom, in this case the 6β-F, confers higher activity than the axial fluorine atom.

glucocorticoid activity and important, but not necessary, for the less structure-specific anti-inflammatory activity) acquires a more acidic character paralleling the electronegativity of the neighbouring halogen atom—an inductive effect seen generally in α-halogenated alcohols. This strongly sterically hindered hydroxyl will then bind more firmly with a proton-acceptor group of the receptor molecule.* In addition, the unfavourable polarization of the equatorial $C-H$ bond renders the dehydrogenation of the carbinol to the ketone more difficult. [This is also seen in the slower chemical CrO_3-oxidation of simple halogenated alcohols (Roček 1960), and was confirmed by Eschenmoser (1964) with 9α-fluoro-11β-hydroxy-steroids in which the removal of the equatorial hydrogen is the rate-limiting step.] Because of this effect the enzymic redox equilibrium in the organism is shifted from the metabolized, biologically inactive 11-oxo-form (58) to the active 11β-hydroxy-form (57) (Bush 1956; Bush and Mahesh 1959, 1964; Bush, Meigs and Hunter 1962). A similar but predominantly stereochemical explanation has been given for the effect of a 2α-methyl group (which stabilizes corticoids also against the 4,5-reductase of the liver) (Bush and Mahesh 1959).

The decreased rates and qualitatively changed patterns observed with other *metabolic inactivations*, such as reductions of the unsaturated system in ring *A* or the oxo-group in the side chain (e.g. Glenn *et al.* 1957; Schriefers, Korus and Dirscherl 1957; see also Schriefers 1961), are not themselves sufficient to explain the high activity of these fluorinated corticoids. They may play a more important role in connection with the progestational properties of progesterones of type (57) in which the higher acidity of the unnecessary and even disadvantageous 11β-hydroxyl is meaningless. As stated before, this activity rises with the size of the halogen inhibiting the approach of enzymes from the α-side. Their corticoid activity, however, does parallel the electronegativity of the halogen.

The induced higher polarity of the 9α-fluorinated 11β-hydroxy corticoids reduces, according to Samuels' 'polarity rule', their *interaction with blood proteins* (see Westphal 1970, 1971). This has been ascertained for the corticoid-binding α-globulin (Florini and Buyske 1961), the so-called transcortin, which has a transport and a storage function and binds cortisol in an inactive complex. A reduction of the binding to serum albumin, increasing from the 9α-bromo- to the 9α-fluoro-substitution, has also been proved (Westphal and Ashley 1959, 1962). These effects may also constitute additional factors for the activity.

More difficult is the explanation of the high activity of 9α-fluoro-derivatives

---

* Indeed, 9α-halogenated 11β-hydroxy progesterones exhibited with model proton-acceptor compounds increased hydrogen-bonding, differing, however, only slightly with the changing electronegativity of the halogen (Devine and Lack 1966).

(59) in the 11-deoxy-progesterone and -mineralocorticoid series, which contrasts with the low activity of such derivatives of androgens and glucocorticoids. In this context can be mentioned—purely hypothetically—the strong hydrogen bonds which a fluorine atom, contrary to other halogens, can form with hydroxyl groups, amino groups and thio groups, possibly those of a receptor; an intramolecular example for this is compound (50).

Halogenation at the 21-position decreases the electron density at C-20. The consequent increase in reactivity of the C-20 carbonyl with a receptor may explain why the fluorine atom in (60) leads to particularly enhanced progestational and antiphlogistic activities.

The situation is complicated in the 6α-fluoro-hormones (61) upon which the fluorine atom confers slightly higher anabolic and considerably higher progestational, glucocorticoid and anti-inflammatory activities; these activities are also higher than those of the corresponding chloro derivatives. These compounds represent vinylogous α-halogen-ketones in which the inductive effect of the halogen destabilizes the conjugated system, thereby enhancing its electrophilic reactivity, for example in a hydride transfer. One consequence of this is that liver enzymes reduce 6α-fluoro- (61) and especially 6β-fluoro-Δ^4-3-ketones more quickly than the unhalogenated compound and this in an abnormal pathway to the allylic alcohols (62) (Ringold, Ramanchandran and Forchielli 1964). Whereas this behaviour may contribute to the low biological activity of the 6β-fluoro-derivatives, it cannot explain the high activity of the 6α-fluoro compounds. Inconsistent with the proposition that electronic properties are the exclusive reason for the enhanced activity of 6α-fluoro steroids is also the very similar biological effect of introducing an electron-releasing group, namely a methyl, into the same position*. Biological effects which parallel those of the fluorination are also seen when a methyl group is introduced at position 16α. Here, the steric hindrance to reactions at the side chain as well as to oxidation of the 11β-hydroxyl (Bush, Meigs and Hunter 1962) can explain the activity of the methylated compounds. In other positions, however, where the electronegativity of the halogen is of decisive importance for the activity, as for example in the 9α- and 12α-positions, corresponding methylations have adverse effects. A special feature of halogenations at C-6 is that they occupy a position at which other major metabolic transformations take place, namely hydroxylations. To my knowledge, the influence of halogenations on these transformations has not been investigated.

---

* As can be anticipated from their electronic properties, both the 6α- and 6β-methyl derivatives are reduced by the liver enzyme more slowly than the unsubstituted and especially the fluorinated compounds.

So much for the altered chemical and physicochemical properties of fluorinated steroid hormones which may explain their specific biological activities. An explanation of their special biochemical interactions with the receptors in the target cells is still lacking. Up till now the biochemistry of steroid hormone action has been studied mainly with natural hormones. However, the inductive effect e.g. of the fluorinated corticoid dexamethasone on the synthesis of an enzyme protein has been studied, and has been shown to be far greater than that of the fluorine-free cortisol (Samuels and Tomkins 1970; Thompson, Tomkins and Curran 1966). This effect may involve the inactivation of a repressor of the messenger RNA-translation step (Tomkins *et al.* 1969; see also Tomkins 1971).

SUMMARY

A short review of chemical methods for introducing fluorine into steroids is given.

Fluorinated steroid hormones are among the biologically most active synthetic analogues of the natural substances. Thus products have been obtained in which, for example, both the anabolic and androgenic or the glucocorticoid and mineralocorticoid activities are considerably enhanced. To reduce the second, undesired effects, additional substituents are introduced into the molecule. These substituents may even be a second or third fluorine atom. Some of these fluorine substitutions also selectively affect the biological spectrum, enhancing the desired and reducing the undesired properties.

Different proposals for explaining the influence of fluorine substitution on biological activity are discussed. Among them are the properties of the 11β-hydroxyl, crucial for the glucocorticoid effects. The acidity of this hydroxyl is strongly enhanced by the electron-withdrawing effect of a vicinal fluorine atom. This causes stabilization of the hydroxyl in the enzymic redox equilibrium, which parallels a decreased chemical oxidation speed. Reduced enzyme interaction in other inactivating metabolic pathways also plays a role. The higher polarity of the derivatives may also positively influence the interaction with the receptor at the specific binding site. In contrast, the higher polarity exerts a negative effect on the binding with proteins and glycoproteins of the blood serum, thus reducing the storage of inactive complexes in favour of free hormonal steroids.

References

ANNER, G., C. MEYSTRE, H. KAUFMANN, V. SCHMIDLIN, H. UEBERWASSER and P. WIELAND (1970) *Abstr. III Int. Congr. Hormonal Steroids*, p. 78, ed. L. Martini. Amsterdam: Excerpta Medica Foundation (Int. Congr. Ser. No. 210).

BARTON, D. H. R. (1969c) Lecture IUPAC Symp., Mexico City.

BARTON, D. H. R., L. J. DANKS, A. K. GANGULY, R. H. HESSE, G. TARZIA and M. M. PECHET (1969a) *Chem. Commun.* 227–228 (cf. Rimac Inc. Belgian Pat. 720'642).

BARTON, D. H. R., A. K. GANGULY, R. H. HESSE, S. N. LOO and M. M. PECHET (1968b) *Chem. Commun.* 806–808.

BARTON, D. H. R., L. S. GODINHO, R. H. HESSE and M. M. PECHET (1968a) *Chem. Commun.* 804–806.

BARTON, D. H. R., R. H. HESSE, G. TARZIA and M. M. PECHET (1969b) *Chem. Commun.* 1497–1498.

BOSWELL, G. A., JR (1965) *Chem. & Ind.* **47**, 1929–1930.

BOSWELL, G. A., JR (1966) *J. Org. Chem.* **31**, 991–1000.

BOSWELL, G. A., JR (1968) *J. Org. Chem.* **33**, 3699–3713.

BOSWELL, G. A., JR, A. L. JOHNSON and J. P. MCDEVITT (1971) *Angew. Chem.* **83**, 116–117.

BOWERS, A. *et al.* (1960) *Tetrahedron Lett.* No. 20, 34–37.

BOWERS, A. *et al.* (1962) *J. Am. Chem. Soc.* **84**, 1050–1053.

BUSH, I. E. (1956) *Experientia* **12**, 325–331.

BUSH, I. E. and V. B. MAHESH (1959) *Biochem. J.* **71**, 718–742.

BUSH, I. E. and V. B. MAHESH (1964) *Biochem. J.* **93**, 236–255.

BUSH, I. E., R. A. MEIGS and S. HUNTER (1962) *J. Endocrinol.* **24**, ii–iii.

BUU-HOÏ, N. P. (1961) *Drug Res.* **3**, 51–62.

CHAMBERLIN, J. W. (1963) In *Steroid Reactions*, pp. 155–178, ed. C. Djerassi. San Francisco: Holden-Day.

CHEN, P. S., JR and P. BORREVANG (1970) *Handb. Exp. Pharmakol.* 20, pt. 2, 193–252.

CIBA A. G. (appl. 1963) German Pat. 1'244'773.

CIBA-GEIGY A. G. (1971) Belg. Pat. 761'363; Dtsche. Pat. Offenlegungsschrift 2'100'324.

COOK, D. L. (1964) In *Methods in Hormone Research*, vol. 3, pt A, pp. 185–225, ed. R. I. Dorfman. New York: Academic Press.

CRABBÉ, P. *et al.* (1961) *Bull. Soc. Chim. Belg.* **70**, 271–284.

DEVINE, A. B. and R. E. LACK (1966) *J. Chem. Soc.* 1902–1907.

DORFMAN, R. I. (ed.) (1965) In *Methods in Hormone Research* vol. 4, pt B, pp. 77–93. New York: Academic Press.

DORFMAN, R. I. (ed.) (1966) In *Methods in Hormone Research* vol. 5, pt C, pp. 235–295. New York: Academic Press.

DORFMAN, R. I. and F. A. KINCL (1966) In *Methods in Hormone Research* vol. 5, pt C, pp. 147–203, ed. R. I. Dorfman. New York: Academic Press.

DRILL, V. A. (1966) *Oral Contraceptives*. New York: McGraw-Hill.

ESCHENMOSER, A. (1964) Quoted in I. E. Bush and V. B. Mahesh (1964) *Biochem. J.* **93**, 253.

FLORINI, J. R. and D. A. BUYSKE (1961) *J. Biol. Chem.* **236**, 247–251.

FRIED, J. and A. BORMAN (1957) *Vitam. & Horm.* **16**, 303–374.

FRIED, J. and E. F. SABO (1953) *J. Am. Chem. Soc.* **75**, 2273–2274.

FRIED, J. and E. F. SABO (1954) *J. Am. Chem. Soc.* **76**, 1455–1456.

GLENN, E. M., R. O. STAFFORD, S. C. LYSTER and B. J. BOWMAN (1957) *Endocrinology* **61**, 128–142.

HAMMOND, G. S. (1971) *Chem. Technol.* **1**, 24–26.

KINCL, F. A. (1965) In *Methods in Hormone Research* vol. 4, pt B, pp. 21–76, ed. R. I. Dorfman. New York: Academic Press.

KOLLONITSCH, J., L. BARASH and G. A. DOLDOURAS (1970) *J. Am. Chem. Soc.* **92**, 7494–7495.

MACKELLER, F. A. and G. SLOMP (1968) *Steroids* **11**, 787–798.
MIYAKE, T. and W. H. ROOKS (1966) In *Methods in Hormone Research*, vol. 5, pt C, pp. 59–145, ed. R. I. Dorfman. New York: Academic Press.
NAKANISHI, S. *et al.* (1959) *J. Am. Chem. Soc.* **81**, 5259–5260.
REGITZ, M. (1967) *Angew. Chem.* **79**, 786–801.
REGITZ, M. and F. MENZ (1968) *Chem. Ber.* **101**, 2622–2632.
REGITZ, M. and J. RÜTER (1968) *Chem. Ber.* **101**, 1263–1270.
RINGOLD, H. J. (1961) In *Mechanism of Action of Steroid Hormones* pp. 200–234, ed. C. A. Villee and L. L. Engel. Oxford: Pergamon Press.
RINGOLD, H. J., S. RAMACHANDRAN and E. FORCHIELLI (1964) *Biochim. & Biophys. Acta* **82**, 143–157.
RINGOLD, H. J. and G. STORK (1958) *J. Am. Chem. Soc.* **80**, 250.
ROČEK, J. (1960) *Collect. Czech. Chem. Commun.* **25**, 1052–1057.
RUDEL, H. W. and F. A. KINCL (1966) *Acta Endocrinol. (Copenh.)* **51**, suppl. 105.
SAMUELS, H. H. and G. M. TOMKINS (1970) *J. Mol. Biol.* **52**, 57–71.
SARETT, L. H. (1959) *Ann. N.Y. Acad. Sci.* **82**, 802–808.
SARETT, L. H. (1965) In *Hormonal Steroids* (Proc. I Int. Congr. Hormonal Steroids) vol. 2, pp. 11–14, ed. L. Martini and A. Pecile. New York: Academic Press.
SCHIEMANN, G. and B. CORNILS (1969) In *Chemie und Technologie cyclischer Fluorverbindungen*, pp. 114–128. Stuttgart: Enke.
SCHRIEFERS, H. (1961) *Hoppe-Seyler's Z. Physiol. Chem.* **324**, 188–196.
SCHRIEFERS, H., W. KORUS and W. DIRSCHERL (1957) *Acta Endocrinol. (Copenh.)* **26**, 331–344.
SPERO, G. B., J. E. PIKE, F. H. LINCOLN and J. L. THOMPSON (1968) *Steroids* **11**, 769–786.
TANABE, M. and D. F. CROWE (1969) *Chem. Commun.* 1498–1499.
THOMPSON, E. B., G. M. TOMKINS and J. F. CURRAN (1966) *Proc. Natl. Acad. Sci. U.S.A.* **56**, 296–303.
TOMKINS, G. M. (1971) In *Schering Workshop on Steroid Hormone Receptors*, p. 154, ed. G. Raspé. Oxford: Pergamon Press.
TOMKINS, G. M., T. D. GELEHRTER, D. GRANNER, D. MARTIN, H. H. SAMUELS and E. B. THOMPSON (1969) *Science* **166**, 1474–1480.
WESTERHOF, P. *et al.* (1965) *Recl. Trav. Chim. Pays-Bas Belg.* **84**, 863–884.
WESTPHAL, U. (1970) In *Biochemical Action of Hormones* vol. 1, pp. 220–265, ed. G. Litwack. New York: Academic Press.
WESTPHAL, U. (1971) *Steroid-Protein Interactions* (Monogr. Endocrinology 4). Berlin: Springer-Verlag.
WESTPHAL, U. and B. D. ASHLEY (1959) *J. Biol. Chem.* **234**, 2847–2851.
WESTPHAL, U. and B. D. ASHLEY (1962) *J. Biol. Chem.* **237**, 2763–2768.

Discussion

Saunders: Could you introduce a fluorine in position 3 in cholesterol?

Wettstein: This has indeed been achieved, for example by reaction with chlorotrifluoroethyl-diethylamine (see Ayer 1962; Knox *et al.* 1964).

Saunders: I think fluorine has been put into stilboestrol and the derivative was many times more active than stilboestrol itself.

Wettstein: Work in the oestrogenic field was directed mainly towards making new compounds which would have lower oestrogenic activity but higher blood-lipid reducing activity or antitumour activities. An example of this is compound 38 (p. 287).

Sharpe: It has long been accepted that fluorine is too electronegative to form oxyacids and that chlorine is not electronegative enough to take part in hydrogen bonding. The dangers of trusting such non-thermodynamic interpretations are well illustrated by the isolation of HOF (Studier and Appelman 1971) and of several compounds containing the HCl_2^- ion, which is strictly analogous to HF_2^- (Waddington 1958; Chang and Westrum 1962; McDaniel and Vallée 1963). (Several salts of organic bases containing the HCl_2^- have, in fact, been known since the turn of the century, but were not previously recognized as such.) Great caution should be exercised in accepting electronic reasons why compounds should not exist; too often 'should not exist' is more accurately rendered 'have not yet been prepared'.

Bergmann: Cholesteryl fluoride and also a few more complex fluoro derivatives of cholesterol and other sterols have an outstanding biological effect on insects: they completely prevent the development and pupation of insect larvae. This is a competitive inhibition which can be overcome by adding cholesterol or other sterols to the food.

Wettstein: It has been said by others during this symposium that fluorine is an excellent leaving group. However, for the organic chemist it is the most stable of the halogen substituents. Can this discrepancy be explained?

Sharpe: I think terms like 'good leaving group' are very dangerous. What are the units in which one measures the 'goodness'? In aqueous media the high hydration energy of F^- (compared with other singly-charged anions) undoubtedly helps in the breaking of a bond to another atom, but several other factors are also involved, as may be seen from consideration of the reaction

$$RF \rightarrow R^+ + F^-$$

in solution.

Wettstein: An interesting example is compound 39 (p. 287) in which the fluorine atom is very stable. The compound is thus not easily converted into the ring A-phenol but nevertheless is an active oestrogen. We have no explanation for this. However, if fluorine were a good leaving group it would be all right.

Sharpe: I think that when one calls something a good leaving group one is really distracting attention from the fact that usually something else helps to pull it off a carbon atom and a third entity replaces it on the carbon atom. The C—F bond is one of the strongest single bonds in chemistry, and I think any truly unimolecular process involving its fission is extremely unlikely. Classifications of leaving groups therefore relate only to specific conditions and reactions.

Barnett: It is possible, nevertheless, to arrange leaving groups in order of their ability to be removed in the presence of the same nucleophilic attacking agent under specified conditions.

Heidelberger: Fluorine atoms have been used successfully to block metabolism.

10-Methyl-1-1,2-benzanthracene is a moderately active carcinogenic hydrocarbon. If a fluorine atom is introduced at the 4-position, the compound is considerably more carcinogenic than the parent hydrocarbon. However, if a fluorine atom is present at the 3-position, the resulting compound is totally inactive as a carcinogen (Miller and Miller 1960). A likely interpretation is that an essential metabolic step involving the interaction of the carcinogen with its target macromolecule occurs in the 3-position; a firmly bound fluorine atom at this position would prevent this essential interaction. The same sort of thing has also been done with fluorine-substituted azo dye carcinogens.

Kent: Professor Wettstein, are any of these very active fluorosteroids available in radioactive form and if so, do these hormone analogues associate with different organelles? For instance, will the 9α-fluoro-11β-hydroxyl moiety which confers high glucocorticoid activity associate, say, with smooth membrane, and the 9α-fluoro (without the 11-hydroxyl) with other parts of the cell?

Wettstein: Nothing specific is known. Another point about the different hormonal activities is that it is assumed that not all the receptors responsible for the different activities approach the molecule from the same side.

Kent: It appears that the fluorine is usually projected axially in the more active forms.

Wettstein: Yes, with the exception of the equatorial 6α-fluorine atom.

Kent: Do such molecules change the conformation of proteins in model systems?

Wettstein: Possibly, but I do not know.

Taylor: To what extent has the stability of the C—F bond in fluorinated steroids been examined in the whole animal? Is fluoride or fluoroacetate formed? This is important with the increasing clinical application of many of these hormonal steroids.

Wettstein: There is no example, as far as I know, of a biological reaction taking out fluorine from the steroid nucleus.

References

AYER, D. E. (1962) *Tetrahedron Lett.* 1065–1069.
CHANG, S. S. and E. F. WESTRUM (1962) *J. Chem. Phys.* **36**, 2571.

KNOX, L. H., E. VERLARDE, S. BERGER, D. CUADRIELLO and A. D. CROSS (1964) *J. Org. Chem.* **29**, 2187–2195.

MCDANIEL, D. H. and R. E. VALLÉE (1963) *Inorg. Chem.* **2**, 996.

MILLER, E. C. and J. A. MILLER (1960) *Cancer Res.* **20**, 133.

STUDIER, M. H. and E. H. APPELMAN (1971) *J. Am. Chem. Soc.* **93**, 2349.

WADDINGTON, T. C. (1958) *J. Chem. Soc.* 1708.

Biochemical effects of fluoroacetate poisoning in rat liver

P. BUFFA, V. GUARRIERA-BOBYLEVA and I. PASQUALI-RONCHETTI

Istituto di Patologia Generale, Università di Modena, Modena

Fluoroacetate ($CH_2F \cdot COO^-$) is converted in the organism to fluorocitrate (Peters *et al.* 1953) which inhibits aconitate hydratase [citrate (isocitrate) hydrolyase; E.C. 4.2.1.3.]—an enzyme of the tricarboxylic acid cycle—thus causing citrate to accumulate in the tissue.

The early finding by Buffa and Peters (1949*a, b*) that fluoroacetate poisoning causes accumulation of citrate in animal tissues has been consistently confirmed (Potter and Busch 1950; DuBois, Cochran and Doull 1951; Lindebaum, White and Schubert 1951; Gál, Peters and Wakelin 1956; Liébecq and Liébecq-Hutter 1958), and elevated concentrations of tissue citrate are now inseparable from fluoroacetate intoxication. In fact, accumulation of citrate has been found in all species of poisoned animals, both vertebrate and invertebrate, so far studied (Buffa and Peters 1949*b*; Potter and Busch 1950; Buffa and Peters 1950, unpublished results; Liébecq and Liébecq-Hutter 1958; Annison *et al.* 1960), and also in explants of chick embryo heart cultured *in vitro* and treated with fluoroacetate (Buffa *et al.* unpublished results).

A typical feature of the phenomenon is the wide range in the amounts of citrate accumulated in different tissues. These seem to be related to the rate of metabolic activity of the individual tissues; thus heart and kidney, which have high respiratory rates, show elevated concentrations of citrate. Similarly, there appears to be a correlation with muscle activity: the diaphragm, which continuously contracts, accumulates much more citrate than the less active thigh muscle (Buffa and Peters 1949*b*). The liver, however, appeared to be a striking exception: it accumulated very low amounts of citrate in spite of its high respiratory and metabolic activity (Buffa and Peters 1949*b*; Potter and Busch 1950).

Subsequent work has revealed that the effect of fluoroacetate on citrate metabolism in the liver is under the influence of the sex hormones and the nutritional state of the animal. DuBois, Cochran and Doull (1951) found that adult female rats poisoned with fluoroacetate accumulated citrate in the liver whereas males did not. In a further study DuBois, Cochran and Zerwic (1951) showed that castrated or adrenalectomized male rats acquired the ability to accumulate citrate in the liver after administration of fluoroacetate, but that castration did not decrease citrate accumulation in the liver of female rats. Fluoroacetate poisoning caused citrate accumulation in the liver of male rats previously treated with the oestrogen, oestradiol dipropionate; conversely, the androgen, testosterone propionate, inhibited the accumulation of citrate in the liver of female rats. Further, young immature female rats which had been injected with fluoroacetate did not accumulate citrate in the liver. Nevertheless, the influence of the sex hormones does not seem to be a determining factor; in fact, Lindebaum, White and Schubert (1951) found that female rats did not accumulate citrate in the liver after administration of fluoroacetate, and Liébecq and Liébecq-Hutter (1958) observed that adult male mice poisoned with fluoroacetate accumulated some citrate in the liver whereas female mice did not.

A factor of more decisive importance for the accumulation of citrate in the hepatic tissue appears to be the nutritional state of the animal. Ord and Stocken (1953) showed that fed rats of both sexes accumulated marked amounts of citrate in the liver, but that animals starved for 24 hours accumulated only minute amounts of the acid after injection of fluoroacetate. However, there are apparently conflicting reports on the effects of the nutritional state on the accumulation of citrate in the liver (Potter, Busch and Bothwell 1951; Cole, Engel and Fredericks 1955; Gál, Peters and Wakelin 1956; Spencer and Lowenstein 1967); but since the nutritional state of the rat must be carefully defined (see Heath and Threlfall 1968) some of the published results cannot be precisely evaluated.

The experiments to be described were done to gain insight into the mode of action of fluoroacetate in the liver. They were prompted by the unique response of the hepatic tissue to fluoroacetate with regard to citrate accumulation, and also by the lack of respiratory inhibition observed in slices of poisoned liver (Buffa *et al.* 1960).

Extensive preliminary tests had shown that our colony of Wistar rats (line Glaxo-2A), under the standard laboratory conditions, did not exhibit sex differences in the accumulation of citrate in the liver during fluoroacetate poisoning: fed adult males and females accumulated citrate to the same extent and starved males and females showed negligible amounts of liver citrate after 60 minutes of an intraperitoneal injection of 10 mg per kg body weight of sodium fluoroacetate.

We used adult rats (150–250 g) of both sexes. They were fed on a commercial balanced diet, made up in pellets, and water, and we took care to use animals in well-defined states of nutrition. The experiments always started at 09.00 hours and rats in two nutritional states were used. (*A*) animals which had food and water available *ad libitum*. These rats were in the 'fed state' according to the definition of Heath and Threlfall (1968), and in this report they will be referred to as *normal fed rats*. (*B*) Animals from which food, but not water, had been withheld for 24 hours. These will be called *starved rats*.

LIVER CITRATE AND GLYCOGEN CHANGES IN FLUOROACETATE POISONING

Four groups of rats, normal fed males, normal fed females, starved males, and starved females were injected with fluoroacetate. The rats were then killed at intervals up to 5 hours and their liver citrate was determined (Fig. 1). Normal

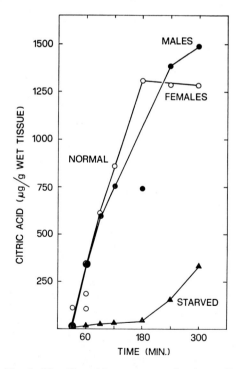

FIG. 1. The effect of fluoroacetate poisoning on liver citrate in normal fed and 24-hour starved rats. Each point represents the mean value from two rats.

The results obtained from starved male and female rats were very close and have been combined; thus each point represents the mean value of four animals.

fed rats accumulated high amounts of citrate for 3–4 hours after the injection; it is apparent that both normal fed males and females accumulated similar amounts of citrate in the liver. Also, the starved animals did not show sex differences in their citrate concentrations. In the starved rats the usual low concentrations of citrate were observed for up to 3 hours of poisoning; after this time the concentration of hepatic citrate increased steadily and after 5 hours it was over 300 μg citric acid per g wet tissue. A similar finding has been reported by Cole, Engel and Fredericks (1955).

Two hours after they were fed, the starved rats were again able to accumulate citrate in the liver in the same way as normal fed animals.

Rats of the Sprague-Dawley strain and commercial non-inbred animals also did not show sex differences in response to fluoroacetate poisoning, and they accumulated high concentrations of citrate in the liver only when they were in the fed state.

Table 1 shows the effect of fluoroacetate and fluorocitrate poisoning on liver

TABLE 1

Effect of fluoroacetate and fluorocitrate poisoning on liver citrate in normal fed and starved rats

Substance injected	Citric acid (μg/g wet tissue)	
(0.1 ml/100 g body weight)	Normal fed rats	Starved rats
Sodium chloride	1	1
(0.13 M)	(8)	(6)
Sodium fluoroacetate	556 ± 41	0
(10 mg/kg body wt/2 hours)	(4)	(2)
Sodium fluorocitrate	1962 ± 355	790 ± 166
(40 mg/kg body wt/2 hours)	(4)	(4)

Results are given as means ±s.e.m. In parentheses are the numbers of rats used.
Synthetic trisodium fluorocitrate contained approximately 4% of the inhibitory isomer. It was treated according to the procedure of Ward and Peters (1961). The substances were dissolved in water and intraperitoneally injected.

citrate in the normal fed and starved rats. It can be seen that fluorocitrate is more active than fluoroacetate in causing accumulation of citrate in the normal fed rats, as was first shown by Gál, Peters and Wakelin (1956). Fluorocitrate, unlike fluoroacetate, also provoked citrate accumulation in the liver of the starved rats. This could mean that either fluoroacetate is not transformed into fluorocitrate in the starved rat, or that the accumulated citrate diffuses out of the hepatic cells. Further, the fact that fluorocitrate caused much less citrate to be accumulated in the starved rats than in the normal fed ones suggests that citrate is synthesized at a slower rate in the liver of starved animals, or that part of it leaks out of the cells. Fawaz, Tutunji and Fawaz (1956) also observed that starvation reduced citrate accumulation in the liver of fluorocitrate-poisoned rats.

Fluoroacetate poisoning also has a profound influence on glucose metabolism: hyperglycaemia has been reported in goats (Foss 1948), rabbits (Mazur, Bodansky and Albaum, quoted by Chenoweth 1949), rats (Elliott and Phillips 1954; Cole, Engel and Hewson 1954; Engel, Hewson and Cole 1954; Cole, Engel and Fredericks 1955) and sheep (Annison *et al.* 1960). Although glucose metabolism has been much studied, glycogen has so far received less attention in spite of the observations by Mazur, Bodansky and Albaum (quoted by Chenoweth 1949) in rabbits and by Annison and co-workers (1960) in sheep that lethal fluoroacetate poisoning causes marked depletions of this polysaccharide in the liver.

In normal fed rats fluoroacetate poisoning causes a rapid fall in the amount of glycogen in the liver; one hour after the injection of a lethal dose the glycogen level is reduced by about 75%, and after two hours is lowered by more than 90%. Fig. 2 illustrates the steep decline in glycogen concentration and the corresponding linear rise in citrate content in a group of normal fed rats injected

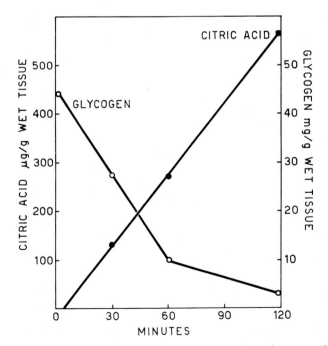

Fig. 2. The effect of fluoroacetate poisoning on liver glycogen and citrate in normal fed rats. Each point represents the mean value of two males and two females.

with fluoroacetate. Starvation in the rat is accompanied by a substantial fall in liver glycogen (Stoner 1958); in 24-hour starved rats it is about 10% of that in normal fed animals. A lethal injection of fluoroacetate in starved rats causes the residual small amount of glycogen present in the liver to disappear completely within an hour.

LIVER ADENOSINE TRIPHOSPHATE

The impairment of the tricarboxylic acid cycle caused by the fluorocitrate which originates from fluoroacetate through the lethal synthesis can be expected to be accompanied by a reduction in the amount of high-energy phosphate compounds which serve as storage reservoirs of phosphate-bond energy in the tissue. The diminutions in temperature of the whole body (Buffa and Filippini-Lera Buffa 1959) and of the liver (Stoner 1956) in fluoroacetate-poisoned rats are in accord with this suggestion. Further, Buffa and co-workers (1960) showed a marked decrease in the amounts of high-energy phosphate compounds in the liver tissue of intoxicated rats.

Adenosine triphosphate (ATP) was measured by the luciferin-luciferase method (Strehler 1965) in the liver tissue of rats injected with a lethal dose of fluoroacetate (Table 2) and, as expected, a marked depletion of this compound

TABLE 2

Effect of *in vivo* fluoroacetate poisoning on rat liver adenosine triphosphate

| *Substance injected* | *ATP (μmole/g wet tissue)* | |
(0.1 ml/100 g body weight)	*Normal fed rats*	*Starved rats*
Sodium chloride	1.64 ± 0.08	1.01 ± 0.05
(0.13 M)	(20)	(12)
Sodium fluoroacetate	0.81 ± 0.11	1.06 ± 0.10
(10 mg/kg body wt/1 hour)	(16)	(8)
Difference	−51%	±0%

Results are given as means ± s.e.m. In parentheses are the numbers of rats used.

was found in the liver of normal fed animals, whereas no decrease in ATP was exhibited by the intoxicated hepatic tissue of starved rats. It is worth noting that starvation by itself is accompanied by about a 40% lowering of liver ATP, which may be significant in view of the dependence on ATP of fluoroacetate activation demonstrated by Brady (1955) in an *in vitro* system.

RESPIRATION AND CITRATE CHANGES IN LIVER SLICES AND HOMOGENATES FROM
FLUOROACETATE-POISONED RATS

The reported results did not unequivocally exclude the occurrence of a block
at the aconitate hydratase stage of the tricarboxylic acid cycle in the liver of
the fluoroacetate-poisoned starved rats. It could be that citrate did not accumulate because it diffused out of the liver cells. The following findings would
support this suggestion: (1) as demonstrated by fluorocitrate poisoning,
synthesis of citrate does occur in the liver of the starved rat; (2) fluoroacetate
poisoning provoked in the liver of the starved rat a small increase of citrate of
the same order as that found in blood (Buffa and Peters 1949*b*; Potter and
Busch 1950); and (3) in experiments carried out on liver slices to test this
suggestion it was found that, under the conditions illustrated in Fig. 5 (right),
at the end of the incubation about 90 % of the accumulated citrate had diffused
out of the cells into the suspending medium.

The experiments with liver slices and whole tissue homogenates were done
to clarify this point and also as a preliminary to the tests with isolated mitochondria. In each experiment one rat was injected with fluoroacetate and
one with physiological saline; after one hour they were killed and their livers

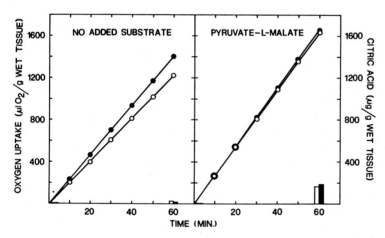

Fig. 3. Respiration rates and citrate contents of liver slices obtained from starved rats poisoned for one hour with 10 mg sodium fluoroacetate per kg body weight. ○—○ and blank
columns: controls; ●—● and black columns: fluoroacetate. Each curve and each column is
the mean from 8 experiments.

The oxygen uptake was measured by conventional Warburg manometry in O_2 at 37 °C.
The incubation medium contained 75 mM-tris-HCl, pH 7.4 and 50 mM-KCl. Pyruvate was
10 mM and L-malate 5 mM. There were 95–150 mg of tissue. The final volume of the reaction
was 3.0 ml.

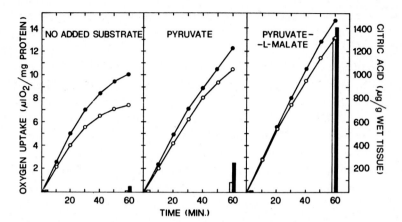

FIG. 4. Oxygen uptake rates and citrate contents of whole liver homogenates obtained from starved rats poisoned for one hour with 10 mg sodium fluoroacetate per kg body weight. ○—○ and blank columns: controls; ●—● and black columns: fluoroacetate. Each curve and each column is the mean from 8 experiments.

The oxygen uptake was measured by conventional Warburg manometry in air at 37 °C. The homogenate was 10 % in 250 mM-sucrose (w/v). The incubation medium contained 50 mM-tris-HCl, pH 7.4, 33 mM-KCl and 84 mM-sucrose. Pyruvate was 10 mM and L-malate 5 mM. There were about 15 mg of protein. The final volume of the reaction was 3.0 ml.

used for preparing slices or homogenates. Citrate determinations were done on the liver tissue soon after the removal of the organ and on the whole suspensions of the slices and homogenates at the end of incubation.

The results are summarized in Figs. 3–6. It is evident from the curves of the oxygen uptake that slices and homogenates from both normal fed and starved rats poisoned with fluoroacetate not only did not show inhibition of respiration but that they consumed oxygen at faster rates than the controls. Similar findings in liver slices obtained from fluoroacetate-poisoned rats had been reported previously (Buffa *et al.* 1960).

As usual very little citrate accumulated in the liver tissue of the starved rats injected with fluoroacetate (Figs. 3 and 4). There was no accumulation of citrate in the slices and homogenates incubated without added substrate. The addition of pyruvate or pyruvate and L-malate caused a slight increase in the amount of citrate with respect to the controls.

Slices and homogenates from normal fed rats poisoned with fluoroacetate incubated without added substrates metabolized the citrate previously accumulated in the liver *in vivo* (Figs. 5 and 6). When pyruvate or pyruvate and L-malate were added they respired at higher rates than the slices and homogenates without added substrates, and they also accumulated marked extra amounts of citrate.

The results of these experiments ruled out the possibility that in starved rats poisoned with fluoroacetate the lack of citrate accumulation in liver is due to a

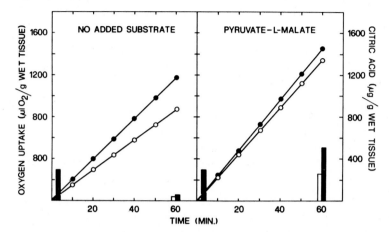

FIG. 5. Respiration rates and citrate contents of liver slices obtained from normal fed rats poisoned for one hour with 10 mg sodium fluoroacetate per kg body weight. ○—○ and blank columns: controls; ●—● and black columns: fluoroacetate. Each curve and each column is the mean from 8 experiments.

The experimental conditions were the same as described in the legend to Fig. 3.

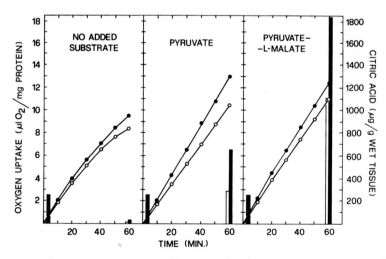

FIG. 6. Oxygen uptake rates and citrate contents of whole liver homogenates obtained from normal fed rats poisoned for one hour with 10 mg sodium fluoroacetate per kg body weight. ○—○ and blank columns: controls; ●—● and black columns: fluoroacetate. Each curve and each column is the mean from 8 experiments.

The experimental conditions were the same as described in the legend to Fig. 4.

leakage of the compound out of the cells into the extracellular fluid and blood. Moreover, the accumulation of similar amounts of citrate in the controls and the poisoned preparations when pyruvate or pyruvate and L-malate were added demonstrates that fluoroacetate intoxication in the starved rats does not cause significant inhibition of aconitate hydratase. Apparently only insignificant lethal synthesis of fluorocitrate occurs in the liver of the starved rat. This is in agreement with the finding of Gál, Drewes and Taylor (1961) with [2-^{14}C] fluoroacetate that very little fluorocitrate is formed in the liver of starved rats.

The results with slices and homogenates from normal fed rats poisoned with fluoroacetate appeared contradictory. Without added substrate, citrate was metabolized at higher rates than it was synthesized. With added pyruvate or pyruvate and L-malate, citrate was metabolized at a lower rate than it was synthesized and accumulated in the suspension. The same was true in control preparations, that is normal liver cells *in vitro* with appropriate substrate-accumulated citrate. However, the preparations from the fluoroacetate-poisoned rats accumulated extra amounts of citrate with respect to the controls, thus revealing the existence of a block at the aconitate hydratase stage of the tricarboxylic acid cycle. But notwithstanding this block, there was no inhibition of respiration.

RESPIRATORY ACTIVITY OF LIVER MITOCHONDRIA ISOLATED FROM FLUOROACETATE-POISONED RATS

The effects of fluoroacetate and fluorocitrate on isolated mitochondria have been investigated by various workers. Gál and Smith (1960) found that 1.1 mM fluoroacetate, with fumarate plus pyruvate as substrate, caused marked inhibition of oxygen uptake in liver mitochondria; whereas Buffa and co-workers (1960), in experiments lasting up to 37 min with pyruvate as the substrate and 6.6 mM-fluoroacetate, did not record any inhibition of oxygen consumption. The apparent discrepancy in the results was possibly due to the absence of fumarate in the system used by the latter workers (Gál 1960).

Fluorocitrate shows a variety of inhibitory effects on intact isolated liver mitochondria. It inhibits the oxidation of fumarate (Fairhurst, Smith and Gál 1959a), pyruvate (Gál and Smith 1960), succinate, glutamate (Fanshier, Gottwald and Kun 1964) and citrate (Guarriera-Bobyleva and Buffa 1969). Some of these inhibitory effects are difficult to reconcile with the known mode of action of fluorocitrate and are likely to originate through mechanisms as yet not clear.

Also, the effects of fluoroacetate and fluorocitrate on oxidative phosphorylation in isolated intact mitochondria have been studied with discordant results. Judah and Rees (1953) showed that fluoroacetate had no effect on oxidative phosphorylation in rat kidney mitochondria, whereas Fairhurst, Smith and Gál (1959a, b) found that fluoroacetate and synthetic fluorocitrate interfered with oxidative phosphorylation of rat liver mitochondria.

In evaluating the effects observed with fluorocitrate it should be remembered that various synthetic and enzymic mixtures of the four isomers of the compound have been used under the name of fluorocitrate. Further, much work has been done with crude preparations of the compound. Thus it cannot be excluded that some of the effects observed were not due to the diastereoisomer, which is active on aconitate hydratase and originates in the organism from fluoroacetate, but to the other isomers or to some impurity present in the preparations used.

Additional information on the biochemical mode of action of fluorocitrate may be obtained by studying the mitochondria isolated from animals poisoned with fluoroacetate. So far studies of this kind have received little attention (Buffa *et al.* 1960; Buffa, Carafoli and Muscatello 1960; Corsi and Granata 1967). Buffa and co-workers (1960) found that intact mitochondria isolated from the liver of fluoroacetate-poisoned rats oxidized citrate and succinate at slower rates than did the controls. The experiments were carried out in conventional Warburg-type respirometers; oxidation of citrate was inhibited by 34% and that of succinate by 24%. The P:O ratios were not affected.

Manometric measurement of oxygen consumption is considered to be the expression of a quasi-steady state reaction condition involving a certain distribution of intermediate metabolites derived from the added substrate. In contrast to the manometric experiments, the recording of oxygen uptake of mitochondrial suspensions by rapid polarographic analysis with an oxygen electrode of the Clark type more closely approaches the initial rate of the enzymic reactions occurring in the mitochondria and is likely to give an insight into the situation existing *in vivo*. The polarographic method was therefore used to study the oxidation of the intermediates of the tricarboxylic acid cycle and of pyruvate and L-glutamate by intact liver mitochondria freshly isolated from rats acutely poisoned with fluoroacetate. A small amount of L-malate was added to pyruvate and L-glutamate to prime the reactions.

The mitochondria from the normal fed poisoned rats showed a reduced ability to oxidize citrate, *cis*-aconitate and oxaloacetate. The inhibition of citrate oxidation was of the order of 50% and never exceeded 60%; the mean inhibition of *cis*-aconitate oxidation was around 15%. The inhibition of oxaloacetate oxidation was 34%; however, this result must be accepted with reservations because of the low oxygen uptake with this substrate. The oxidation of pyruvate, isocitrate, α-oxoglutarate, fumarate, L-malate and L-glutamate was not affected. The oxidation of succinate was significantly increased; in 16 experiments a mean increment of 24% was recorded.

The respiratory control index (acceptor control) was markedly reduced for citrate, *cis*-aconitate and oxaloacetate and the ADP:O ratios were slightly lowered. However, with all the other substrates the respiratory control indexes and ADP:O ratios were unchanged with respect to the controls.

The mitochondria from the starved rats were much less affected by fluoro-acetate poisoning than the ones from normal fed rats. They showed a slight inhibition in the oxidation of pyruvate and citrate and no change in the ability to oxidize all the other substrates, except succinate whose oxidation was increased by an average of 23% in 10 experiments. Figs. 7 and 8 are the polaro-graphic tracings of two typical experiments with liver mitochondria from normal fed rats, with citrate and *cis*-aconitate as substrates. It can be seen that state 4 respiration (Chance and Williams 1956) was unchanged in poisoned mitochondria and that the lowering of respiratory control was due to a decrease of state 3 coupled respiration when citrate was the substrate.

The results with intact mitochondria isolated from starved rats are in agreement with the findings *in vivo* and *in vitro* on liver slices and homogenates. Aconitate hydratase is only slightly inhibited and this explains the low citrate increase in the liver of fluoroacetate-poisoned starved rats. The results with mitochondria isolated from poisoned normal fed rats are also in keeping with

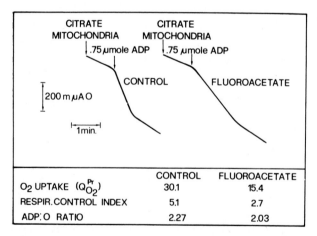

FIG. 7. Polarographic tracings of the oxygen uptakes by intact liver mitochondria isolated from a normal fed control rat and a normal fed rat poisoned for one hour with 10 mg sodium fluoroacetate per kg body weight. The substrate was 2 mM-citrate.

The experimental conditions were as described by Guarriera-Bobyleva and Buffa (1969).

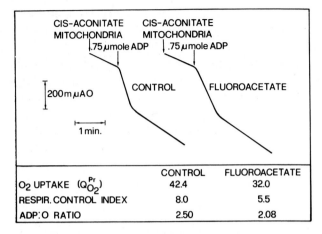

FIG. 8. Polarographic tracings of the oxygen uptake by intact liver mitochondria isolated from a normal fed control rat and a normal fed rat poisoned for one hour with 10 mg sodium fluoroacetate per kg body weight. The substrate was 2 mM-*cis*-aconitate.

The experimental conditions were as described by Guarriera-Bobyleva and Buffa (1969).

the effects on cell citrate. The inhibition of mitochondrial aconitate hydratase at the first stage of the reaction catalysed by the enzyme is the cause of citrate accumulation found *in vivo* and *in vitro*.

The marked inhibition of citrate oxidation and the slight inhibition of *cis*-aconitate oxidation confirm the findings with fluorocitrate *in vitro* of Fanshier, Gottwald and Kun (1964) on sonically disrupted rat kidney mitochondria, and of Guarriera-Bobyleva and Buffa (1969) on intact rat liver mitochondria. In fact, the latter found that 6.7 μM-fluorocitrate (containing 4% of the active isomer) caused 94% inhibition of citrate oxidation, whereas 1.0 mM-fluorocitrate was necessary to provoke the same degree of inhibition of *cis*-aconitate oxidation. Apparently the citrate part of the mitochondrial aconitate hydratase reaction is some 150 times more sensitive to the inhibitory action of fluorocitrate than is the isocitrate part of the reaction.

The ADP:O ratios and the respiratory control indexes recorded show that fluorocitrate does not affect the mechanism of oxidative phosphorylation. The present results confirm the previous findings by Buffa and co-workers (1960) on rat liver and muscle mitochondria, by Buffa, Carafoli and Muscatello (1960) on muscle mitochondria, and by Corsi and Granata (1967) on heart, kidney and brain mitochondria studied by Warburg manometry. The slight lowering of the ADP:O ratios and the depression of the respiratory control indexes associated with citrate, *cis*-aconitate and oxaloacetate oxidation are dependent on the decreased activity of the respiratory chain with these inhibited substrates.

The finding that succinate oxidation is increased is at variance with the results obtained by Fanshier, Gottwald and Kun (1964) with fluorocitrate *in vitro,* and by Buffa and co-workers (1960) with fluoroacetate *in vivo.*

To analyse the point further, we studied the effect of fluorocitrate on the oxidation of pyruvate, citrate, *cis*-aconitate and succinate by intact mitochondria isolated from the liver of the normal fed rat. The inhibitor was added to the mitochondrial suspension either during respiratory state 4 or during respiratory state 3. The results clearly show (Table 3) an inhibitory effect on pyruvate oxidation, and no effect at all on succinate oxidation. The effects of fluorocitrate upon citrate and *cis*-aconitate confirmed the findings *in vitro* by Guarriera-Bobyleva and Buffa (1969) and the present ones on mitochondria poisoned by fluoroacetate *in vivo.*

The recorded effects of fluorocitrate on succinate dehydrogenase appear somewhat conflicting and more work is necessary to clarify this problem.

TABLE 3

Effects of fluorocitrate on respiratory activity of isolated rat liver mitochondria

Substrate	Oxygen uptake ($\mu lO_2/mg\ protein/h$)		Effect of inhibitor (%)	Respiratory control index	ADP:O ratio
	State 4	State 3			
Pyruvate + L-malate	7.6	37.8	—	4.9	2.42
+ FCit in state 4	6.9	32.3	−16	4.7	2.39
+ FCit in state 3	7.9	41.5→37.8	−11	4.7	2.43
Citrate	6.5	32.0.	—	4.9	2.39
+ FCit in state 4	6.5	6.5	−100	1	—
+ FCit in state 3	6.5	29.4→8.7	−90	1.3	—
cis-Aconitate	7.6	54.2	—	7.1	2.30
+ FCit in state 4	8.0	47.6	−16	5.9	2.30
+ FCit in state 3	7.6	47.1→47.1	0	6.2	2.29
Succinate	11.6	56.0	—	4.8	1.53
+ FCit in state 4	12.7	62.9	+14	5.0	1.41
+ FCit in state 3	10.5	52.4→52.4	0	5.0	1.50

The experimental conditions were the same as described by Guarriera-Bobyleva and Buffa (1969). Substrate concentrations were 1.5 mM-pyruvate and 0.5-mM-L-malate. Fluorocitrate was 0.1 mM.

A POSSIBLE FUNCTION FOR EXTRAMITOCHONDRIAL ACONITATE HYDRATASE

The experiments with liver slices and homogenates from fluoroacetate-poisoned rats have shown that the endogenous respiration of these preparations is not inhibited and that it is paralleled by the disappearance of the citrate previously accumulated in the tissue *in vivo*. It has been shown by Schneider, Striebich and Hogeboom (1956) that approximately 70% of the accumulated citrate in the liver of fluoroacetate-poisoned rats is associated with mitochondria. It has also been shown that citrate diffuses out passively from the mitochondria into the cytosol (Max and Purvis 1965; Lowenstein 1968).

The distribution of citrate amongst the liver cell fractions obtained by conventional differential centrifugation was re-investigated and the results are summarized in Table 4. It can be seen that in fluoroacetate-poisoned liver cells more than 43% of the accumulated citrate was retained in the mitochondria and over 32% was found in the extramitochondrial cytoplasm. The question now arises as to why the endogenous respiration of the liver slices and homogenates is not inhibited in spite of the block to the tricarboxylic acid cycle and whether the respiration is supported by the citrate accumulated in the cells. Dickman and Speyer (1954) demonstrated that aconitate hydratase has a dual distribution in the rat liver cell: a large proportion of the enzyme is found in the soluble extramitochondrial cytoplasm and some is tightly bound to the mitochondria. Guarriera-Bobyleva and Buffa (1969) confirmed this finding and showed that approximately 70% of the enzyme is extramitochondrial and that in fluoroacetate poisoning this fraction of aconitate hydratase is virtually uninhibited. Apparently the endogenous fluorocitrate is bound to the mitochondria of liver cells and does not diffuse out into the cytosol.

Other enzymes of the tricarboxylic acid cycle are found in the mitochondria and in the extramitochondrial cytoplasm. Some intermediates of the cycle have been found to shuttle between these mitochondrial and extramitochondrial enzymes (Greville 1966).

TABLE 4

Distribution of citric acid in the liver cell of the normal fed rat poisoned with fluoroacetate

Liver fraction	Citric acid		
	$\mu g/g$ wet tissue	*Per cent*	*$\mu g/mg$ N*
Whole homogenate	724 ± 126	100	20.5 ± 3.1
Nuclei and cell debris	184 ± 37	25.4	18.7 ± 3.1
Mitochondria	314 ± 44	43.4	38.9 ± 4.3
Postmitochondrial fraction	235 ± 69	32.4	13.2 ± 3.3
Recovery	—	101.2	—

Mean values from 5 experiments $\pm$ S.E.M.
The rats were injected with sodium fluoroacetate (10 mg per kg body weight for two hours). The liver tissue was homogenized and the fractions prepared according to the description given by Guarriera-Bobyleva and Buffa (1969).

An experiment was done to see whether the addition of soluble cytoplasm would reactivate the oxidation of citrate by isolated mitochondria inhibited by fluoroacetate *in vivo*. The soluble cytoplasm contains aconitate hydratase and isocitrate dehydrogenase [L_s-isocitrate:NADP oxidoreductase (decarboxylating); E.C.1.1.1.42]; these would transform citrate into *cis*-aconitate, isocitrate and α-oxoglutarate, and either isocitrate or α-oxoglutarate or both would penetrate the mitochondria and enter the tricarboxylic acid cycle, thus by-passing the block. NADP was added to fortify the diluted cytosol and radioactive citrate was used to ensure that it was the substrate that was oxidized and not other substrates present in the cytosol. The results (Table 5) show that the soluble cytoplasm markedly reactivated the oxidation of added citrate by fluoroacetate-poisoned mitochondria.

TABLE 5

Effect of the addition of soluble cytoplasm on the oxidation of citrate by liver mitochondria isolated from the normal fed rat poisoned with fluoroacetate

Additions	Oxygen uptake in state 3 ($\mu l\ O_2/mg$ protein/h)	Δ (%)	$^{14}CO_2$ Production (Counts/mg protein/h)	Δ (%)
[1,5-^{14}C] Citrate	12.0 ± 0.5	—	410 ± 32	—
Citrate + Soluble cytoplasm	18.3 ± 1.1	+ 53	506 ± 32	+ 23
Citrate + NADP	14.1 ± 0.6	+ 18	419 ± 34	± 0
Citrate + Soluble cytoplasm + NADP	19.0 ± 1.1	+ 58	673 ± 54	+ 64

Mean values from 5 experiments ± S.E.M.
The rats were injected with sodium fluoroacetate (10 mg per kg body weight for one hour) and the liver mitochondria were prepared and tested as described by Guarriera-Bobyleva and Buffa (1969). The soluble cytoplasm was prepared from a portion of the homogenate used for the preparation of mitochondria by centrifugation at 139,000 g_{av} for 45 min. The clear supernatant was collected and used. It did not contain detectable cytochrome oxidase activity. The oxygen uptake and $^{14}CO_2$ production were measured simultaneously in two separate cuvettes, 1 and 2 respectively. Additions: soluble cytoplasm fraction, 0.5 ml; NADP, 0.1 mg; ADP, 1 μmole in cuvette 1 and 5 μmole in cuvette 2. In cuvette 2, respiration was continued for 5 min; $^{14}CO_2$ was collected in ethanolamine-methanol and measured by liquid scintillation.

ULTRASTRUCTURAL CHANGES IN THE LIVER OF THE NORMAL FED RAT POISONED WITH FLUOROACETATE

Fluoroacetate poisoning causes ultrastructural changes in the rat liver (Figs. 9 and 10). One hour after the injection of a lethal dose of the inhibitor the

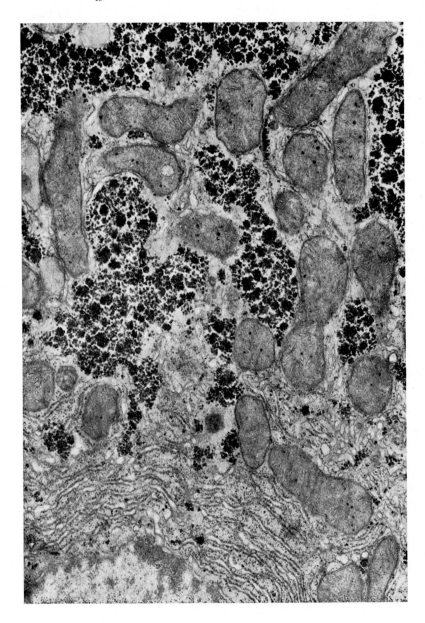

FIG. 9. Parenchymal liver cell from a normal fed control rat. × 24 000.
Fixation in glutaraldehyde and postfixation in osmium tetroxide in phosphate buffer; dehydration in acetone; tissue blocks stained with uranyl acetate in absolute acetone. Embedding in Durcupan; sections stained with lead citrate.

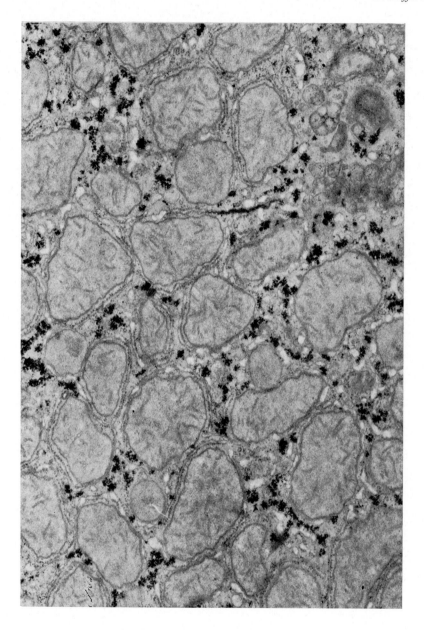

FIG. 10. Parenchymal liver cell from a normal fed rat poisoned for 3 hours with 10 mg sodium fluoroacetate per kg body weight. The mitochondria are swollen and the limiting membranes have a wrinkled profile. The clear matrix does not contain electron dense granules. The cristae are disintegrated. The smooth endoplasmic reticulum shows some small dilatations; the glycogen is much reduced. × 24 000.

mitochondria of normal fed rats show pronounced modifications; the number of elongated forms is decreased and the mitochondria are mainly rounded and expanded. The matrix is rather transparent to the electrons and the cristae are disintegrated. Three hours after poisoning, the mitochondria are still more expanded and relaxed and the limiting membranes show an irregular profile; the matrix is very clear and the cristae further damaged. No dense granules can be seen in the matrix. The glycogen is much reduced and the smooth endoplasmic reticulum shows some small dilatations. It may be relevant to note that the changes involve primarily the mitochondria and that only minor modifications occur in other cell constituents, except for the reduction in the amount of glycogen. Changes of a similar appearance have been observed in the mitochondria of chick embryo explants cultured *in vitro* and treated with fluoroacetate (Buffa *et al.* 1964).

DISCUSSION

The results of this work confirmed that the nutritional state of the rat is a determining factor in the response of the liver to fluoroacetate intoxication (Ord and Stocken 1953). Fed animals accumulated marked amounts of citrate, whereas starved rats showed barely significant increases of citrate in the first 2–3 hours of intoxication (Fig. 1). The late rise in liver citrate concentrations in starved fluoroacetate-poisoned rats (Fig. 1) had not been investigated before, though it was observed by Cole, Engel and Fredericks (1955) together with a parallel increase in the amount of blood glucose. Increased amounts of citrate have been recorded in the liver of rats made diabetic with alloxan (see Spencer and Lowenstein 1967). Hence, the late accumulation of citrate in the liver of the starved fluoroacetate-poisoned rats might be due either to a failure in insulin production secondary to the effect of fluoroacetate on β-cells of the pancreatic islets, or to the restoration in the liver, by the increased blood glucose, of a state similar to that which causes citrate accumulation in the fed rat.

With regard to the cause of depletion of liver glycogen in the normal fed fluoroacetate-poisoned rat (Fig. 2), it is likely that the orthophosphate that becomes available in the cell as a consequence of the reduction in ATP synthesis (Table 2) plays an important role in the rapid decrease in the amount of that polysaccharide.

The experiments with liver slices, homogenates and intact mitochondria prepared from starved rats poisoned with fluoroacetate (Figs. 3 and 4) demonstrated that in this nutritional state little synthesis of fluorocitrate takes place in the liver. This is in accord with the results of Gál, Drewes and Taylor (1961) who found that [2-^{14}C] fluoroacetate was very poorly converted to fluorocitrate in the rat liver *in vivo*.

The results with liver slices, homogenates and intact mitochondria obtained from normal fed rats poisoned with fluoroacetate showed that the endogenous fluorocitrate specifically inhibits the first of the two reactions catalysed by mitochondrial aconitate hydratase, namely the dehydration of citrate to *cis*-aconitate (Figs. 7 and 8). The inhibition of this tricarboxylic acid cycle enzyme, localized in the mitochondrial matrix, does not affect the mechanism of oxidative phosphorylation in the respiratory chain. Also, the ADP:O ratios, as measured in isolated mitochondria, remain at normal levels, although less ATP is formed in the tissue because of the impairment of the cycle (Fig. 7 and Table 2). The available evidence with intact mitochondria isolated from different tissues of fluoroacetate-poisoned rats indicates that *in vivo* oxidative phosphorylation is not inhibited by the biochemical injury caused by this agent in the cells (Buffa *et al.* 1960; Buffa, Carafoli and Muscatello 1960; Corsi and Granata 1967). Further, the inhibition of citrate oxidation caused by synthetic fluorocitrate in intact liver mitochondria *in vitro* is not associated with uncoupling of oxidative phosphorylation (Table 3).

The increased endogenous respiration of liver slices obtained from fluoro-acetate-poisoned rats and the disappearance from the incubated slices of the citrate previously accumulated *in vivo* appeared to be in contrast to the marked inhibition of citrate oxidation observed in intact isolated mitochondria. Also, the respiration *in vitro* of the diaphragm of the fluoroacetate-poisoned rat (Buffa, Carafoli and Muscatello 1960) and that of explants of chick embryo heart cultured *in vitro* and treated with fluoroacetate (Buffa, unpublished observation) is not inhibited in spite of the block in the tricarboxylic acid cycle revealed by accumulations of citrate in the cells.

Citrate originates in the animal cell in the mitochondria by the condensation of acetyl-CoA and oxaloacetate catalysed by citrate synthase. Then, according to the need of the cell, some of it is metabolized via the mitochondrial aconitate hydratase through the tricarboxylic acid cycle and some of it diffuses out of the mitochondria and is split into acetyl-CoA and oxaloacetate by the citrate cleavage enzyme [citrate oxaloacetate-lyase (CoA acetylating and ATP dephosphorylating); E.C. 4.1.3.8] (Srere and Bhaduri 1962; Atkinson 1968). The two extramitochondrial enzymes, aconitate hydratase and NADP-linked isocitrate dehydrogenase, may represent another alternative pathway for the metabolism of cytosolic citrate. This possibility was suggested by Racker (1961).

Previous work has shown that in the liver of the fluoroacetate-poisoned rats the endogenous fluorocitrate leaves the extramitochondrial aconitate hydratase virtually unaffected (Guarriera-Bobyleva and Buffa 1969); thus the enzyme remains efficient and available for function.

The citrate cleavage enzyme is also localized in the soluble cytoplasm (Srere and Bhaduri 1962); however, in all probability it was inactive in the system used because of the low amounts of ATP and the presence of ADP, which strongly inhibits it (Atkinson 1968). Thus it seems very unlikely that the $^{14}CO_2$ derived from compounds such as oxaloacetate and acetoacetate originated in the system from cleaved citrate. Conversely, the activating effect of added NADP indicates that the $^{14}CO_2$ formed in the reaction was derived from the oxidation of citrate. In conclusion, the evidence obtained (Tables 4 and 5) is consistent with the suggestion that extramitochondrial aconitate hydratase and NADP-linked isocitrate dehydrogenase represent an alternative pathway for the oxidation of citrate in the liver cell.

The mode of action of fluoroacetate in the rat liver is depicted in Fig. 11. In the liver of normal fed rats the tricarboxylic acid cycle runs mostly on pyruvate. Under these conditions the injected fluoroacetate is activated and

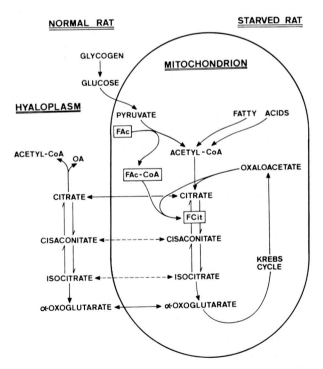

FIG. 11. Schematic representation of the mode of action of fluoroacetate in the liver of a normal fed rat and in that of a rat starved for 24 hours.

converted into fluorocitrate which specifically inhibits the first reaction of the mitochondrial aconitate hydratase (Figs. 7 and 8), thus causing accumulation of citrate (Fig. 1) and lowering of the amount of ATP (Table 2). Some citrate diffuses out of the mitochondria into the cytosol and part of it is metabolized through the extramitochondrial aconitate hydratase and NADP-linked isocitrate dehydrogenase, thus by-passing the block in the tricarboxylic acid cycle. Hence, cell respiration may be maintained in spite of the inhibition in the cycle, but the energy production is decreased. Concomitant with the biochemical lesion, ultrastructural changes occur in the mitochondria, starting from the matrix where the inhibited enzyme is localized (Figs. 9 and 10).

In the 24-hour starved rat concentrations of liver glycogen are very low, the amounts of ATP and pyruvate are much decreased and the tricarboxylic acid cycle runs mostly on fat. Under these conditions very little fluorocitrate is formed, mitochondrial aconitate hydratase is hardly inhibited and only small amounts of citrate are found in the liver tissue.

The results of this work also show that in the rat liver the basic biochemical lesion induced by fluoroacetate poisoning is the inhibition of mitochondrial aconitate hydratase by the fluorocitrate which originates in the cells through a lethal synthesis (Peters 1952). However, other biochemical effects have occasionally been observed in isolated liver mitochondria poisoned *in vivo* with fluoroacetate (Buffa *et al.* 1960) or *in vitro* with fluorocitrate (Fanshier, Gottwald and Kun 1964); namely the inhibition of succinate and glutamate oxidation. Also, inhibition of the oxidation of various tricarboxylic acid cycle intermediates has been observed in mitochondria isolated from heart, kidney and muscle of fluoroacetate-poisoned rats (Buffa *et al.* 1960; Buffa, Carafoli and Muscatello 1960; Corsi and Granata 1967). The problem as to the mechanisms of these multifarious effects of fluoroacetate poisoning remains open. Gál, Drewes and Taylor (1961) injected [2-^{14}C] fluoroacetate into rats and found that in liver, apart from fluorocitrate, activity was also incorporated into fatty acids and cholesterol. Hence it seems likely that in the organism fluoroacetate gives rise to a variety of compounds endowed with different biological activities.

SUMMARY

The mode of action of fluoroacetate in the liver of the rat has been studied.

The response of the liver tissue to fluoroacetate poisoning depends in a decisive manner on the nutritional state of the animal. In the fed rat a lethal dose of fluoroacetate causes marked accumulation of citrate, rapid fall in the amount of glycogen and decrease of tissue ATP. In the rat starved for 24 hours the concentration of liver glycogen is very low, the amount of ATP is reduced by about 40%, and fluoroacetate poisoning causes only a slight increase in liver citrate. In the fed rat the tricarboxylic acid cycle runs mostly on pyruvate, and fluoroacetate is converted to fluorocitrate; when the tricarboxylic acid cycle runs mostly on fatty acids, as is the case in the starved rat, very little fluoroacetate is converted to fluorocitrate in the liver.

In the liver slices obtained from fluoroacetate-poisoned fed rats the endogenous respiration is not inhibited and the citrate previously accumulated is metabolized. The mitochondria isolated from the poisoned liver are inhibited in their ability to oxidize added citrate (by over 50%), *cis*-aconitate (by about 15%) and oxaloacetate (by about 30%), whereas they oxidize the other intermediates of the cycle and pyruvate and glutamate at normal rates. Oxidative phosphorylation is not affected.

Extramitochondrial aconitate hydratase is virtually unaffected in the liver cell poisoned with fluoroacetate and, in conjunction with extramitochondrial NADP-linked isocitrate dehydrogenase, it provides an alternative metabolic pathway for the citrate which has diffused out of the mitochondria into the cytosol.

ACKNOWLEDGMENTS

We are grateful to Sir Rudolph Peters, F. R. S. for generous gifts of sodium fluoroacetate and synthetic trisodium fluorocitrate. We thank Mrs Maddalena Gualtieri and Mr Giuseppe Mori for excellent assistance.

This work was supported by grants from the Consiglio Nazionale delle Ricerche of Italy.

References

ANNISON, E. F., K. J. HILL, D. B. LINDSAY and R. A. PETERS (1960) *J. Comp. Pathol. & Ther.* **70**, 145–155.

ATKINSON, D. E. (1968) In *The Metabolic Roles of Citrate*, pp. 23–39, ed. T. W. Goodwin. London: Academic Press.

BRADY, R. O. (1955) *J. Biol. Chem.* **217**, 213–224.

BUFFA, P., G. F. AZZONE, E. CARAFOLI and U. MUSCATELLO (1960) *Sperimentale* **110**, 79–107.

BUFFA, P., E. CARAFOLI and U. MUSCATELLO (1960) *Rend. Accad. Naz. Lincei* **28**, ser. III, 929–936.

BUFFA, P. and E. FILIPPINI-LERA BUFFA (1959) *Atti Soc. Ital. Patol.* **6**, pt 2, 551–556.

BUFFA, P., U. MUSCATELLO, G. GODINA and A. BARASA (1964) *Proc. III Eur. Regional Conf. Int. Fed. Soc. Electron Microscopy* (Prague 1964), vol. B, p. 125, ed. M. Titlbach. Prague: Publishing House of the Czechoslovak Academy of Sciences.

BUFFA, P. and R. A. PETERS (1949*a*) *Nature (Lond.)* **163**, 914.
BUFFA, P. and R. A. PETERS (1949*b*) *J. Physiol. (Lond.)* **110**, 488–500.
CHANCE, B. and G. R. WILLIAMS (1956) *Adv. Enzymol.* **17**, 65–134.
CHENOWETH, M. B. (1949) *Pharmacol. Rev.* **1**, 383–424.
COLE, B. T., F. L. ENGEL and J. FREDERICKS (1955) *Endocrinology* **56**, 675–683.
COLE, B. T., F. L. ENGEL and K. HEWSON (1954) *Fed. Proc. Fed. Am. Soc. Exp. Biol.* **13**, 28.
CORSI, A. and A. L. GRANATA (1967) *Biochem. Pharmacol.* **16**, 1083–1089.
DICKMAN, S. R. and J. F. SPEYER (1954) *J. Biol. Chem.* **206**, 65–75.
DuBois, K. P., K. W. COCHRAN and J. DOULL (1951) *Proc. Soc. Exp. Biol. & Med.* **76**, 422–427.
DuBois, K. P., K. W. COCHRAN and M. M. ZERWIC (1951) *Proc. Soc. Exp. Biol. & Med.* **78**, 452–455.
ELLIOTT, W. B. and A. H. PHILLIPS (1954) *Arch. Biochem.* **49**, 389–395.
ENGEL, F. L., K. HEWSON and B. T. COLE (1954) *Am. J. Physiol.* **179**, 325–332.
FAIRHURST, A. S., R. E. SMITH and E. M. GÁL (1959*a*) *Biochem. Pharmacol.* **1**, 273–279.
FAIRHURST, A. S., R. E. SMITH and E. M. GÁL (1959*b*) *Biochem. Pharmacol.* **1**, 280–287.
FANSHIER, D. W., L. K. GOTTWALD and E. KUN (1964) *J. Biol. Chem.* **239**, 425–434.
FAWAZ, E. N., B. TUTUNJI and G. FAWAZ (1956) *Proc. Soc. Exp. Biol. & Med.* **92**, 311–313.
FOSS, G. L. (1948) *Br. J. Pharmacol. & Chemother.* **3**, 118–127.
GÁL, E. M. (1960) *Arch. Biochem. & Biophys.* **90**, 278–287.
GÁL, E. M., P. A. DREWES and N. F. TAYLOR (1961) *Arch. Biochem. & Biophys.* **93**, 1–14.
GÁL, E. M., R. A. PETERS and R. W. WAKELIN (1956) *Biochem. J.* **64**, 161–168.
GÁL, E. M. and R. E. SMITH (1960) *Proc. Soc. Exp. Biol. & Med.* **103**, 401–404.
GREVILLE, G. D. (1966) In *Regulation of Metabolic Processes in Mitochondria*, pp. 86–107, ed. J. M. Tager, S. Papa, E. Quagliariello and E. C. Slater. Amsterdam: Elsevier.
GUARRIERA-BOBYLEVA, V. and P. BUFFA (1969) *Biochem. J.* **113**, 853–860.
HEATH, D. F. and C. J. THRELFALL (1968) *Biochem. J.* **110**, 337–362.
JUDAH, J. D. and K. R. REES (1953) *Biochem. J.* **55**, 664–668.
LIÉBECQ, C. and S. LIÉBECQ-HUTTER (1958) *Experientia* **14**, 216–217.
LINDEBAUM, A., M. R. WHITE and J. SCHUBERT (1951) *J. Biol. Chem.* **190**, 585–593.
LOWENSTEIN, J. M. (1968) In *The Metabolic Roles of Citrate*, pp. 61–86, ed. T. W. Goodwin. London: Academic Press.
MAX, S. R. and J. L. PURVIS (1965) *Biochem. & Biophys. Res. Commun.* **21**, 587–594.
ORD, M. G. and L. A. STOCKEN (1953) *Proc. Soc. Exp. Biol. & Med.* **83**, 695–697.
PETERS, R. A. (1952) *Proc. R. Soc. B* **139**, 143–170.
PETERS, R. A., R. W. WAKELIN, P. BUFFA and L. C. THOMAS (1953) *Proc. R. Soc. B* **140**, 497–507.
POTTER, V. R. and H. BUSCH (1950) *Cancer Res.* **10**, 353–356.
POTTER, V. R., H. BUSCH and J. BOTHWELL (1951) *Proc. Soc. Exp. Biol. & Med.* **76**, 38–41.
RACKER, E. (1961) *Adv. Enzymol.* **23**, 323–399.
SCHNEIDER, W. C., M. J. STRIEBICH and G. H. HOGEBOOM (1956) *J. Biol. Chem.* **222**, 969–977.
SPENCER, A. F. and J. M. LOWENSTEIN (1967) *Biochem. J.* **103**, 342–348.
SRERE, P. A. and A. BHADURI (1962) *Biochim. & Biophys. Acta* **59**, 487–489.
STONER, H. B. (1956) *Br. J. Exp. Pathol.* **37**, 176–198.
STONER, H. B. (1958) *Br. J. Exp. Pathol.* **39**, 635–651.
STREHLER, B. L. (1965) In *Methods of Enzymatic Analysis*, pp. 559–567, ed. H. U. Bergmeyer. New York: Academic Press.
WARD, P. F. V. and R. A. PETERS (1961) *Biochem. J.* **78**, 661–668.

Discussion

Gál: In the starved animals one has to consider the possibility of increased activation of acetyl-CoA carboxylase. If citrate and fluorocitrate concentrations are not low the citric acid cycle does not operate well and there is increased activation of the carboxylase due to its allosteric stimulation by citrates. The activation of the carboxylase and the diminution of existing ATP in the presence of fluorocitrate will be regulatory to the condensing enzyme. If ATP can be turned into ADP and inorganic phosphate, both the acetoacetate synthesis and the inhibition of citrate formation may decrease. We have seen such effects with ATP, hexokinase and glucose in the kidney mitochondria (Gál 1960).

Buffa: We have not done any experiments on those lines. Does the hexokinase complicate things?

Gál: In 1960 we had no idea of the inhibition of the condensing enzyme by ATP. Intuitively we added potassium fluoride, hexokinase and glucose to the kidney mitochondria, and so altered the levels of citrate.

Buffa: All our experiments on isolated mitochondria were done polarographically with an oxygen electrode in order to measure the early rates of enzymic reactions.

Kun: Your experiments with substrate couples, Professor Buffa, shed considerable light on the mechanism of fluorocitrate toxicity in the mitochondrial system. Skilleter, Dummel and Kun (1972) found similar results to yours. Chappell's (1968) work shows that malate is an activator of the tricarboxylate carrier. Malate can be replaced by fluoromalate which does not penetrate the mitochondrial metabolic compartment. If fluorocitrate, malate or fluoromalate are added simultaneously, the inhibitory effect of fluorocitrate is less pronounced because the activator of the carrier competes with fluorocitrate for binding to the carrier. However, if the mitochondria are preincubated with fluorocitrate, the fluorocitrate binds irreversibly to the carrier. Then it does not matter which substrate couple is added. That pyruvate–malate or malate–citrate can be used as the substrate pair in your experiments appears to be in agreement with the proposed type of mechanism.

Peters: Professor Gál, have you further investigated your discovery that the so-called 'acetone bodies' are excreted in the urine in the presence of fluorocitrate (Gál, Peters and Wakelin 1956)?

Gál: No. Fluorocitrate and citrate lead to activation of the CoA carboxylase which leads to an active polymer of the enzyme. Indeed acetoacetate *in vivo* is being made immediately.

Professor Kun, did you add radioactive fluorocitrate to your system to see whether you got fluorocitrate poisoning?

Kun: No, we have not done that experiment.

Gál: Is the transfer block in the intramitochondrial phase?

Kun: It is assumed that the carrier is in the inner membrane. I suppose that part of the carrier molecule must communicate with the intermembrane compartment.

Gál: Does it point the outer surface rather than the intramitochondrial matrix towards the aconitase?

Kun: As far as I know active citrate transfer is directed towards uptake, and there is a passive diffusion of citrate outwards. Did you observe a passive outward diffusion of citrate and an energy-coupled inward movement of citrate?

Gál: Yes. But some of these carriers are localized in certain places and have an outer/inner phase and an inner/matrix phase. Do you know the location of this carrier?

Kun: Not at the moment. If one assumes that the carrier macromolecule interacts with citrate and 'carries' it inwards, then I cannot see how the macromolecule could be located exclusively on the outside of the inner membrane. In molecular terms, citrate has a large distance to travel to cross the membrane. So I would imagine that the carrier macromolecule must have both an inward and an outward phase.

Gál: I am asking for the location because we can now separate the outer/inner membrane.

Kun: According to general opinion, the carrier is in the inner mitochondrial membrane.

Gál: Can you see where the attachment is?

Kun: On isolation of the inner membrane, the orientation of the membrane is lost. The only way to prepare an inner membrane preparation that retains its orientation is to isolate a combined inner membrane plus matrix preparation. This has been done by Schnaitman and Greenawalt (1968).

Gál: Could the fluorocitrate be covalently bound to the membrane?

Kun: Possibly, but I think it could be chelated.

Buffa: In our laboratory Drs V. Guarriera-Bobyleva and R. Costa-Tiozzo have done some experiments (personal communication) on the distribution of calcium, magnesium and potassium in the liver cell components of the fed rat poisoned with fluoroacetate. The liver tissue was homogenized in 0.25 M-sucrose and the following fractions were obtained by differential centrifugation: nuclei and cell debris; heavy mitochondria; light mitochondria; microsomes; and supernatant. Total calcium, magnesium and potassium were determined on each fraction by atomic absorption spectrophotometry. No significant changes were found in the liver of the fluoroacetate-poisoned rat with respect to the control when the results were referred to the wet tissue. However, if the results were referred to the protein there was an apparent increase of total calcium in the mitochondria. This may be explained by the fact that the mitochondria in the fluoroacetate-poisoned liver are swollen (Fig. 10 p. 320) and their protein content is diminished with respect to the controls.

Peters: Do you think the starved rat is metabolizing fat under certain circumstances?

Buffa: In the liver of the 24-hour starved rat the concentration of glycogen is very low, little pyruvate is formed and fatty acids are actively oxidized.

Peters: Have you considered that fluoroacetone may be being made from fluoroacetate and excreted in the urine?

Buffa: We have not looked for ketone bodies in the liver or urine of the fluoroacetate-poisoned rat.

Peters: We tried the Rothera keto-bodies reaction and got a positive result with fluoroacetone but not with the urine from rats which had been dosed with fluoroacetate.

Buffa: Were these results from fed or starved rats? The nutritional state of the animal is most important. Rats feed at night, so one has to keep them under standardized conditions, always start experiments at the same time and carefully control the illumination of the animal house.

Gál: There is an active transport of calcium, but I agree with you about the effect *in vivo* with respect to brain mitochondria. One finds that in the mitochondria from fluorocitrate-poisoned animals the affinity of binding for calcium changes and the calcium comes out much faster.

Buffa: What was the citrate content of these mitochondria?

Gál: We did not measure that, but fluorocitrate was present.

Buffa: Dr E. Carafoli (personal communication) in our laboratory did some experiments on the energy-dependent accumulation of calcium and the discharge of the accumulated calcium caused by uncouplers of oxidation phosphorylation in intact liver mitochondria isolated from fed rats poisoned with fluoroacetate. He did not find any difference between the normal and poisoned mitochondria.

Kun: It is extremely difficult to evaluate the data on calcium, magnesium and potassium content because the bivalent ions exist in a bound form in the inner membrane and intermembrane spaces. Direct analysis on subfractions of mitochondria shows that about 70% of magnesium in the liver mitochondria is in a non-diffusible form. Therefore one may only observe passive magnesium movement in the remaining 20–30%. And even more variable results are obtained with calcium where the binding is directly related to the energy state of mitochondria. Furthermore, your results for the potassium content seem to be out of line with the results of direct analysis of mitochondrial fractions. The mitochondria in the energized form contain about 200 mM-potassium, if one calculates per gramme of protein (Lee, Widemann and Kun 1971). It seems difficult to follow your results because they refer to a very complicated mixture of water and protein.

Buffa: The elements calcium, magnesium and potassium were determined on freshly prepared fractions suspended in 0.25 M-sucrose; hence the mitochondria were not in any of the five respiratory states described by Chance and Williams (1956). The concentrations of potassium in the mitochondrial fractions were in the range of 200–270 nmol/mg protein.

Professor Kun, have you any evidence that in mitochondria there are different carriers for *cis*-aconitate and isocitrate?

Kun: No. Binding studies would have to be carried out on a carrier protein which has not yet been isolated.

Gál: If butylmalonate and SH poisons are coupled, they will be as effective as ethylcitrate for citric acid. This implies that there is really no difference in the carriers.

Kun: Yes, but that seems to be another question: obviously all carriers are inhibited if large quantitites of inhibitors are used. I think the question is: which inhibitor reacts with a particular carrier?

Gál: You mean which penetrates fastest?

Kun: Yes, and also which inhibitor has higher affinity for the carrier.

Gál: That would also depend on the state of utilization.

Kun: The transfer of substrates would have to be measured under conditions in which metabolic utilization was excluded.

References

CHANCE, B. and G. R. WILLIAMS (1956) *Adv. Enzymol.* **17**, 65–134.
CHAPPELL, J. B. (1968) *Br. Med. Bull.* **24**, 150.
GÁL, E. M. (1960) *Arch. Biochem. & Biophys.* **90**, 278–287.
GÁL, E. M., R. A. PETERS and R. W. WAKELIN (1956) *Biochem. J.* **64**, 161–168.
LEE, N. M., I. WIDEMANN and E. KUN (1971) *Biochem. & Biophys. Res. Commun.* **42**, 1030.
SCHNAITMAN, C. and J. N. GREENAWALT (1968) *J. Cell. Biol.* **38**, 158.
SKILLETER, D. N., R. J. DUMMEL and E. KUN (1972) *Mol. Pharmacol.* in press.

General discussion III

Heidelberger: What is the mechanism of fluorocitrate poisoning? Is it connected with binding to this carrier protein Professor Kun mentioned (p. 327)?

Gál: I think it is concerned with both aconitase and the carrier protein.

Peters: People have shown a fall in ATP. Dr Koenig and his colleagues (Patel and Koenig 1968) originally thought that the ammonia increased before the convulsions. Now, Koenig thinks magnesium is chelating in a way that reduces the efficiency of the ATPase system, and that this causes the increased excitability in the animal.

Gál: We do not see this magnesium shift. I think it is more probable that as a result of fluorocitrate poisoning there may be a change at a synaptosomal site of calcium binding. In view of the proposals of Nachmansohn (1968), this may change the conformation of the synaptosomal membrane and affect permeability and acetylcholine breakdown.

Peters: Do we agree that there is a fall in ATP?

Gál: We can agree on that, not because there is reduced ATPase activity, as you suggested, but rather because there may be an increase of intramitochondrial ATPase activity.

Peters: Professor Hastings and I (Hastings, Peters and Wakelin 1953) put fluorocitrate into the subarachnoid space of the pigeon and tried to cure the convulsions with magnesium salts of isocitrate and all the other compounds that are in the citric acid cycle after citrate; and we got absolutely no reversal of poisoning. So any hypothesis has to explain why substances given beyond the block in the citric acid cycle do not cure the animal.

Kun: If the carrier is irreversibly inhibited, a major route of substrate transfer is blocked and eventually the mitochondria are starved of an energy source.

Peters: It may be that the Golgi apparatus is out of action. Perhaps the lysosomes break up and liberate their thiamine pyrophosphatase. If that is bound up with your permeability factor, then the situation may be similar to that with mustard gas where so many compounds are alkylated that there is no hope of reversing the poisoning.

Gál: Recently Cooper and Pincus (1967) implicated thiamine pyrophosphatase in nerve conduction. So this enzyme could be involved here also.

Peters: It is thought that thiamine pyrophosphatase is in the Golgi apparatus. I thought that the Golgi apparatus was mainly concerned with sugar metabolism.

Kent: There is substantial evidence that some sugar transferases in glycoprotein formation occur in the Golgi apparatus.

Bergmann: One can make house flies resistant to several times the lethal dose of fluoroacetate (Takori 1963), and there is an interesting difference between the citrate biochemistry of susceptible and resistant strains: the latter does not accumulate citrate, whereas the susceptible strain does (300% of the normal amount). Also, both the susceptible and the resistant strains accumulate pyruvate; one has the impression that fluoroacetate interferes with the oxidation of pyruvate at a stage unrelated to aconitase. Furthermore, fluoroacetate-resistant house flies are cross-resistant to DDT, which is surprising, but may be connected with the hypothesis that DDT interferes with the pentose phosphate shunt. I think more work should be done on the effects of fluoro compounds on insects, because these systems are relatively easy to analyse.

Gál: Both types are dependent on pyruvate metabolism, and fluoroacetate indeed interferes with pyruvate. So how is it that one is resistant and the other is not?

Bergmann: I do not know.

Buffa: We did some experiments (unpublished) on chick embryo heart explants cultured *in vitro*. The cells of these explants grow very well and if given fluoroacetate accumulate citrate, thus showing that the biochemical lesion is at the level of aconitate hydratase. These cells can be grown in the presence of fluoroacetate and they are just as viable as the untreated cells. However, there are differences. In the control explants the rate of beating of the heart is higher than in the fluoroacetate-treated ones. Both lines of cells have the same rates of respiration; however, if an uncoupler of oxidative phosphorylation is added to the cultures, there is an increase in the respiratory rate in the controls, but not in the treated cells. In the electron microscope the cells grown in fluoroacetate appear profoundly altered and the mitochondria, especially, are very badly damaged.

Kun: Perhaps one of the activating enzymes is missing in the fluoroacetate-resistant house flies.

Goldman: The DDT resistance of house flies can be related to increased levels of DDT dehydrochlorinase (Sternburg, Vinson and Kearns 1953). Perhaps either fluoroacetate or fluorocitrate is degraded by that enzyme or some of the related dehalogenating enzymes found in flies.

Cain: Professor Bergmann, how is fluoroacetate administered to house flies? Are they allowed to feed on it or is it injected?

Bergmann: Increasing amounts of fluoroacetate are fed to the house flies, and a certain percentage of them become resistant to it.

Cain: Do you select your resistant ones by training?

Bergmann: Yes. There is a considerable increase in resistance over a number of generations.

Heidelberger: Have any breeding experiments been done to see how this property segregates genetically?

Bergmann: Very few.

References

COOPER, J. R. and J. PINCUS (1967) *Ciba Found. Study Group* 28, *Thiamine Deficiency*, pp. 112–121. London: Churchill.

HASTINGS, A. B., R. A. PETERS and R. W. WAKELIN (1953) *J. Physiol. (Lond.)* **120**, 50P.

NACHMANSOHN, D. (1968) *Science* **160**, 440–441.

PATEL, A. and H. KOENIG (1968) *Neurology (Minneap.)* **18**, 296.

STERNBURG, J., E. B. VINSON and C. W. KEARNS (1953) *J. Econ. Entomol.* **46**, 513–515.

TAKORI, A. S. (1963) *J. Econ. Entomol.* **56**, 69.

The use of microorganisms in the study of fluorinated compounds

PETER GOLDMAN

National Institute of Arthritis and Metabolic Diseases, National Institutes of Health, Bethesda, Maryland

Initial insight into phenomena later found to be of general interest in biology has often been gained from studies of microorganisms. In biochemistry, for example, the elucidation of the pathways of glycolysis in yeast and of fatty acid synthesis in *Clostridium kluyverii* demonstrated principles which were found later to be applicable to the metabolism of these compounds in other organisms. The value of the microorganisms in these examples was related to the great emphasis that the particular organism placed on the metabolic pathway under study. This background suggests that it is reasonable to consider the metabolism of fluorinated compounds by selected bacteria as a means of gaining insight into properties of these compounds that may be of general interest in biology. Emphasis is then not on the use of fluorinated compounds to elucidate cellular function but rather on the microorganism as a means of studying biochemistry relating to the compound itself.

Enrichment culture (Hayaishi 1955) has been particularly valuable as a means of providing bacteria for the type of study to be considered. By this method microorganisms are selected from soil on the basis of their ability to utilize a particular compound as a source of carbon or nitrogen. An organism relying on a compound in this way must necessarily be capable of metabolizing the compound; rapid growth of the organism indicates the presence of active enzyme systems for the metabolism of the compound. By enrichment culture it has been possible to isolate an organism capable of rapid growth with fluoroacetate as a sole source of carbon and energy (Goldman 1965).

An aerobic organism which utilizes a potential Krebs' cycle inhibitor as a source of energy and carbon can understandably be regarded either with scepticism or amusement. Nevertheless, it is reasonable from teleological considerations that fluoroacetate, a natural product (Marais 1944), should not accumulate indefinitely and that bacteria might be found in soil which are capable of degrading it. Without this kind of degradation, natural products might be expected to accumulate in the environment and this would have led long ago to pollution of the earth with natural products in the way the 'unnatural' insecticides and detergents are now polluting it. Soil organisms capable of adapting to the varied ecology to be found in soil offer a diverse source of microorganisms from which to select one with the desired metabolic potential. An enrichment culture might be regarded as an extension to the laboratory of a selection which has already favoured the survival in the soil of those organisms capable of making profitable use of the carbon, nitrogen or energy to be gained from compounds in their environment.

Fluoroacetate in particular might be expected to undergo microbial degradation since, because of its chemical stability, it is unlikely to undergo significant degradation under conditions generally found on the earth's surface.

It might be expected that an organism capable of degrading fluoroacetate would only be found in the soil around fluoroacetate-containing plants and that such soil might have to be used as the source of the organism. Nevertheless, we isolated a suitable organism from soil in a creek in Maryland, an area where, at least as far as we know, there are no plants which produce fluoroacetate. Organisms apparently related to the one isolated in our laboratory are widely distributed, since other bacteria capable of growth on fluoroacetate have been isolated in Japan (Horiuchi 1961; Tonomura *et al.* 1965) and the United Kingdom (Kelly 1965).

The biological aspects of enrichment culture have been reviewed elsewhere (Schlegel and Jannasch 1967), but even the fragmentary characterization of the fluoroacetate-degrading organisms suggests how selection conditions may influence the properties of the organism isolated. The pseudomonad found in our laboratory (Goldman 1965) was obtained by repeated transfers in liquid medium, conditions which favour the selection of an organism capable of rapid growth. Tonomura and co-workers (1965) apparently used similar enrichment techniques and the organisms isolated in both laboratories contain an inducible enzyme capable of cleaving fluoroacetate. Indeed, the properties of the two enzymes appear almost identical. An organism which grows more slowly on fluoroacetate was obtained by isolation on agar plates containing fluoroacetate (Kelly 1965). As discussed below, this organism has different properties from the one isolated in liquid culture, and it is possible that the defluorination mechanism may also be different.

Before the studies on fluoroacetate, a stock culture of *Vibrio* had been shown to release fluoride from 4-fluorobenzoate (Ali, Callely and Hayes 1962); and enzymic cleavage of the carbon-fluorine bond had been observed in other compounds in which fluorine was substituted on an aromatic ring (Hughes and

Saunders 1954; Kaufman 1961). These reactions might be considered adventitious since the enzymes involved were in no sense particularly adapted to cleavage of the carbon-fluorine bond. We were led to the selection of an organism capable of rapid growth on fluoroacetate in the hope of finding in that organism an enzyme more specifically adapted to cleavage of the carbon-fluorine bond.

At the outset of this study it seemed likely that an enzyme selected on this basis might provide several opportunities for observing biologically significant properties of the carbon-fluorine bond. Since the energy of this bond is among the highest found in natural products (Pauling 1960), an enzyme cleaving such a bond might have unique properties or might require a novel co-factor. If, however, fluoroacetate were metabolized to another compound before the carbon-fluorine bond was cleaved, the organism might offer the opportunity to study a new pathway of fluoroacetate metabolism. Although none of these were found, they are mentioned to show how enrichment culture offers a number of possibilities for providing biologically useful information on the compound selected for study.

The organism found in our laboratory yielded an enzyme which cleaves fluoroacetate according to the equation

$$FCH_2COO^- + HO^- \rightarrow HOCH_2COO^- + F^-$$

In addition the enzyme catalyses the analogous release of the halide from chloroacetate and iodoacetate (bromoacetate was not tested). Nevertheless, fluoroacetate is the preferred substrate for the enzyme, as indicated by the kinetic constants of the various substrates (Table 1). In addition to these compounds a variety of others containing the carbon-fluorine bond were tested with the enzyme and no significant release of fluoride could be detected. Among the compounds examined were several which are structurally quite similar to fluoroacetate, for example difluoroacetate, 2-fluoropropionate and 3-fluoropropionate. It then appears that the specificity of the enzyme is for the acetate

TABLE 1

Kinetic constants for substrates of haloacetate halidohydrolase

Substrate	V_{max}†	K_m
		mM
Fluoroacetate	100	2.4
Chloroacetate	15	20
Iodoacetate	0.8*	

† Relative to fluoroacetate taken as 100.
* Rate measured with substrate at 10^{-1} M.

portion of the molecule rather than for the carbon-fluorine bond; on this basis the enzyme was named haloacetate halidohydrolase (Goldman 1965).

When the defluorination reaction was carried out in $H_2^{18}O$ it was found that the hydroxyl group of glycollate had the same ^{18}O enrichment as the water of the reaction mixture (Goldman and Milne 1966). In the reaction of phenyl-alanine hydroxylase with p-fluorophenylalanine, fluoride is also released from the substrate and replaced with a hydroxyl group (Kaufman 1961). But any similarity between the two reactions is only apparent since the hydroxyl group of tyrosine in this reaction is derived from the NADPH-dependent reduction of molecular oxygen and not from water (Kaufman et al. 1962).

The reversibility of the haloacetate halidohydrolase reaction was also examined in an experiment where glycollate and fluoride were incubated with the enzyme in the presence of $H_2^{18}O$. As indicated schematically in Fig. 1, if a small amount of fluoroacetate were formed, even if it could not be detected chemically, it would on hydrolysis lead to the introduction of ^{18}O into glycol-late. Mass spectrometry of the glycollate isolated after the incubation failed to show any ^{18}O incorporation (Goldman and Milne 1966).

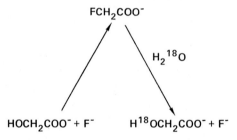

FIG. 1. Diagram to illustrate how a reversible reaction leading to the formation of fluoroace-tate from glycollate and fluoride would, in the presence of $H_2^{18}O$, lead to the incorporation of ^{18}O into glycollate.

The reversibility of the reaction was also examined by an experiment in which the substrate was chloroacetate instead of fluoroacetate. In this case $Na^{36}Cl$, a non-competitive inhibitor of the defluorination reaction, was added to a reaction mixture in which chloroacetate was being hydrolysed. If the reaction were reversible it might have been possible to detect the presence of ^{36}Cl in the chloroacetate remaining after the reaction had gone only partially to completion. This possibility was tested by adding $Na^{36}Cl$ at several con-centrations to the reaction mixture, but no evidence of $^{36}ClCH_2COO^-$ could be detected in the chloroacetate which had been partially hydrolysed by the enzyme. Hence reversibility of this reaction could not be demonstrated (Gold-man and Milne 1966).

The removal of all halides from carbon-2 of a substituted acetic acid by an apparently similar mechanism suggested that the mechanism might be studied further with other enzymes of this type. Accordingly we considered two additional halidohydrolases (I and II) which have been obtained from a pseudomonad isolated on dichloroacetate as a carbon source (Goldman, Milne and Keister 1968). Halidohydrolases I and II have different physical properties and are separable by gel filtration on Sephadex G-100. Both enzymes are reactive with the same substrates but at different rates. The substrates and products of these reactions with halidohydrolases I and II are summarized in Table 2;

TABLE 2

Compounds tested with halidohydrolases I and II

Substrate	Product
Dichloroacetate	Glyoxylate
Chloroacetate	Glycollate
L-2-Chloropropionate	D-2-Hydroxypropionate
L-2-Chlorobutyrate	D-2-Hydroxybutyrate
Iodoacetate	Glycollate
Not reactive	
D-2-Chloropropionate	Trichloroacetate
D-2-Chlorobutyrate	Chlorodifluoroacetate
2,2-Dichloropropionate	2-Chloro-2-methylpropionate

also shown are related compounds towards which no reactions of the halidohydrolases were detected. With regard to substrate specificity, halidohydrolases I and II differ from haloacetate halidohydrolase in their inability to react with fluoroacetate and their greater ability to tolerate substituents at carbon-2 of the substrate. The three enzymes have a pH optimum of about 9.2.

On the basis of the reactions summarized in Table 2 it can be seen that halidohydrolases I and II catalyse reactions of the general form

$$\text{L-RCHXCOO}^- + \text{HO}^- \rightarrow \text{D-RCHOHCOO}^- + \text{X}^-$$

where $X = Cl$ or I
and $R = H$, CH_3 or CH_3CH_2

The apparent exception to the rule of a simple bimolecular nucleophilic displacement of the halide by the hydroxide ion is the reaction with dichloroacetate in which the product is glyoxylate. Actually, this reaction is also in accord with the general equation since the expected product of the reaction with dichloroacetate would be HOClCHCOO^- and this would readily undergo a subsequent transformation to glyoxylate. At least two mechanisms are possible for this transformation of HOClCHCOO^-. One would involve the spontaneous elimination of HCl to form glyoxylate. The other would involve a second enzymic dehalogenation by the enzyme to yield $(HO)_2CHCOO^-$, which would be expected to eliminate water spontaneously to form glyoxylate. It is not possible to distinguish between the mechanisms. As with haloacetate halidohydrolase, ^{18}O from water is incorporated into carbon-2 during the halide displacement catalysed by halidohydrolases I and II.

A number of different mechanisms for the halidohydrolase reaction are shown in Fig. 2. To be acceptable a mechanism must explain both the ^{18}O labelling data and the inversion of configuration seen in the reactions of L-2-chloro-propionate and L-2-chlorobutyrate. The simplest possibility (scheme A) is a displacement of the halide by a hydroxyl ion, resulting in the observed incorporation of ^{18}O at carbon-2 and the required inversion of configuration. In scheme B the halide is displaced by the formation of an α-lactone. Hydrolysis of this lactone by nucleophilic displacement of the hydroxide ion at carbon-1, as illustrated, would transfer the oxygen atom originally at the carboxyl group to carbon-2. In such a scheme ^{18}O would be found at carbon-1 instead of

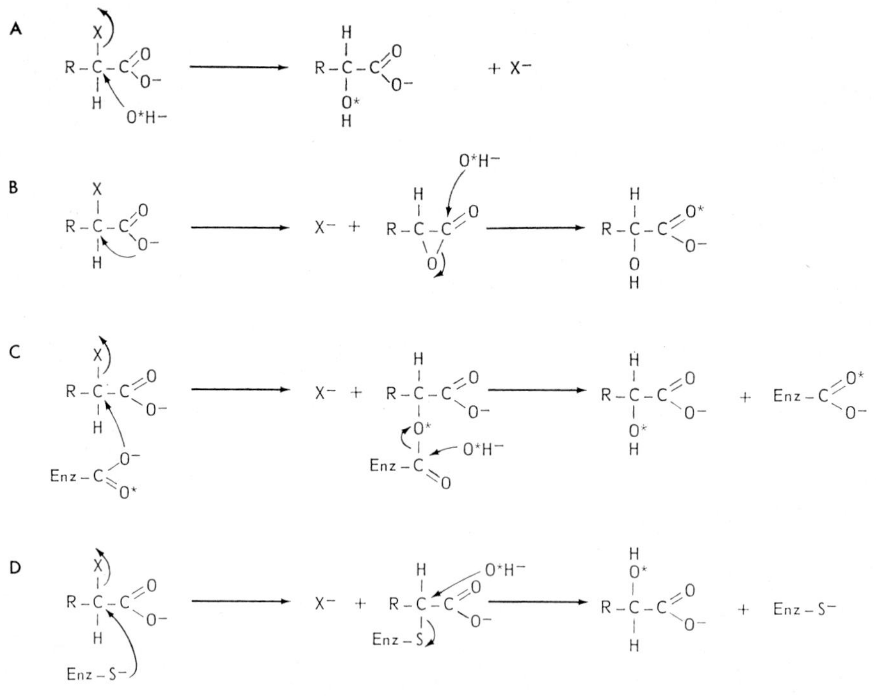

FIG. 2. Diagram for possible reaction mechanisms of the halidohydrolases. For discussion, see text. The position of ^{18}O in each mechanism is indicated by an asterisk. After the enzyme reacts once according to mechanism C, ^{18}O would be incorporated into the reactive carboxyl group on the enzyme; hence, ^{18}O labelling of this group is already indicated.

carbon-2. Since this is not in accord with the experimental results, this mechanism need not be considered further. The lactone formed as in scheme B could also be hydrolysed by the attack of the hydroxide atom at carbon-2. This would lead to the observed incorporation of ^{18}O at carbon-2, but this mechanism must be rejected because it involves a double displacement and hence would require that the configuration of carbon-2 be retained during the reaction. Scheme C is one in which the carboxyl group of the enzyme acts as a nucleophil to displace the halide. This mechanism, a special case of scheme A, accounts for the observed position of the ^{18}O and the observed inversion of configuration. A detailed study of the enzyme would be required to examine further the validity of this mechanism. A fourth possibility, the displacement of the halide by a thiol group on the enzyme and the subsequent hydrolysis of the thiol ether, is shown in scheme D. This double displacement mechanism, although in accord with the available data on haloacetate halidohydrolase (Goldman 1965; Goldman and Milne 1966), is not suitable for halidohydrolases I and II, where the configuration at carbon-2 has been shown to undergo inversion during the reaction. Schemes A and C therefore remain as possible mechanisms which are in accord with the observations on labelling and configuration.

An examination of the analogues of chloroacetate which react with the enzyme (Table 2) indicates that the reaction can proceed only if the halide has the correct stereochemical orientation. However, an examination of the analogues which are not attacked by the enzyme indicates that in addition to the correct position of the halide, the substrate also requires the right position for a proton on carbon-2. That is, compounds like trichloroacetate or 2,2-dichloropropionate, which have the halide in the correct position, can be thought of as being unreactive because of the lack of a proton on carbon-2. From this point of view the lack of reaction with D-2-chloropropionate and D-2-chlorobutyrate can be attributed to an inappropriate position of the proton as well as to an inappropriately positioned halogen. These considerations led us to examine the possibility that the proton on carbon-2 undergoes exchange during the reaction. We did this by carrying out the dechlorination of chloroacetate in 98% D_2O and examining the reaction product for the incorporation of deuterium. No deuterium was found in the glycollate formed in this manner. Conditions used for the isolation of glycollate from the reaction mixture would have allowed the exchange of deuterium in the hydroxyl group but not from a bond with the carbon-2, if indeed one had been formed during the enzymic reaction. Presumably the proton is required on carbon-2 for steric reasons, the other substituents being so bulky as to impede the catalytic function of the enzyme.

Another pseudomonad has been isolated from an enrichment culture on 2-fluorobenzoic acid. This organism metabolizes 2-fluorobenzoic acid by two pathways. One pathway, analogous to that for the metabolism of benzoic acid, proceeds through 3-fluorocatechol to 2-fluoromuconic acid; fluoromuconic acid does not seem to be further metabolized by the organism. This pathway accounts for 15% of the 2-fluorobenzoate utilized by the organism. The remaining 85% of the 2-fluorobenzoic acid undergoes defluorination to produce catechol (Goldman, Milne and Pignataro 1967; Milne, Goldman and Holtzman 1968) and it is from this pathway that the organism is provided with carbon and energy by reactions that are well known.

The defluorination reaction in this case takes place by the introduction of molecular oxygen in a one-step process shown in Fig. 3 (Milne, Goldman and Holtzman 1968). For this experiment, 2-fluorobenzoic acid was allowed to react with a suspension of this pseudomonad in an atmosphere that was 50% $^{18}O_2$ and 50% $^{16}O_2$. As indicated in the figure, the catechol recovered from this incubation mixture contained either two atoms of ^{16}O or two atoms of ^{18}O. This finding could only be explained by a one-step reaction with a single molecule of molecular oxygen. An alternative mechanism involving the successive addition of oxygen from different molecules of oxygen would lead to

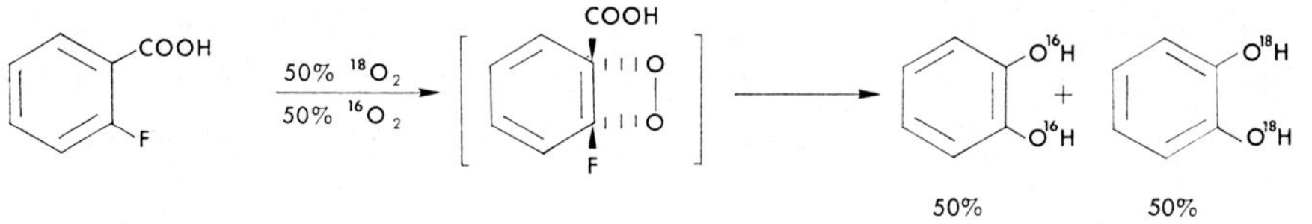

FIG. 3. The formation of catechol (III) from 2-fluorobenzoate (I) catalysed by *Pseudomonas* species in an atmosphere containing equal quantities of $^{16}O_2$ and $^{18}O_2$. The proposed cyclic peroxide intermediate (II) is in accord with the oxygen labelling found in the catechol.

the formation of sizable amounts of a catechol containing one atom of ^{16}O and one of ^{18}O. If one of the catechol oxygens had come from either the starting compound or water, there would be no catechol doubly labelled with ^{18}O. Although other mechanisms are possible to explain these findings, it seemed most attractive to postulate the *cis*-addition of molecular oxygen to form the cyclic peroxide intermediate shown in Fig. 3.

The oxygen labelling pattern of these experiments is comparable to that found previously in the analogous elimination of NH_3 and CO_2 which occurs during the conversion of anthranilic acid to catechol (Kobayashi *et al.* 1964). A single molecule of oxygen also provides the two oxygen atoms in the 3-fluorocatechol formed from 2-fluorobenzoic acid (Milne, Goldman and Holtzman 1968). Thus a basically similar mechanism of oxygenation is involved in three separate reactions in which a catechol is formed from a 2-substituted derivative of benzoic acid. Depending on the reaction, the leaving group can be F, NH_2 or H.

Enrichment culture has also been used in our laboratory to isolate an organism capable of the selective degradation of one enantiomer of the racemate containing the inhibitory isomer of fluorocitrate (Kirk and Goldman 1970). In this study the racemate was the sole source of carbon and it was hoped that organisms could be isolated which would selectively degrade one or the other of the two enantiomers. Evidence consistent with this expectation was obtained with a pseudomonad which was capable of releasing, as fluoride, 50% of the organic fluorine in the racemate. The significance of this observation was clarified when, after the cessation of fluoride release, it was possible to isolate almost quantitatively only one of the original enantiomers from the growth medium. In contrast to the original racemate, the isolated enantiomer had no inhibitory activity against aconitase. In addition, the enantiomer gave an optical rotation opposite to that of the active biosynthetic isomer (Pattison and Peters 1966) and the same as that of the inactive isomer obtained by chemical resolution (Dummel and Kun 1969).

It is interesting to speculate how various pseudomonads can flourish in media containing either 10^{-3} M-fluorocitrate or 10^{-3} M-fluoroacetate. Fluoroacetate seems to be toxic to organisms of the genus *Pseudomonas* and the mechanism of toxicity is apparently the classical one of fluorocitrate synthesis and the consequent inhibiton of aconitase (Peters 1957). This is supported by the accumulation of citrate in the presence of fluoroacetate (Jayasuriya 1956) and the formation of fluorocitrate from fluoroacetate (Behrman and Stanier 1957).

Like many other pseudomonads, the one capable of degrading fluoroacetate that was isolated in our laboratory can use as a carbon source several compounds that are metabolized via the tricarboxylic acid cycle. However, acetate, which is converted to citrate by most of these organisms, cannot support growth of the fluoroacetate-utilizing organism. Presumably the inability to utilize acetate is related to the capacity to survive the lethal synthesis ordinarily expected in the presence of fluoroacetate. Underlying both phenomena could be either a defect in the transport system or activating enzyme for acetate or a lack of citrate synthase.

Protective mechanisms of this kind, however, cannot be invoked to explain the observations of Kelly (1965) on a bacterium which can grow both on fluoroacetate and acetate. An organism with this capacity would probably synthesize fluorocitrate from fluoroacetate and thus would face the problem of fluorocitrate intoxication that is faced by the pseudomonad which degrades the active isomer of fluorocitrate. Kelly's suggestion of a fluorocitrate-resistant aconitase is a convenient explanation for the survival of both of these bacteria.

We were never successful in isolating an organism capable of growth on the purified enantiomer of the active isomer of fluorocitrate. Many explanations for this are possible, but the failure to isolate such an organism might be considered consistent with the absence in soil organisms of an enzyme capable of an appropriate interaction with the inactive isomer.

The mechanism of lethal synthesis of fluorocitrate from fluoroacetate, itself a natural product, might be considered proof that the inhibitory isomer of fluorocitrate is also a natural product. It is therefore not surprising that enzymes which can degrade this compound are present in soil organisms. The enantiomer of the inhibitory isomer, on the other hand, has not been found in nature; it has only been made by the organic chemist or by the biochemist using citrate synthase to condense acetate and fluoro-oxaloacetate (Fanshier, Gottwald and Kun 1962). It therefore might be expected that there has been no selective pressure to develop an enzyme capable of attacking this compound.

It may seem that I am suggesting the generalization that natural products can be used to select organisms by enrichment culture and that compounds not found in nature will not serve as a source of carbon or energy in this type of selection. In a sense this may be a useful concept, but the difficulties of using such a guideline must be considered. One difficulty is that it is not the compound alone which determines whether an organism capable of degrading it can be isolated. Perhaps an organism exists in soil which is capable of degrading the compound but it cannot be cultivated under laboratory conditions because of the lack of some other factor such as a nutritional requirement or a satisfactory pH in the medium. For this reason failure to isolate an organism may reflect nothing more than an inability to establish the appropriate ecological conditions in the culture medium.

Another difficulty is that of deciding whether a compound is a natural product. Fluoroacetate, for example, would not have been considered a natural product when it was synthesized in 1896; but in 1944 Marais showed that it is the toxic component of *Dichapetalum cymosum*, and since then it has been found in a number of different plants. Similarly dichloroacetate would not have been considered a natural product until the structure of chloramphenicol was elucidated. Other problems along these lines are compounds which are probably not natural products but which may undergo degradation on the basis of their similarity to a natural product. For example, the folate analogue amethopterin can serve as a carbon and nitrogen source for a pseudomonad which is also capable of growth on folate (Levy and Goldman 1967). Then there are a number of enzymes which are capable of catalysing reactions that apparently have little resemblance to their normal function. The release of fluoride from fluorophenylalanine by phenylalanine hydroxylase (Kaufman 1961) or from glycosylfluorides by appropriate glycosidases (Barnett, Jarvis and Munday 1967) might be considered purely accidental in terms of the normal role of these enzymes. Other examples of the adventitious cleavage of the carbon-halogen bond have recently been collected (Goldman 1972). The defluorination of the fluorobenzoic acids (Ali, Callely and Hayes 1962; Goldman, Milne and Pignataro 1967) may also be examples of reactions by enzymes whose primary function lies elsewhere.

So far this discussion has emphasized the degradative reactions carried out by microorganisms. Sometimes microbial transformations also yield compounds which might otherwise be difficult to synthesize. For example, among the products of the metabolism of *p*-fluorophenylacetic acid, Harper and Blakley (1971) have isolated D(+)-fluorosuccinate, which they point out is not easy to purify from a racemic mixture formed by organic synthesis. A fluorine substituent in an aromatic ring is frequently retained during metabolism; this enables fluorinated analogues of intermediates in the metabolism of aromatic compounds to be isolated from the growth media (Goldman, Milne and Pignataro 1967; Cain, Tranter and Darrah 1968; Harper and Blakley 1971). The incorporation of fluoroacetate into the uridine nucleotides of *Staphylococcus aureus* (Viriyanondha and Baxter 1970) is an interesting example of another biosynthetic reaction where acetate can be replaced by fluoroacetate.

Fluorine-containing natural products also may be synthesized by microorganisms. An example is the antibiotic nucleocidin produced by a *Streptomyces* (Morton *et al.* 1969). Failure to find fluorine in the initial structural determinations of this compound illustrates a difficulty which should be overcome more readily in the future because of the availability of techniques such as mass spectroscopy and nuclear magnetic resonance.

Recently attention has been given to some of the properties of macromolecules formed from various analogues. It has been shown, for example, that several fluorinated amino acids can be incorporated into protein and that 5-fluorouracil can be incorporated into RNA. Bacteria are useful for this type of study since large numbers of cells can be grown rapidly and it is often possible to select growth conditions which favour the incorporation of the analogue.

A considerable proportion of phenylalanine can be replaced by *p*-fluorophenylalanine in the protein of *E. coli* when the organism is grown in the presence of the analogue (Munier and Cohen 1959). However, in some cases, as with γ-fluoroglutamate, simply adding the analogue to the growth medium of growing cells does not lead to significant replacement of the normal amino acid (J. C. Unkeless and P. Goldman, unpublished). Where auxotrophs for the amino acid are available they can be valuable in assuring that replacement of the normal amino acid is complete. Such an approach, however, is only possible when the analogue will support growth. These conditions can be met in *E. coli* auxotrophs for phenylalanine, tyrosine and tryptophan where it has been shown that the normal amino acids can be replaced respectively by *o*-, *m*- or *p*-fluorophenylalanine, *m*-fluorotyrosine and 4-, 5- or 6-fluorotryptophan (Munier 1960; Munier and Sarrazin 1966; Munier, Drappier and Thommegay 1967; Browne, Kenyon and Hegeman 1970). In an auxotroph for leucine, trifluoroleucine will only support growth and replace leucine after the organism has undergone a period of adaptation to the analogue (Rennert and Anker 1963).

Amino acid replacements lead to extensive changes in the bacteria which are manifest in their morphology and physiology as recently reviewed by Marquis (1970). Nevertheless, it has been possible to define some specific changes that occur in isolated proteins whose properties have been altered by the replacement of the normal amino acid. The specific changes observed depend on the type of microorganism, the amino acid replaced and the particular protein being examined. A fluorophenylalanine substitution can leave enzyme activity unaffected in one case (Richmond 1963), change it in other circumstances (Munier 1960), or essentially eliminate it in others (Richmond 1960). Two enzymes whose synthesis is controlled from the same operon can undergo different responses in their activity to this type of substitution (Kepes 1967). Altered immunological properties (Richmond 1960) and reduced temperature stability (Munier and Sarrazin 1966; Kepes 1967) are among the changes that have been found in the properties of protein as the result of amino acid substitution. The increased heat-lability that has been observed with isolated proteins is in accord with observations on the function of certain bacterial proteins *in vivo*. For example, an *E. coli* mutant has a repressor for alkaline phosphatase which is active at 30 °C but inactive at 37 °C. When the bacteria are grown in the presence of *p*-fluorophenylalanine, the repressor becomes inactive at 30 °C and the temperature must be lowered to 22 °C for its activity to be restored. Presumably the additional heat-lability of the mutant repressor protein can be attributed to further conformational changes brought about by the analogue (Kang and Markovitz 1967).

Interesting observations have also been made on the physical and biochemical properties of bacterial RNA when 5-fluorouracil is added to the growth medium (Lowrie and Bergquist 1968; Kaiser 1969). Under these conditions 5-fluorouracil can largely replace uracil in transfer RNA. The introduction of 5-fluorouracil alters the properties of the transfer RNA and enables the fluorouracil-containing tRNA to be separated from normal tRNA by ion exchange chromatography (Kaiser 1969). The introduction of 5-fluorouracil also changes the spectral absorbance (Kaiser 1969) and heat stability (Lowrie and Bergquist 1968) of the transfer RNA. Some of the evidence on the effect of 5-fluorouracil in altering the ability of the RNA to accept amino acids is contradictory. It is not clear that changes of this kind are direct effects of the incorporation of 5-fluorouracil. They might be indirect effects due to changes in the relative amounts of the types of transfer RNA which are formed due to alterations of bacterial physiology caused by the presence of the analogue. Questions of this sort can only be answered when the transfer RNA can be purified and the exact nature of the substituents characterized.

It seems likely that macromolecules with substituents of the kind we have discussed will offer an opportunity to correlate specific changes in chemical and physical properties of these molecules with their biological function. Their great adaptability should make bacteria particularly valuable for this kind of study.

SUMMARY

Appropriate bacteria can often provide a convenient system for demonstrating biochemical properties of carbon-fluorine compounds. A pseudomonad selected from soil for its ability to utilize fluoroacetate as a sole carbon source has yielded an enzyme which cleaves the carbon-fluorine bond according to the equation

$$FCH_2COO^- + H_2O \rightarrow HOCH_2COO^- + HF$$

Analogous reactions with chloroacetate and iodoacetate are catalysed by the same enzyme which has been named haloacetate halidohydrolase. The mechanism of this reaction and similar ones catalysed by other enzymes have been investigated. Another example of enzymic cleavage of the carbon-fluorine bond is found in an organism selected for its ability to utilize 2-fluorobenzoate as a carbon source. The defluorination in this case yields catechol in which the oxygen of the hydroxyl groups is derived from molecular oxygen rather than from water. Enrichment culture has also been used to obtain a pseudomonad which can selectively degrade the inhibitory isomer of fluorocitrate in a mixture of the isomer and its enantiomer. Bacteria can also serve in a biosynthetic role by incorporating fluorine-containing analogues into biologically interesting molecules such as nucleic acids and proteins.

References

ALI, D. A., A. G. CALLELY and M. HAYES (1962) *Nature (Lond.)* **196**, 194–195.
BARNETT, J. E. G., W. T. S. JARVIS and K. A. MUNDAY (1967) *Biochem. J.* **106**, 211–227.
BEHRMAN, E. J. and R. Y. STANIER (1957) *J. Biol. Chem.* **228**, 947–953.
BROWNE, D. T., G. L. KENYON and G. L. HEGEMAN (1970) *Biochem. & Biophys. Res. Commun.* **39**, 13–19.
CAIN, R. B., E. K. TRANTER and J. A. DARRAH (1968) *Biochem. J.* **106**, 211–227
DUMMEL, R. J. and E. KUN (1969) *J. Biol. Chem.* **244**, 2966–2969.
FANSHIER, D. W., L. K. GOTTWALD and E. KUN (1962) *J. Biol. Chem.* **237**, 3588–3596.
GOLDMAN, P. (1965) *J. Biol. Chem.* **240**, 3434–3438.
GOLDMAN, P. (1972) In *Degradation of Synthetic Organic Molecules*, ed. S. Dagley and W. W. Kilgore. Washington: National Research Council (in press).
GOLDMAN, P. and W. G. A. MILNE (1966) *J. Biol. Chem.* **241**, 5557–5559.
GOLDMAN, P., W. G. A. MILNE and D. B. KEISTER (1968) *J. Biol. Chem.* **243**, 428–434.
GOLDMAN, P., W. G. A. MILNE and M. T. PIGNATARO (1967) *Arch. Biochem. & Biophys.* **118**, 178–184.
HARPER, D. B. and E. R. BLAKLEY (1971) *Can. J. Microbiol.* **17**, 635–644.
HAYAISHI, O. (1955) In *Methods in Enzymology*, vol. 1, pp. 126–137, ed. S. P. Colowick and N. O. Kaplan. New York: Academic Press.
HORIUCHI, N. (1961) *Nippon Nogei Kagaku Kaishi* **35**, 870–873.
HUGHES, G. M. K. and B. C. SAUNDERS (1954) *J. Chem. Soc.* 4630–4634.
JAYASURIYA, G. C. N. (1956) *Biochem. J.* **64**, 469–477.
KAISER, I. I. (1969) *Biochemistry* **8**, 231–237.
KANG, S. and A. MARKOVITZ (1967) *J. Bacteriol.* **94**, 87–91.
KAUFMAN, S. (1961) *Biochim. & Biophys. Acta* **51**, 619–621.
KAUFMAN, S., W. F. BRIDGERS, F. EISENBERG and S. FRIEDMAN (1962) *Biochem. & Biophys. Res. Commun.* **9**, 497–502.
KELLY, M. (1965) *Nature (Lond.)* **208**, 809–810.
KEPES, A. (1967) *Biochim. & Biophys. Acta* **138**, 107–123.
KIRK, K. and P. GOLDMAN (1970) *Biochem. J.* **117**, 409–410.
KOBAYASHI, S., S. KUNO, N. ITADA, O. HAYAISHI, S. KOZUKA and S. OAE (1964) *Biochem. & Biophys. Res. Commun.* **16**, 556–561.
LEVY, C. C. and P. GOLDMAN (1967) *J. Biol. Chem.* **242**, 2933–2938.
LOWRIE, R. J. and P. L. BERGQUIST (1968) *Biochemistry* **7**, 1761–1770.
MARAIS, J. S. C. (1944) *Onderstepoort J. Vet. Sci. Anim. Ind.* **20**, 67–73.
MARQUIS, R. E. (1970) *Handb. Exp. Pharmakol.* **20**, pt. 2, 166.
MILNE, G. W. A., P. GOLDMAN and J. L. HOLTZMAN (1968) *J. Biol. Chem.* **243**, 5374–5376.
MORTON, G. O., J. E. LANCASTER, W. FULMOR VAN LEAR and W. E. MEYER (1969) *J. Am. Chem. Soc.* **91**, 1535.
MUNIER, R. L. (1960) *C. R. Hebd. Séance Acad. Sci. Paris* **250**, 3524–3526.
MUNIER, R. and G. N. COHEN (1959) *Biochim. & Biophys. Acta* **31**, 378–391.
MUNIER, R. L., A.-M. DRAPPIER and C. THOMMEGAY (1967) *C. R. Hebd. Séance Acad. Sci. Sér. D, Paris* **265**, 1429–1432.
MUNIER, R. L. and G. SARRAZIN (1966) *C. R. Hebd. Séance Acad. Sci. Sér. C, Paris* **262**, 1029–1032.
PATTISON, F. L. M. and R. A. PETERS (1966) *Handb. Exp. Pharmakol.* **20**, pt. 1, 393.
PAULING, L. (1960) *The Theory of the Chemical Bond*, p. 85, Ithaca, N. Y.: Cornell University Press.
PETERS, R. A. (1957) *Adv. Enzymol.* **17**, 113.
RENNERT, O. M. and H. S. ANKER (1963) *Biochemistry* **2**, 471–476.
RICHMOND, M. H. (1960) *Biochem. J.* **77**, 121–135.

RICHMOND, M. H. (1963) *J. Mol. Biol.* **8**, 284–294.
SCHLEGEL, H. G. and H. W. JANNASCH (1967) *Annu. Rev. Microbiol.* **21**, 49.
TONOMURA, K., F. FUTAI, O. TANABE and T. YAMAOKA (1965) *Agric. & Biol. Chem. (Jap.)* **29**, 124–128.
VIRIYANONDHA, P. and R. M. BAXTER (1970) *Biochim. & Biophys. Acta* **201**, 495–496.

Discussion

Shorthouse: We tried an extract of the fluorocitrate-degrading pseudomonad on particles of guinea-pig kidney in an attempt to stop the fluorocitrate forming from fluoroacetate. We had no success.

Goldman: Several explanations are possible for this observation. First of all the fluorocitrate-degrading organism is not very active. In our experiments (Kirk and Goldman 1970) there seemed to be degradation of approximately 1 millimole of fluorocitrate in 24 hours by about a gramme of cells. Thus a small amount of extract might not have been capable of eliminating the fluorocitrate produced under the conditions of your experiment. Then too we only used the intact bacteria to degrade fluorocitrate; difficulty in making active extracts or dilution of necessary co-factors might be responsible for your observations. Another problem might be the effect of other components in your reaction mixture. Then one must also consider the relative affinity of fluorocitrate for the two reactions. The inhibition of the guinea-pig kidney particles might occur at a concentration of fluorocitrate which is much lower than the Michaelis constant for the degradative reaction.

Cain: Professor E. F. Gale has said that 'it is probably not unscientific to suggest that somewhere or other some organism exists which can, under suitable conditions, oxidize any substance which is theoretically capable of being oxidized' (Gale 1952). This concept of 'microbial infallibility', as Alexander (1965) has called it, has recently been challenged by the obvious recalcitrance of many synthetic organic compounds in the environment. The synthetic plastics, the chlorinated hydrocarbon insecticides and many herbicides persist for very long periods under conditions where organic compounds of biosynthetic origin rapidly disappear. It is recognized that in many of our agricultural and industrial processes, we depend upon the ultimate disappearance of noxious, and frequently toxic, material by biological degradation. Synthetic detergents from both industrial and domestic cleaning processes, cyanides from electroplating plants and organo-nitro compounds from the manufacture of explosives, as well as insecticide formulations containing fluoroacetate, are known to undergo such breakdown by microorganisms.

The fact remains, however, that these microorganisms are *not* ubiquitous. We have at Newcastle attempted several times by the usual enrichment procedures to isolate fluoroacetate-decomposing species from the river Tyne, the lower reaches of which must be among the most grossly polluted stretches of fresh water in Europe and from which one might reasonably expect to isolate bacteria capable of anything! The total microflora count is extremely high, but we have never obtained a fluoroacetate-decomposer. Further, Mr R. J. Hall and I have recently reported the presence of significant quantities of fluoroacetate in soil collected from around the roots of *Dichapetalum toxicarium* (Hall and Cain 1972). If it is assumed that this fluoroacetate is derived from the ω-fluoro fatty acids known to occur in the leaves and fruits of this perennial genus (Peters *et al*. 1960), then it might reasonably be supposed that a suitably adapted microflora of fluoroacetate-degraders would have developed over several seasons and would result in the immediate breakdown of any fluoroacetate reaching the soil from the plant or its remains. The fact that fluoroacetate *is* present suggests that such a microflora must be absent or ineffective.

This particular study has pointed out two other features that are perhaps worth emphasizing: (*i*) fluoroacetate can only rarely be removed from soils by acidification and extraction with diethyl ether. The anion seems to bind tightly to soil colloids and we have found that good recoveries of fluoroacetate can only be obtained after electrodialysis (Hall 1970). (*ii*) Identification of an organofluorine compound has frequently been based, at least in part, on the presence of the strong absorption band in the infrared region of the spectrum (around 1000–1100 cm^{-1}) due to the C—F bond. We have observed such a band in the infrared spectra of soils containing organofluorine compounds, and also in soils containing kaolinite clays, and no trace of fluorine at all. Indeed the spectrum of china clay itself is remarkably similar to that of a soil containing fluoroacetate in small quantities (Hall and Cain 1972). Clearly, then, chemical identification of the C—F material must be used to substantiate indications by physical methods.

One of the most promising uses to which fluorine can be, and has been, put is as a probe in delineating metabolic pathways in microorganisms. Although there are only a few species which can rapidly cleave the C—F bond (Kelly 1965; Goldman 1965), many soil bacteria, particularly the ubiquitous pseudomonads and actinomycetes, can oxidize the fluorinated analogues of substrates upon which they have been grown, and keep the C—F bond intact right through the metabolic sequence until the derivatives enter the terminal oxidation sequences —at which stage they accumulate and can be identified (Behrman and Stanier 1957; Smith, Tranter and Cain 1968; Goldman, Milne and Pignataro 1967).

In some examples from our own studies, for instance, we found that certain actinomycetes grown on *p*-nitrobenzoate would oxidize both this substrate and the 2- and 3-fluoro-derivatives via the corresponding fluoro-3,4-dihydroxy-benzoates. Only 2-fluoronitrobenzoate gave rise to fluoroacetate as an end-product, thus indicating the type of ring cleavage and subsequent lactonization step that this microorganism adopted (Cain, Tranter and Darrah 1968).

Cain and Farr (1968) believed that the ability of *Pseudomonas aeruginosa* to oxidize fluorobenzenesulphonate after growth on benzenesulphonate was due to the hydroxylation of the ring at the 1- and 2-positions, which presumably resulted in the formation of the corresponding diol, 4-fluorocatechol. The corresponding 2-fluorobenzenesulphonate was also extensively oxidized (Cain, unpublished results) because the 2-position can be avoided by hydroxylation at C-1 and C-6 to form the corresponding 3-fluorocatechol.

A very similar picture is revealed in some recent work Mr G. K. Watson has been doing in my laboratory. He has found that a bacterium that grew with pyridine as the sole carbon and nitrogen source could not oxidize 2-fluoropyridine, which implicated the 2-position as the site of an early metabolic step. He has subsequently shown that this organism cleaves the ring of pyridine to give succinic semi-aldehyde and formamide, probably from positions C-3 to C-6 and N-1 and C-2 of the ring, respectively (Watson and Cain 1971). Clearly, the presence of fluorine at C-2 hinders ring cleavage and the release of the (fluoro)-formamide. No fluoride was produced above that formed in the absence of the cell suspension.

Saunders: The Japanese claim that the enzyme, fluorohydrolase, has an optimum activity at pH 9 and does not affect β-fluoropropionate or fluoro-acetamide (Horiuchi 1962; Tonomura *et al.* 1965). This is surprising because β-fluoropropionate is unstable at this pH, in my experience.

Goldman: We also saw that fluoride is liberated slowly at pH 9 from β-fluoro-propionate. When a substrate or potential substrate is to be tested with an enzyme in conditions where the compound is unstable, it is customary to com-pare the rate in the presence of enzyme to the rate in a similar incubation mix-ture containing either no enzyme or heat-inactivated enzyme. Under such circumstances one is testing for an enzyme-catalysed enhancement of the spontaneous rate. The spontaneous rate of fluoride release from β-fluoropro-pionate is not so great as to make this type of experiment unfeasible.

Fowden: Dr Goldman, I suspect you performed your enrichment culture ex-periments at pH values above 7.0 At lower pH, the growth of fungi would be increasingly enhanced. Have you any evidence to suggest that fungi capable of degrading fluoroacetate can be obtained in this way?

Goldman: No, we have not done this. We have concentrated on isolating bacteria, feeling that they would be better and more convenient sources of enzymes.

Fowden: Certainly, bacteria would grow more uniformly and therefore would be easier to handle. However, it is often not difficult to obtain active enzyme preparations from fungal mycelia.

Cain: This can certainly be done; we now have fungi that degrade detergents and related compounds quite rapidly at pH values below 7.0. These fungi, some of them originally isolated as hydrocarbon-utilizers from dirty aviation fuel, grow upon compounds like benzenesulphonate and alkylbenzenesulphonate and will also oxidize the 2-fluoro derivative of the former at pH 5.5 to 6.0 to form fluoroacetate.

Wettstein: In connection with stereospecific enzymic transformations it is appropriate to cite some practical and even industrially important examples from the steroid series. In the production of steroid hormones by total synthesis we have the problem of finally obtaining the correct enantiomer rather than the racemate. We solved this problem (Vischer, Schmidlin and Wettstein 1956) initially by introducing one of the desired substituents by microbial enzymes which attacked, in this case, only the 'natural' D-enantiomer contained in the racemate, leaving the 'unnatural' L-enantiomer untouched. With this method the natural D-aldosterone has been obtained for the first time. Similar transformations have also been observed with fluorinated steroids. More appropriate is the recent method of Gibian and co-workers (Gibian *et al.* 1966; Kosmol *et al.* 1967; Rufer, Schröder and Gibian 1967; Rufer *et al.* 1967) and Bellet, Nominé and Mathieu (1966) who introduced the first asymmetric centre into symmetrical, optically inactive precursors by means of microbial enzymes, thus obtaining 'natural' enantiomer.

Armstrong: I was interested, but disappointed, Dr Goldman, in your finding that the enzyme you have studied which hydrolyses fluoroacetate to fluoride ion and glycollic acid is not capable of reversing this reaction. An attractive hypothesis is that fluoroacetate is produced, in those plants that contain it, from fluoride ion and glycollic acid. Your experiments were, of course, carried out in solutions. Is it possible that the equilibrium of the hydrolytic reaction is so far to the right, in favour of hydrolysis, that a small reversibility of the reaction cannot be observed? Further, is it possible that fluoroacetate, when synthesized in plants, is sequestered or compartmentalized in such a way as to remove it from solution, thus allowing fluoroacetate to accumulate due to the feeble reversibility of the reaction you have studied? On this point, is fluoroacetate of plant tissues directly soluble in water?

Peters: A lot is water-soluble. L. Murray thought that in *Acacia georginae* fluoroacetate was combined with trigonelline. Dr Ward's work, as already mentioned (p. 120), shows that the fluoroacetate can combine with a sulphur compound with elimination of inorganic fluoride. This is different from a combined fluoroacetate.

Shorthouse: The *Acacia georginae* is believed, by farmers anyway, to be less toxic after rain. The trees also give off pungent sulphurous odours detectable over large distances. They may well be emitting organosulphur fluoride compounds. We have noticed the distinctive odour from the roots of our plants grown in pots.

Armstrong: Is the principle of the proposition I have just asked about plausible?

Kun: Incorporation of fluorine into organic molecules could occur by the activation of fluorine in the form of a high energy intermediate. The reversal of of ATPase by phosphorylation of ADP provides such an example.

Harper: At the Prairie Regional Laboratory in Canada we isolated, by the enrichment culture technique, a bacterium capable of using *p*-fluorophenylacetic acid as the sole carbon source (Harper and Blakley 1971*a*, *b*). We showed that this compound was degraded through 3-fluoro-3-hexenedioic acid. Lactonization of this acid can take two courses, depending on whether lactonization occurs through the hydrogen-bearing carbon or the fluorine-bearing carbon of the double bond. If the 4-fluorolactone is formed, delactonization will result in the spontaneous elimination of hydrogen fluoride and the formation of β-ketoadipic acid, since 3-hydroxy-3-fluoroadipic acid is highly unstable.

Alternatively if the 3-fluorolactone is formed, delactonization will give 3-hydroxy-4-fluoroadipic acid, which can be oxidized to the 3-keto-4-fluoro compound. This is subsequently cleaved in a similar manner to β-ketoadipic acid, to give acetate and monofluorosuccinic acid. The latter compound we isolated from the medium as the D($+$)-form and found that it was metabolized via fluorofumarate. The action of fumarase on fluorofumarate results in the formation of 2-fluoromalate as has been shown by Teipel, Hass and Hill (1968) and Clarke, Nicklas and Palumbo (1968). This inherently unstable compound undergoes spontaneous non-enzymic elimination of hydrogen fluoride to give oxaloacetate. Thus it appears that in this case organic fluorine is eliminated by a mechanism which requires no special enzymic complement or specificity. This also appears to be the case in the degradation of *p*-fluorobenzoic acid (Harper and Blakley 1971*c*). The failure of this bacterium to grow on *p*-chlorophenylacetic acid may well be related to the fact that, whereas D-fluorosuccinate is a substrate for succinic dehydrogenase, D-chlorosuccinate is an inhibitor (Gawron *et al.* 1962) and thus may effectively block the tricarboxylic acid cycle.

Heidelberger: What is the first stage in the degradation of *p*-fluorophenylacetic acid?

Harper: We did not discover the mechanism of ring cleavage although we assume from the formation of 3-fluoro-3-hexenedioic acid that cleavage takes place between C-1 and C-2 of the aromatic ring. Both 2- and 3-hydroxy-4-fluorophenylacetic acid were isolated from the culture medium but neither appeared to be metabolized by the bacterium. We theorize that they were probably formed by a constitutive hydroxylase playing no part in breakdown. We concluded that partial reduction of the aromatic ring followed by oxidative cleavage between C-1 and C-2 must occur. This would result in the formation of a β-keto acid which would undergo cleavage to acetate and 3-fluoro-3-hexenedioic acid. This is strictly speculation however.

Kun: Could defluorination occur by a peroxide radical mechanism?

Harper: It is possible—but a reduction of the ring must also occur. Another difficult fact to account for is the *trans*-configuration of the double bond in 3-fluoro-3-hexenedioic acid.

Barnett: Dr Goldman, have you considered that the mechanism of fluoride displacement, for which you show an attack by sulphur giving an intermediate enzyme $\cdot S\cdot CH_2\cdot COOH$, might involve a sulphonium ion of the enzyme type $\cdot \overset{+}{S}(R)\cdot CH_2\cdot COOH$? I would think that the thiol ether structure you showed would be too stable for further reaction, since it is the type of derivative one tries to make when one wishes to inactivate an enzyme.

Goldman: A sulphonium ion would, as you suggest, facilitate the hydroxide ion attack that we proposed for the second stage of the defluorination reaction and it seems likely that the enzyme would have to make an adjustment, perhaps of the kind you propose, to break the thiol ether bond. This of course gets us into an area of paper chemistry and speculation that is not susceptible to experimental testing. We proposed this model simply to tie together all the available evidence and perhaps to offer a stimulus for further experiments.

Barnett: I would have thought that the thiol ether intermediate would have been kinetically very stable. The intermediate you propose ($C-S-CH_2COOH$) seems to me to be more stable than either starting material or product. It would be very stable to attack by hydroxyl ion. A sulphonium ion intermediate would be less stable, but direct displacement of the fluoride ion by the hydroxyl ion may be a preferable mechanism. This would lead to inversion.

For the general halidohydrolase, which hydrolyses α-chloropropionic acid as well as the halogenoacetates, you were able to show that the reaction went with inversion. This was not possible with haloacetate halidohydrolase because it would only accept fluoroacetate as substrate. Have you considered using specifically deuterated *2R* or *2S*-fluoroacetic acid, which you could relate to the *2R* and *2S* glycollates of Rose (1958) (Fig. 1)?

Goldman: It would probably be easier to prepare specifically tritiated bromo-

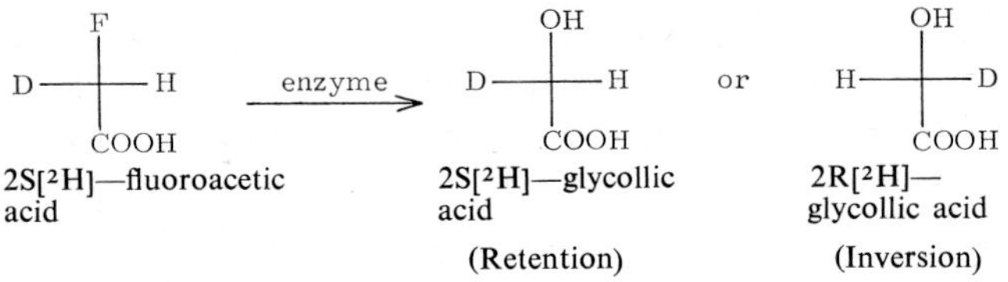

F   OH   OH

D————H   enzyme   D————H   or   H————D

COOH   COOH   COOH

2S[^{2}H]—fluoroacetic acid   2S[^{2}H]—glycollic acid   2R[^{2}H]— glycollic acid

(Retention)   (Inversion)

FIG. 1 (Barnett).

acetate or chloroacetate, which are also substrates for the enzyme. We had thought of doing this by preparing specifically tritiated glycollate by the stereospecific reduction of glyoxylate. Tosylation of glycollate followed by halide displacement of the tosyl group would lead to a specifically tritiated substrate for the enzyme. However, Dr Paul Meloche has suggested that an easier route to specifically tritiated bromoacetate might be through the formation of specifically tritiated bromopyruvate, which he has been able to form using the enzyme 2-keto-3-deoxy-6-phosphogluconic aldolase. We hope soon to be able to do the experiment using the substrate which he can prepare.

Bergmann: Do you have any phosphate in your enzymic system?

Goldman: No.

Bergmann: I ask this because it would be likely, in view of your results, that glycollic acid is the precursor of fluoroacetic acid in plants. Our experiments with tissue cultures have failed to show this, but we believe that the phosphate ester of glycollic acid may be a likely precursor.

Goldman: The phosphate ester of glycollic acid that is present in plants does indeed seem to be attractive as a precursor of fluoroacetate. It meets the requirements in terms of energy that Professor Kun has indicated are important. Of course this mechanism would require the phosphate ester to undergo cleavage of the oxygen-carbon bond instead of the oxygen-phosphorus bond, as is the case in ester hydrolysis.

Cain: Glycollate and fluoride were found as the products of fluoroacetate hydrolysis by your purified bacterial enzyme, Dr Goldman, which you have reminded us could not utilize fluoroacetamide. The soil bacterium originally isolated by Kelly (1965) released fluoride from both fluoroacetate and fluoroacetamide. Have you yourself, or has anyone else, studied the enzymology of Kelly's organism to see whether the enzymic mechanism or its specificity is significantly different from yours?

Goldman: No.

Taylor: I am not sure that I understand your use of the terms 'natural' and 'unnatural' substrates as far as microorganisms are concerned.

Goldman: I agree there is a difficulty here. Fluoroacetate is an example of a natural product but only since Marais (1943) discovered it in a plant. Before that, it would have been considered as a synthetic 'unnatural' compound.

Taylor: I believe the problem is largely one of semantics. 3-Deoxy-3-fluoro-glucose is now part of the environment and metabolized by certain bacteria into new fluorinated products. Perhaps all chemical transformations should be considered natural. After all, the molecules were present and undergoing transformation before the appearance of bacteria or man.

References

ALEXANDER, M. (1965) *Adv. Appl. Microbiol.* **7**, 35–80.

BEHRMAN, E. J. and R. Y. STANIER (1957) *J. Biol. Chem.* **228**, 947–953.

BELLET, P., G. NOMINÉ and J. MATHIEU (1966) *C.R. Hebd. Séances Acad. Sci., Paris, Sér. C.* **263**, 88–89.

CAIN, R. B. and D. R. FARR (1968) *Biochem. J.* **106**, 859–877.

CAIN, R. B., E. K. TRANTER and J. A. DARRAH (1968) *Biochem. J.* **106**, 211–217.

CLARKE, D. D., W. J. NICKLAS and J. PALUMBO (1968) *Arch. Biochem. Biophys.* **123**, 205.

GALE, E. F. (1952) *The Chemical Activities of Bacteria*, p. 5. New York: Academic Press.

GAWRON, O., A. J. GLAID, T. P. FONDY and M. M. BECHTOLD (1962) *J. Am. Chem. Soc.* **84**, 3877.

GIBIAN, H., K. KIESLICH, H. J. KOCH, H. KOSMOL, C. RUFER, E. SCHRÖDER and R. VÖSSING (1966) *Tetrahedron Lett.* 2321–2330.

GOLDMAN, P. (1965) *J. Biol. Chem.* **240**, 3434–3438.

GOLDMAN, P., G. W. A. MILNE and M. T. PIGNATARO (1967) *Arch. Biochem Biophys.* **118**, 178–184.

HALL, R. J. (1970) M.Sc. Thesis. University of Newcastle-upon-Tyne.

HALL, R. J. and R. B. CAIN (1972) *New Phytologist* in press.

HARPER, D. B. and E. R. BLAKLEY (1971*a*) *Can. J. Microbiol.* **17**, 635–644.

HARPER, D. B. and E. R. BLAKLEY (1971*b*) *Can. J. Microbiol.* **17**, 645–650.

HARPER, D. B. and E. R. BLAKLEY (1971*c*) *Can. J. Microbiol.* **17**, 1015–1023.

HORIUCHI, N. (1962) *J. Jap. Biochem. Soc.* **34**, 92–98.

KELLY, M. (1965) *Nature (Lond.)* **208**, 809–810.

KIRK, K. and P. GOLDMAN (1970) *Biochem. J.* **117**, 409–410.

KOSMOL, H., K. KIESLICH, R. VÖSSING, H-J. KOCH, K. PETZOLDT and H. GIBIAN (1967) *Justus Liebigs Ann. Chem.* **701**, 198–205.

MARAIS, J. S. C. (1943) *Onderstepoort J. Vet. Sci. Anim. Ind.* **18**, 203–206.

PETERS, R. A., R. J. HALL, P. F. V. WARD and N. SHEPPARD (1960) *Biochem. J.* **77**, 17–23.

ROSE, I. A. (1958) *J. Am. Chem. Soc.* **80**, 5835.

RUFER, C., H. KOSMOL, E. SCHRÖDER, K. KIESLICH and H. GIBIAN (1967) *Justus Liebigs Ann. Chem.* **702**, 141–148.

RUFER, C., E. SCHRÖDER and H. GIBIAN (1967) *Justus Liebigs Ann. Chem.* **701**, 206–216.

SMITH, A., E. K. TRANTER and R. B. CAIN (1968) *Biochem. J.* **106**, 203–210.

TEIPEL, J. W., G. M. HASS and R. L. HILL (1968) *J. Biol. Chem.* **243**, 5684.

TONOMURA, K., F. FUTAI, O. TANABE and T. YAMAOKA (1965) *Agric. & Biol. Chem. (Jap.)* **29**, 124–128.

VISCHER, E., J. SCHMIDLIN and A. WETTSTEIN (1956) *Experientia* **12**, 50–52.

WATSON, G. K. and R. B. CAIN (1971) *Biochem. J.* **127**, 44P.

Clinical and pathological effects of fluoride toxicity in animals

JAMES L. SHUPE

Utah State University, Logan, Utah

Fluorine rarely occurs in its free state in nature, but combines chemically to form fluorides. Fluorides are widely distributed in nature and are universally present in various amounts in soils, water, the atmosphere, vegetation, and animal tissues. In nature fluorine occurs most commonly as fluorite (CaF_2) and fluorapatite ($CaF_2 \cdot 3Ca_3(PO_4)_2$). Compounds with fluorine bound to carbon (organic fluorides) do not present a significant environmental health hazard to animals.

Animals normally ingest some fluoride with no adverse effects. Although such amounts of fluoride may have beneficial effects (Bernstein and Cohen 1967; Shaw 1954; WHO 1970; Hodge and Smith 1965; Smith 1962; Taves 1970; Taves *et al.* 1968), adverse or toxic effects may occur when excessive amounts (Table 1) are ingested (Agate *et al.* 1949; Allcroft, Burns and Hebert 1965; Alther 1961; Blakemore, Bosworth and Green 1948; Burns and Allcroft 1964; Hupka and Luy 1929; Mitchell and Edman 1952; Phillips *et al.* 1955; Phillips, Hart and Bohstedt 1934; Roholm 1937; Shupe *et al.* 1963; Slagsvold 1934; Velu 1932).

Various sources may contribute to the total fluoride intake of animals. The most frequently encountered sources of excessive fluoride in the diet are: (1) forages subjected to air-borne contamination in areas near industrial operations that emit fluorides; (2) water with a naturally high fluoride content; (3) feed supplements and mineral mixtures with excessive fluoride content; (4) forages contaminated by soils with a high fluoride content; or (5) any combination of the above. Calcium fluoride and compounds of similar solubility are less toxic than sodium fluoride and hay atmospherically contaminated by fluoride effluents from an industrial operation (Fig. 1) (Shupe *et al.* 1962).

With the increase and expansion of certain industries into agricultural areas, especially during and shortly after World War II, fluoride toxicosis in animals

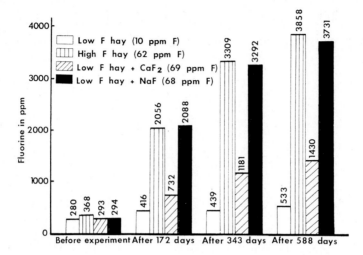

FIG. 1. Average fluoride content of biopsied ribs (dry, fat-free basis) from dairy heifers receiving different sources of fluoride*

TABLE 1

Tolerance* of animals for fluoride (concentration in dry matter in ration)

Species	Breeding or lactating animals (p.p.m. fluorine)	Finishing animals to be sold for slaughter with average feeding period (p.p.m. fluorine)
Dairy and beef heifers	30	100
Dairy cows	30	100
Beef cows	40	100
Steers	N.A.	100
Sheep	50	160
Horses	60	N.A.
Swine	70	N.D.
Turkeys	N.D.	100
Chickens	N.D.	150

* Tolerances based on sodium fluoride or other fluorides of similar toxicity.
N.A. = Not applicable. N.D. = Not determined. (From Shupe and Olson 1971.)

Average	Age (years)	Normal conditions	No adverse effects	Chronic fluoride toxicosis			Acute fluoride toxicosis
				Threshold	Moderate	Severe	
F in moisture-free diet (p.p.m.)	2	up to 15	15–30	30–40	40–60	60–109	over 250
	4	up to 15	15–30	30–40	40–60	60–109	over 250
	6	up to 15	15–30	30–40	40–60	60–109	over 250
Teeth classification* (incisors)	2	0–1	0–2	2–3	3–4	4–5	
	4	0–1	0–2	2–3	3–4	4–5	
	6	0–1	0–2	2–3	3–4	4–5	
Teeth classification* (molars)	2	0–1	0–1	0–1	0–1	0–3	
	4	0–1	0–1	0–1	1–2	1–4	
	6	0–1	0–1	1–2	1–3	1–5	
F in bone (p.p.m.)	2	401–714	714–1605	1605–2130	2130–3027	3027–4206	
	4	706–1138	1138–2379	2379–3138	3138–4504	4504–6620	
	6	653–1221	1221–2794	2794–3788	3788–5622	5622–8676	
F in urine (p.p.m.)	2	2.27–3.78	3.78–8.04	8.04–10.54	10.54–14.71	14.71–19.86	
	4	3.54–5.3	5.3–10.32	10.32–13.31	13.31–18.49	18.49–25.63	
	6	3.51–6.03	6.03–11.29	11.29–14.78	14.78–20.96	20.96–30.09	
F in milk (p.p.m.)	2	up to 0.12	up to 0.12	0.08–0.15	0.15–0.25	0.15 and above	
	4	up to 0.12	up to 0.12	0.08–0.15	0.15–0.25	0.15 and above	
	6	up to 0.12	up to 0.12	0.08–0.15	0.15–0.25	0.15 and above	
F in blood (p.p.m.)	2	up to 0.30	up to 0.30	0.15–0.40	0.30–0.50	0.50 and above	
	4	up to 0.30	up to 0.30	0.15–0.40	0.30–0.50	0.50 and above	
	6	up to 0.30	up to 0.30	0.15–0.40	0.30–0.50	0.50 and above	
F in soft tissue (p.p.m.)	2	up to 1.20	up to 1.20	up to 1.20	up to 1.20	up to 1.20	
	4	up to 1.20	up to 1.20	up to 1.20	up to 1.20	up to 1.20	
	6	up to 1.20	up to 1.20	up to 1.20	up to 1.20	up to 1.20	
Periosteal hyperostosis**	2	0	0–1	0–1	0–2	0–3	
	4	0	0–1	0–1	0–3	0–4	
	6	0	0–1	0–2	0–4	0–5	
Secondary changes may occur***	all	absent	absent	occasionally noticed	present	present	

Acute fluoride toxicosis — Characterized by: excitement, high fluorine content of blood and urine, stiffness, anorexia, reduced milk production, excessive salivation, nausea, vomiting, incontinence of urine and faeces, clonic convulsions, necrosis of mucosa of digestive tract, weakness, severe depression in some cases, and cardiac failure.

These data are based on controlled experiments, but can be correlated with numerous field cases that have been extensively studied and evaluated.

* Average classification of full mouth: Incisor teeth classification: (Degree of fluoride effects)
0-normal 1-questionable 2-slight 3-moderate 4-marked 5-excessive.
Molar teeth classification: (degree of wear)
0-normal 1-questionable 2-slight 3-moderate 4-marked 5-excessive.

** Periosteal hyperostosis: 0-normal 1-questionable 2-slight 3-moderate 4-marked 5-excessive.

*** Stiffness & lameness—loss of body weight—reduced feed intake—rough hair coat—unpliable skin—reduced milk production.

† Revised from Shupe and co-workers 1963.

has become an important toxicological problem in many areas of the United States and some other countries. Incorporation of corrective processing procedures and preventive measures into the industrial processes have eliminated or reduced the problems in some areas (Shupe 1970a). Extensive epidemiological and experimental studies have established that the biological responses of animals to fluorides are related to dosage and other factors that influence the animals' physiological and anatomical responses (Allcroft, Burns and Hebert 1965; Burns and Allcroft 1964; Greenwood *et al.* 1964; Hobbs and Merriman 1962; Phillips, Suttie and Zebrowski 1963; Roholm 1937; Schmidt, Newell and Rand 1954; Shupe *et al.* 1963; Suttie and Phillips 1959).

FLUORIDE TOXICITY

Excessive ingestion of fluoride (Tables 1 and 2) can induce either an acute toxicity or a debilitating chronic disease (Shupe and Alther 1966). The latter condition has been called chronic fluoride toxicity, fluoride toxicosis, or fluorosis. Although specific conditions such as osteofluorosis and dental fluorosis are definable, the general term fluorosis does not lend itself to a clear-cut definition and the term fluoride toxicity will be used in this presentation to describe the conditions that result when excess fluorides are ingested. An understanding of the clinical manifestations and a thorough knowledge of the pathogenesis and lesions of fluoride toxicity are essential for correct diagnosis and accurate evaluation of the disease (Shupe 1966b).

ACUTE FLUORIDE TOXICITY

Acute fluoride toxicity has not been studied as extensively nor elucidated as well as chronic fluoride toxicity because it is relatively rare (Allcroft and Jones 1969; Allcroft *et al.* 1969; WHO 1970; Phillips *et al.* 1955; Shupe and Alther 1966). There is need for additional evaluation of acute and sub-acute toxic effects (WHO 1970; Shupe 1970a). Acute fluoride toxicity has resulted most frequently from accidental ingestion of large amounts of fluoride compounds (Krug 1927; Leone, Geever and Moran 1956; Shupe and Alther 1966) such as sodium fluorosilicate and sodium fluoride. Accidentally and experimentally induced acute responses have been reviewed by Cass (1961).

Depending on a number of factors, different types of toxicological responses may occur. The rapidity with which the symptoms may appear depends on the amount of fluoride ingested. The following symptoms and changes are usually observed: high fluoride content of blood and urine, restlessness, stiffness, anorexia, reduced milk production, excessive salivation, nausea, vomiting, incontinence of urine and faeces, clonic convulsions, necrosis of the mucosa of the digestive tract, weakness, severe depression and cardiac failure. In one of the best documented cases of acute toxicity, 18 dairy cows died within 12 to 14 hours after ingesting sodium fluorosilicate (Na_2SiF_6) which had been erroneously added to the diet in place of a mineral mixture. Convulsions, exaggerated chewing motions, and hyperaemia were observed (Krug 1927).

A distinction between acute and chronic toxicity is not always possible. Severe but non-lethal effects have been observed in animals ingesting diets containing high concentrations of fluoride. Extreme inanition occurred 18 days after feeding trials began in pregnant beef heifers fed 1200, 900 and 600 p.p.m. fluoride (NaF). The symptoms observed in these animals were not typical of those seen in either acute or chronic fluoride toxicity, however (Hobbs and Merriman 1962).

CHRONIC FLUORIDE TOXICITY

Chronic fluoride toxicity is the type of fluoride poisoning most often observed in livestock. The development and onset of this condition is insidious. Some of the symptoms of the disease may be confused with other chronic debilitating diseases such as osteoarthritis and some trace element poisonings or deficiencies (Shupe 1961). It is difficult to define a precise point at which fluoride ingestion becomes detrimental to the animal. This point may vary in individual cases and can be influenced by the following factors: (1) amount of fluoride ingested; (2) duration of fluoride ingestion; (3) fluctuations in amounts of fluoride ingested; (4) solubility of fluoride ingested; (5) species; (6) age at time of ingestion; (7) nutritional status; (8) stress factors; and (9) individual biological response.

The varying interval between ingestion of elevated concentrations of fluorides and manifestation of the clinical symptoms of chronic fluoride toxicity may in some cases complicate the clinical syndrome.

No single criterion should be relied upon in diagnosing and evaluating fluoride toxicity. All clinical observations, necropsy findings and chemical evidence must be carefully evaluated and interrelated before a definite diagnosis and evaluation of fluoride toxicity is made. The following symptoms, lesions and analytical determinations are of particular diagnostic importance: (1) degree of dental fluorosis; (2) degree of osteofluorosis; (3) intermittent lameness; and (4) the amount of fluoride in the bone, urine and components of the diet.

DENTAL FLUOROSIS

Developing teeth are extremely sensitive to ingestion of excessive fluoride. Fluorotic lesions in permanent dentition are one of the most obvious symptoms of excessive fluoride ingestion during the period of tooth formation (Garlick 1955; Shupe *et al.* 1963).

Histological changes in teeth induced by fluoride have been described as follows: the ameloblasts are prematurely reduced in size. The epithelial papillae are abnormal and the entire enamel epithelium has a series of interruptions in its smooth surface in the form of foldings or dips that extend into the enamel. This abnormal enamel matrix fails to achieve normal calcification, and the defect is later present in the mature teeth (Schour and Smith 1934).

Gross fluorotic lesions of the incisor enamel are generally described as chalkiness (dull white chalk-like appearance), mottling (white chalk-like striations or patches in the enamel), hypoplasia (defective enamel), and hypo-calcification (defective calcification). Teeth affected to a moderate, marked, or excessive degree (Table 2) are subject to more rapid attrition and in some cases to an erosion of the enamel from the dentin (Shupe, Miner and Greenwood 1964).

The period of crown formation of the permanent teeth of cattle has been elucidated (Brown *et al.* 1960). Because the enamel matrix formation precedes the calcification process, the period during which the developing teeth in cattle are sensitive to excessive fluoride is from 6 months to approximately 3 years of age. Thus cattle not exposed to fluoride until they are more than 3 years old will not develop typical fluoride-induced dental lesions when excessive fluoride is ingested (Brown *et al.* 1960; Garlick 1955; Shupe 1966a; Shupe 1970b).

For the purpose of classifying the degrees of dental fluorosis the following categories and standards have been developed:

0. Normal: smooth, translucent, glossy white appearance of enamel; tooth normal shovel shape.

1. Questionable effect: slight deviation from normal, unable to determine exact cause; may have enamel flecks but no mottling.

2. Slight effect: slight mottling of enamel; best observed as horizontal striations with transmitted light; may have slight staining but no increase in normal rate of attrition.

3. Moderate effect: definite mottling; large areas of chalky enamel or generalized mottling of entire tooth; teeth may show a slightly increased rate of attrition and may be stained.

4. Marked effect: definite mottling, hypoplasia, and hypocalcification; may have pitting of enamel; with use, tooth will have increased rate of attrition;

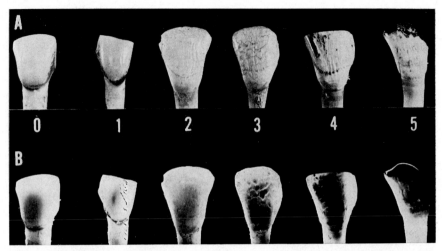

FIG. 2. Classification of representative incisor teeth from cattle is from 0 to 5 reading left to right in *A* and *B*. *A* is with direct lighting. *B* is with transmitted light (from Shupe *et al.* 1963).

enamel may be stained or discoloured.

5. *Excessive effect*: definite mottling, hypoplasia and hypocalcification; with use, tooth will have excessive increase in rate of attrition and may have erosion or pitting of enamel. Tooth may be stained or discoloured (Fig. 2).

Dental fluorosis in animals is usually diagnosed by examining the incisor teeth. Cheek (premolar and molar) teeth are more difficult to examine in the live animal because of problems with proper restraint of the animal being

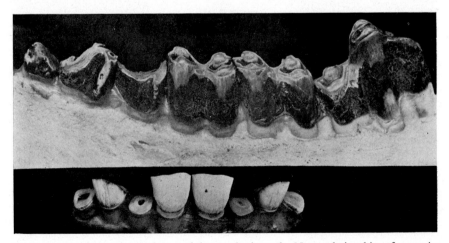

FIG. 3. Permanent bovine incisor and lower cheek teeth. Note relationship of excessive abrasion of certain permanent incisor and cheek teeth.

examined, poor illumination of the teeth, the presence of the tongue and un-swallowed food, and discoloration caused by vegetative matter. When one is diagnosing and evaluating chronic fluoride toxicity in animals, the cheek teeth should be examined and the findings correlated with changes in the incisors and other fluoride-induced symptoms and lesions. The criteria used in diag-nosing and evaluating incisor lesions are not the same as those used in diag-nosing and evaluating cheek teeth for dental fluorosis (Shupe 1966*b*; Shupe and Olson 1971). Fluorosis of premolars and molars is estimated on the degree of selective abrasion and correlation with fluoride lesions in incisor teeth. A corre-lation exists between the degree of dental fluorosis of certain incisor teeth and the degree of abrasion of certain cheek teeth (Fig. 3). Abrasion of the premolars and molars appears to be somewhat delayed beyond that of incisor teeth. Abraded cheek teeth are protected by adjacent good teeth that are not adversely

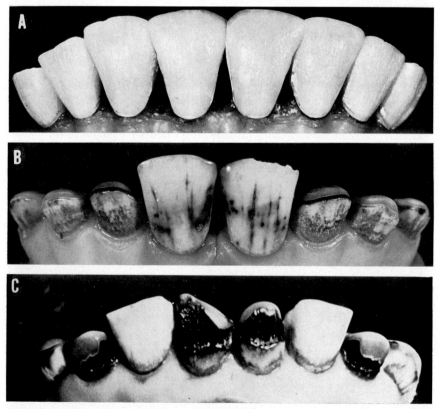

Fig. 4. Permanent bovine incisor teeth. *A*: Normal; *B*: Dental fluorosis due to continuous ingestion of excessive fluoride; *C*: Dental fluorosis due to intermittent ingestion of excessive fluoride.

affected and are not abraded. Abrasive feeds such as coarse, fibrous and tough roughage will increase the rate of dental abrasion (Shupe 1966*b*).

The degree of dental lesion induced by fluoride under controlled experimental conditions has been correlated with the amount of fluoride ingested, the age of the animal, duration of excessive fluoride ingestion, the severity of osteofluorosis and other factors influencing the biological response to fluorides (Shupe *et al.* 1963). The dental lesions, therefore, are useful in diagnosing chronic fluoride toxicity clinically (Fig. 4). To be most meaningful, however, the fluoride dental lesions should be evaluated with other tissue changes and symptoms induced by excessive ingestion of fluoride. In other words, dental lesions should not be used as the sole criterion when evaluating the degree of chronic fluoride toxicity (Shupe 1966*b*, 1970*b*).

OSTEOFLUOROSIS

The amount of fluoride stored in bone can increase over a period of time without evidence of any demonstrable changes in bone structure and function. If the amounts of fluoride ingested are sufficiently higher than normal for an appreciable length of time, structural bone changes will become evident (Johnson 1965; Shupe and Alther 1966; Shupe *et al.* 1963).

In livestock, the first clinically palpable bone lesions usually occur on the medial surface of the proximal third of the metatarsal bones and are bilateral. Subsequently, palpable bone lesions occur on the mandible, metacarpals and ribs. The severity of osteofluorotic lesions appears related in some degree to the stress and strain imposed on various bones and to the structure and function of the bones. This is exemplified by the difference in the osteofluorotic lesions of the ribs, mandible and metaphyseal areas of the metatarsal bones as compared to the less severe lesions and lower fluoride content of the diaphyseal areas of the metatarsal and metacarpal bones (Shupe *et al.* 1963). Experiments with dairy heifers and cows showed an increase in bone alkaline phosphatase in response to increased dietary fluoride (Miller and Shupe 1962; Shupe, Miner and Greenwood 1964).

Grossly, bones that are severely affected by fluoride appear chalky white with a roughened irregular periosteal surface and are larger in diameter and heavier than normal (Fig. 5).

The type of bone changes depends on the amount and duration of fluoride ingestion and one or more of the following conditions may occur: osteoporosis, osteosclerosis, hyperostosis, osteophytosis, or osteomalacia.

Characteristic histological changes are associated with the various degrees

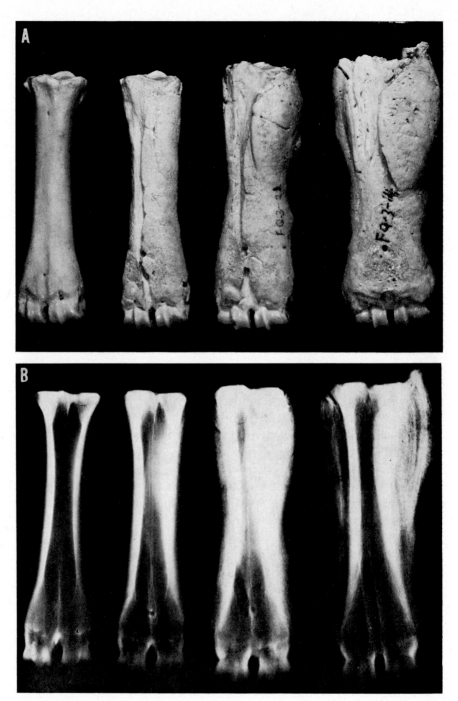

FIG. 5. Metatarsal bones from four cows of the same breed, size and age depicting various degrees of fluoride-induced periosteal hyperostosis. Left to right: normal to excessive osteo-fluorosis. Note that the articular surfaces appear normal. *A*: Gross appearance; *B*: Radiographic appearance (from Shupe 1963).

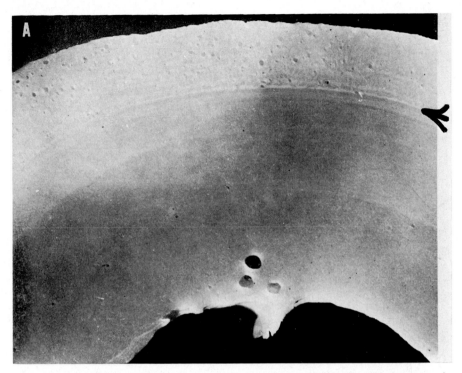

FIG. 6. Cross sections of bovine metatarsal bones depicting: (*A*) Osteosclerosis—cortex is thicker than normal with a zone of hyperostotic bone on periosteal surface. The animal had ingested slightly elevated concentrations of fluoride for a prolonged period of time (×6); (*B*) (p. 368) Osteoporosis—massive, lacy, periosteal new bone and marked porosis of the endosteal portion of the original cortex. The animal had ingested high concentrations of fluoride for a prolonged period of time (6×).

of osteofluorosis. These have been elucidated, described and illustrated (Johnson 1965; Shupe and Alther 1966).

The more consistent and major changes are abnormal bone formation on the premorbid periosteal surface (Fig. 6); thickened cortex due to periosteal hyperostosis; irregular bone pattern and mineralization; coarse, haphazardly arranged collagenous fibres; islands of calcified cartilage in the cortex; and excessive osteoid tissue. Also characteristic are: fibrillar bone; deposits of mineral salts on surfaces of trabeculae, increased osteoclastic reabsorption on endosteal surfaces and of the premorbid bone; fibrous marrow in some cases; hypervascularization adjacent to resorptive cavities; and irregular shapes and sizes of osteones with irregular and abnormal distribution of osteocytes in osteones. Some osteones have osteocytes clumped together or located at the

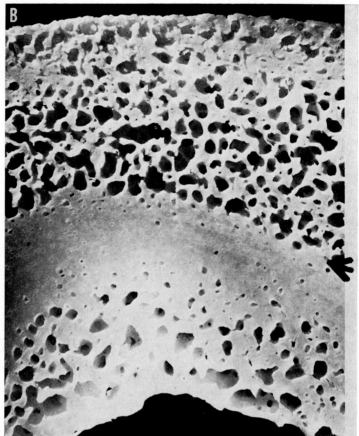

Fig.6*B*

periphery of the osteone with very few osteocytes in the remainder of the osteone (Fig. 7). Some osteocytes are enlarged and some lacunae are empty (Johnson 1965; Shupe and Alther 1966). Impairment and disturbance of normal blood supply to the bone occurs in cows with advanced marked osteofluorosis with excessive periosteal hyperostosis (Shupe 1966*b*).

A slight increase in the fluoride content of bone above the normal concentration (Table 2) can occur without detectable change in bone structure or function. A slight to moderate increase in the fluoride content causes only slight changes in the microscopic bone structure; there is no detectable alteration of function. With increasing amounts of bone fluoride, progressive structural changes occur that result in definite bone lesions (Johnson 1965; Shupe and Alther 1966).

Fɪɢ. 7. Microradiographs of bovine metatarsal bones. *A*: Normal (×11); *B* (p. 370): Osteo-fluorosis (×7.6); *C*: Normal—an enlargement of section *A* (×80); *D* (p. 371): Osteofluoro-sis—an enlargement of section *B* (×80); *E*: Osteofluorosis due to intermittent ingestion of excessive fluoride (×6). Note uneven distribution of large, irregular osteocytes and interstitial changes in *B*, *D*, and *E* (from Shupe *et al.* 1963).

The degree of progressive structural bone changes due to excessive fluoride (Tables 1 and 2) intake and their patterns of combinations are governed by the factors that influence the manifestations of fluoride toxicity. Some of the gross and histological changes induced by fluoride toxicity may resemble bone lesions and alterations that are associated with other bone diseases. Therefore, the lesions observed must be correlated carefully with other lesions and symptoms in making a definitive diagnosis of fluoride toxicity (Shupe 1970*b*; Shupe and Olson 1971).

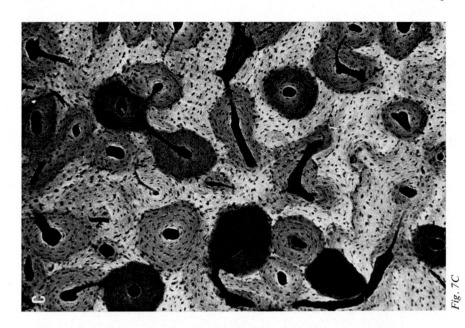

Fig. 7C

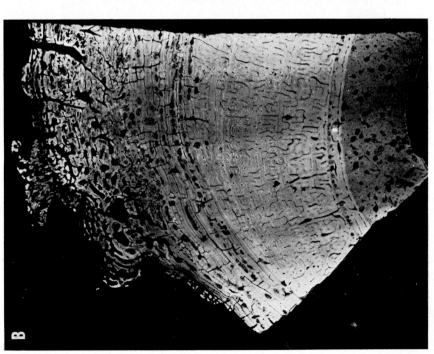

Fig. 7B

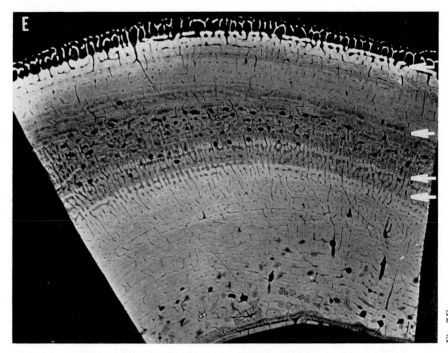

Fig. 7E

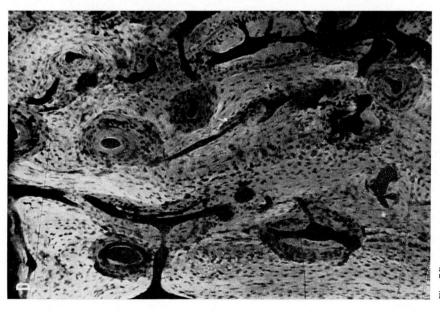

Fig. 7D

LAMENESS AND STIFFNESS

Stiffness and lameness are observed rather frequently in livestock. Without supporting evidences, however, they are inconclusive measures of fluoride toxicity.

The impaired movement observed in fluoride toxicity is more frequently a non-specific rather than a specific lameness and stiffness. In some severe cases, animals become progressively worse and may eventually refuse to stand and instead move around on their knees (Fig. 8). In most cases, however, periods of stiffness and lameness are intermittent.

FIG. 8. Seven-year-old Holstein cow showing effects of severe fluoride toxicity. Note posture, which is due to extreme lameness. This cow had severe osteofluorosis.

In advanced cases of fluoride toxicity where the animals have marked periosteal hyperostosis, spurring and bridging of the joints may occur. This condition eventually causes marked rigidity of the limbs and spine (Shupe, Miner and Greenwood 1964; Shupe 1966*b*).

The stiffness and lameness that are seen in more advanced cases of fluoride toxicity appear to be associated with the osteofluorotic lesions and the mineralization of periarticular structures and tendon insertions. The specific lesions responsible for the lameness and stiffness are not known and can deter the affected animals from grazing or standing at the feeder, and subsequently the reduced feed intake may adversely reduce their standards of performance (Stoddard *et al.* 1963).

It is diagnostically important to remember that the initial stages of the bone lesions associated with fluoride toxicity are not intra-articular. In contrast, lesions associated with osteoarthritis are intra-articular and secondarily may become peri-articular (Shupe 1961).

Crepitation, a symptom of osteoarthritis, is not detectable in uncomplicated cases of fluoride toxicity. Varying degrees of osteoarthritis have been diagnosed in some field cases that were severely affected by fluorides. However, such arthritic changes and the symptoms of acute inflammatory arthritis, such as elevated body temperature, hot swollen joints and suppressed ruminations, are not associated with the characteristic intermittent stiffness and lameness of advanced fluoride toxicity (Shupe 1961).

In England a distinct lameness associated with the fracture of the distal phalanx has been reported to be a characteristic lesion of fluoride toxicity (Burns and Allcroft 1964). This condition has not often been observed in other geographical areas or reported in experimental studies.

Radiology

Radiographically, sclerosis, porosis, malacia, hyperostosis, osteophytosis, or any combination of these conditions have been observed in animals that have ingested excessive amounts of fluoride over long periods of time (Leone *et al.* 1955; Shupe and Alther 1966; Shupe *et al.* 1963). The radiographic findings vary depending on the interaction of factors influencing fluoride toxicity.

FLUORIDE RETENTION IN TISSUES

Bone

The most definitive evidence that animals are ingesting increased amounts of fluoride in their diet comes from chemical analysis of the forage consumed and an increase in the fluoride content of urine and body tissues. Excretion and retention are dependent on a number of factors such as: the fluoride content of the diet, duration of ingestion, solubility of the fluoride, age of the animal, composition of the diet, amount of fluoride in the bone and management practices (Shupe *et al.* 1963).

In a number of long-term experimental studies with dairy or beef cattle where as much as 100 p.p.m. of sodium fluoride was added to the diet, the skeletal retention of fluoride was approximately proportional to the amount ingested. The rate of increase of skeletal uptake of fluoride decreases with time, and this was readily apparent in experiments on dairy cattle in which serial biopsies were taken (Shupe *et al.* 1963).

The amount of fluoride in different bones of the bovine skeleton varies appreciably, with cancellous bones—such as those of the pelvis, head, ribs and vertebrae—containing higher concentrations of fluoride than the more compact metacarpal and metatarsal bones. There are also large variations in the fluoride contents of different areas within some bones, such as the metacarpal or metatarsal; the diaphyseal portion has a lower fluoride content than the metaphyseal portion (Shupe *et al.* 1963). It is, therefore, extremely important that analysis of such bones be accompanied by a description of the method of sampling, and that the method of expressing the data (in terms of ash weight or dry fat-free weight) be clearly indicated.

In experimental conditions where there is a constant and invariable intake of fluoride, the amount of fluoride in the diet can be reasonably estimated if the amount of fluoride in the bone and the period of exposure are known (Shupe and Alther 1966; Shupe *et al.* 1963). This cannot be assumed in field cases as differing degrees of dental damage are indicated in range cattle with similar concentrations of fluoride in their bones (Mortenson *et al.* 1964; Shupe 1966*b*, 1970*b*). Experiments do indicate, however, that bone analysis is a good indication of total fluoride intake, but not necessarily of the concentration of fluoride in the present diet (Shupe *et al.* 1962, 1963; Suttie 1969; Suttie, Miller and Phillips 1957).

Urine

A knowledge of the concentration of fluoride in the urine can be a useful diagnostic aid and it can to some extent be correlated with dietary intake (Shupe *et al.* 1963). It is, however, affected by a number of important variables, including the duration of fluoride ingestion, time of day sampling is conducted and total urinary output. Because of variations in urinary output, results of urinary fluoride analysis are usually expressed on a common specific gravity basis. Individual random samples of urine have little value for indicating the fluoride intake of a herd of animals. Sufficient uncontaminated samples should be taken at various intervals to ensure that they are representative of the entire herd (Shupe *et al.* 1963). Urinary fluoride concentrations will remain high for some time after animals are taken off a high fluoride ration (Allcroft, Burns and Hebert 1965; Blakemore, Bosworth and Green 1948; Phillips, Suttie and Zebrowski 1963) and they are therefore not the best indicators of current exposure.

Other tissues

The ingestion of excessive amounts of fluoride by livestock will result in increased concentrations of fluoride in blood and some soft tissues. Recent advances in analytical methods have made it possible to measure the fluoride in the blood plasma with sufficient accuracy to demonstrate a direct correlation between fluoride intake and plasma fluoride concentration. This does not appear to be a practical diagnostic aid, however.

RECOVERY FROM FLUORIDE EXPOSURE

There have been only limited investigations of the response of animals after their removal from a diet containing excessive amounts of fluoride. In these investigations, a decline in the amount of skeletal fluoride has been found, which varied from as low as 10% in 2 years to as much as 50% in a 4–5 year period. Studies are in progress on Hereford cattle which had accumulated 11 000 p.p.m. of fluoride in the bone; when the ration of these animals was changed to a low fluoride control one, 42% of bone fluoride was lost in 6 months and 66% in 24 months. It appears that the percentage loss varies inversely with the time it takes to reach a particular skeletal level and also with the degree of osteofluorosis. Animals ingesting a high concentration of fluoride for a short time will be able to eliminate more fluoride during a subsequent recovery period than animals which accumulate the same concentration in the bone over a longer time at a lower level of intake.

Teeth that are formed after animals are removed from a high fluoride diet do not have dental fluorosis (Brown *et al.* 1960; Garlick 1955; Shupe 1966*b*, 1970*b*). However, teeth which have marked to excessive fluoride lesions will continue to have an increased rate of attrition in subsequent years (Shupe and Olson 1971).

Plasma fluoride concentrations drop to near normal a few days after cattle are removed from a high fluoride diet. Urinary fluoride concentrations decrease at a slower rate than plasma concentrations. There are periods during which urinary fluoride concentrations do not reflect either current intake or bone fluoride concentrations, and their use in diagnosis could be very misleading unless properly interpreted.

The response of animals when changed from an excessive to a low fluoride intake depends on several factors. Each case should be individually evaluated and the potential rate of recovery considered. It is not possible to make a specific prognosis concerning the subsequent performance of animals which have ingested damaging amounts of fluoride. When feed intake and performance are impaired as a result of molar abrasion, there may be limited improvement after the animals have been taken off a high fluoride diet (Shupe 1966b, 1970b; Shupe and Olson 1971). If, however, the decreased intake of food was due to some metabolic effect, it will usually improve as the fluoride intake is decreased.

OTHER EFFECTS OF FLUORIDE TOXICITY

General condition

Borderline cases of fluoride toxicity may not be specific and thus are difficult to diagnose and evaluate. Certain symptoms and lesions due to other causes may mimic some of the non-specific symptoms and lesions of fluoride toxicity (Shupe 1961).

Appetite impairment resulting in general unthriftiness and loss of condition can be the result of excessive fluoride ingestion. Since many toxicities and deficiencies also result in impairment of feed intake, this criterion is not in itself a good diagnostic aid. Abnormal and excessive molar abrasion can also affect feed intake and utilization. A generalized unthriftiness characterized by dry hair and thick non-pliable skin has been observed in animals with other unequivocal symptoms and lesions of fluoride toxicity (Shupe et al. 1963). Reports of high incidences of diarrhoea (Udall and Keller 1952) have not been confirmed in controlled experimental studies or careful field investigations (Burns and Allcroft 1964; Greenwood et al. 1964; Shupe et al. 1963; Suttie and Phillips 1959).

Reproduction

As measured by services required per conception, prolonged ingestion of excessive fluoride under controlled experimental conditions had no effect on reproduction. Conception rates of experimental cows that received up to 93 p.p.m. of sodium fluoride for as long as $7\frac{1}{2}$ years were normal (Shupe et al. 1963). Severe fluoride toxicity can indirectly influence reproduction by its generalized effects (Shupe 1966b, 1970b).

Placental transfer

Examinations of calves from cows that had received various concentrations of sodium fluoride have demonstrated that fluoride passed into the placenta. The amount of bone fluoride in the offspring varied with the amount of fluoride ingested by the dam. The amount of fluoride transmitted across the placenta does not increase in subsequent gestations regardless of the duration of fluoride ingestion (Shupe *et al.* 1963). The concentrations of fluoride in foetal bone are much lower than those in comparable maternal tissues.

Milk production

There is no evidence that milk production is directly affected by ingestion of low concentrations of fluoride. In cases of marked fluoride toxicity, many effects on milk production are probably secondary to major symptoms and lesions induced by fluoride rather than due to direct interference with lactogenesis (Stoddard *et al.* 1963).

In general, well-fed dairy cows can consume 40 to 50 p.p.m. of fluoride in their ration for two or three lactations without any measurable effect on milk production, even when ingestion of such amounts is started at four months of age. Later lactations of some animals maintained on such an intake of fluoride, however, were adversely affected. But in all cases the uptake of fluoride by the body and the subsequent clinical manifestations of the primary symptoms and lesions normally associated with fluoride toxicity will usually precede any alteration in milk production (Greenwood *et al.* 1964; Stoddard *et al.* 1963).

Milk production of cows fed 93 p.p.m. of sodium fluoride apparently was adversely affected in the second lactation and was definitely reduced in the third and subsequent lactations. Milk production of some cows on 49 p.p.m. of sodium fluoride was adversely affected in the fourth and subsequent lactations. The amounts of fluoride in the milk correlated with the amounts of fluoride ingested and did not exceed 0.2 p.p.m. of fluoride even when the cows ingested 93 p.p.m. of fluoride. Experiments and tests demonstrated that the mammary gland was a minor means of fluoride excretion from the body and that the concentrations of fluoride in milk were low and within safe levels for human consumption (Greenwood *et al.* 1964; Stoddard *et al.* 1963).

Blood

Cows receiving up to 93 p.p.m. of sodium fluoride for $7\frac{1}{2}$ years showed no change in blood morphology or adverse effects on the haemopoietic system. More variation could be found within a single treatment group than between different treatment groups (Hoogstratten *et al.* 1965; Shupe *et al.* 1963). No correlation was found between the amount of fluoride in the diet and the concentration of calcium, inorganic phosphorus, or alkaline phosphatase in the blood serum. The amount of fluoride in the blood correlated with the amount of fluoride in the diet (Greenwood *et al.* 1964; Hoogstratten *et al.* 1965; Shupe *et al.* 1963).

Hair and skin

Some cows receiving 93 p.p.m. or 49 p.p.m. of sodium fluoride for $7\frac{1}{2}$ years had dry hair and thick non-pliable skin. These conditions were manifested after osteofluorosis and intermittent lameness and stiffness were apparent (Greenwood *et al.* 1964; Shupe *et al.* 1963).

Hoofs

Adverse effects of fluorides on hoofs had been reported (Udall and Keller 1952). This condition has not been observed in other geographical areas or reported in experimental studies (Hobbs and Merriman 1962; Shupe *et al.* 1963; Suttie and Phillips 1959).

No adverse effect on the growth and shape of the cows' hoofs could be associated with ingestion of 27 p.p.m., 49 p.p.m., or 93 p.p.m. of sodium fluoride for $7\frac{1}{2}$ years. More variation occurred within a single fluoride treatment group than between groups (Greenwood *et al.* 1964; Shupe *et al.* 1963). Extensive field studies of clinical fluoride toxicity cases did not reveal any correlation between fluoride intake and shape and growth of hoofs (Shupe and Alther 1966; Shupe 1966*b*, 1970*b*).

Soft tissues

Only small amounts of fluoride (less than 2.5 p.p.m.) were retained in the soft tissues of cattle receiving up to 93 p.p.m. sodium fluoride for $7\frac{1}{2}$ years. The kidneys routinely had a higher content of fluoride than any other soft tissue, probably because of their role in eliminating fluorides from the body. No significant gross or histological changes were noted in the brain, pituitary, liver, kidneys, adrenals, spleen, pancreas, thyroid, ovaries, mammary glands, uterus, rumen, reticulum, omasum, abomasum, or intestines. In enzyme studies of some of the above selected soft tissues, there were no significant effects related to fluoride ingestion (Greenwood *et al.* 1964; Shupe *et al.* 1963). In field cases of fluoride toxicity, interpretation of the pathological changes in the soft tissues may be difficult because of the association of conditions unrelated to the ingestion of fluoride. To date, no unequivocal histological or significant functional changes in the soft tissues of cattle can be correlated with high (up to 93 p.p.m.) prolonged ($7\frac{1}{2}$ years) fluoride ingestion. According to the brom-sulphthalein (BSP) liver function tests, no correlation existed between fluoride treatments and BSP clearance (Greenwood *et al.* 1964; Shupe *et al.* 1963).

DIAGNOSTIC AIDS

Symptoms and visible lesions are sometimes not definitive enough to warrant an unequivocal diagnosis of fluoride toxicity. Its subtle and insidious effects may in some cases interfere with normal performance and productivity. In such instances, or to substantiate a reasonably certain diagnosis, additional verification can be sought in several ways. When properly interpreted, urine analyses are a useful diagnostic aid. Bone radiographs can supply valuable information. Biopsies or necropsies of properly selected cases can be used to obtain tissues for gross and histopathological evaluations and for chemical analyses for fluoride content. Correlation of symptoms, lesions and results of chemical analyses of tissues with the ascertained fluoride contents of animals' water and forage sources can often help substantiate suspected instances of fluoride toxicity.

METHODS OF ALLEVIATING FLUORIDE TOXICITY

No substances are known to prevent completely the toxic effects of high concentrations of ingested fluorides. Some products, however, can counteract and lessen the damage that fluorides can induce.

Aluminium sulphate, aluminium chloride, calcium aluminate, calcium carbonate, and defluorinated phosphate have reduced the toxicity of fluoride. Heifers ingesting aluminium sulphate as an inhibitor of fluoride toxicity deposited 30–42% less fluoride in their ribs than did heifers ingesting comparable diets without aluminium sulphate (Greenwood *et al.* 1964; Shupe and Alther 1966; Shupe 1970*b*). If animals must be fed hay high in fluoride, it should at least be combined with hay of low fluoride content. If the drinking water available to the animals has an unavoidably high fluoride content, then it is advisable to feed roughage of low fluoride content. Land with a high fluoride content can be planted with cereals or non-roughage crops because grain does not accumulate fluoride as readily as do forage crops. In such cases, the necessary hay or roughage can be imported from nearby areas of low fluoride content. All mineral mixtures fed to animals should be made from defluorinated phosphate.

EXTRAPOLATION OF CONTROLLED EXPERIMENTAL FINDINGS TO CLINICAL FIELD CONDITIONS

Individuals diagnosing and evaluating fluoride toxicity in animals should be properly trained, knowledgeable and experienced. People assisting with fluoride-induced or suspected fluoride-induced problems should have access to all available records, data and findings that relate to the animals involved.

The diagnosticians should make detailed clinical examinations of all animals suspected of being adversely affected. Regardless of monitoring and sampling data related to the source(s) of fluoride and analyses of fluoride values in the vegetation, each individual animal must be evaluated on its own expression of fluoride toxicity (Shupe 1966*b*, 1970*b*; Shupe and Olson 1971).

It is advantageous to compare animals under approximately similar exposure and management practices with animals having different degrees of exposure and varying management practices. Important factors may be missed if one does not also observe animals in adjacent areas. Some abnormal symptoms thought to be due to excessive fluoride may not be fluoride-induced, and some animals thought to be normal may be adversely affected. The variability that can occur between individual animals within and between herds must also be considered.

Under field conditions the amounts of fluoride ingested, the duration of excessive fluoride ingestion and management practices are usually subject to wide variations. In some instances, extremely high or low amounts of fluoride may be ingested intermittently. Management practices such as preventive medical programmes and housing facilities, type, quantity and quality of feed, breeding programme, and routine animal care can influence manifestations of fluoride-induced symptoms and lesions.

The relationship of the age of the animal to excessive fluoride intake and lesions is also important (Shupe *et al.* 1963). For example, molar abrasion usually occurs with severely affected first, second, or third pairs of incisor teeth (I_1, I_2, I_3); but molar abrasion with severely affected corner incisor teeth (I_4) has a different significance because when there are severe fluoride lesions in these incisors only, there is usually little or no molar abrasion. Additional factors that should be considered are the type of operation, such as dairy or beef, and whether the enterprise is a fattening, replacement, or cow and calf operation. All findings must be properly considered and interrelated before a definite and final evaluation of fluoride toxicity is made (Shupe and Alther 1966; Shupe *et al.* 1963; Shupe 1966*b*: Shupe and Olson 1971).

Based on short and long-term controlled experiments, clinical examinations of more than 85 825 animals in enzootic fluorotic areas and 655 detailed necropsy findings from animals of different ages showing various degrees of fluoride effects, Tables 1 and 2 were developed and consolidated to facilitate the diagnosis and evaluation of fluoride toxicity.

SUMMARY

With the expansion of certain types of industrial operations that emit excessive fluorides into the atmosphere in agricultural areas, fluoride toxicity in animals has been an important toxicological problem in the United States and many other parts of the world. Animals normally ingest low concentrations of fluoride with no adverse effects. Small amounts of fluoride may have beneficial effects, but when excessive amounts are ingested for prolonged periods of time adverse effects are induced. Many sources may contribute to the total fluoride intake of animals. Various factors influence biological responses of animals to excessive ingested fluorides. Symptoms and lesions of fluoride toxicity in animals have been characterized. Fluoride standards and a comprehensive guide for their use in diagnosing and evaluating fluoride toxicity in livestock have been compiled. Prevention and control of fluoride toxicity in livestock can be achieved when the complexity of the disease is realized and the pathogenesis, symptomatology and lesions are properly correlated, interpreted and evaluated, and the source(s) of excessive fluorides are eliminated.

References

AGATE, J. N., G. H. BELL, G. F. BODDIE, R. G. BOWLER, M. BUCKELL, E. A. CHEESEMAN, T. H. J. DOUGLAS, H. A. DRUETT, J. GARRAD, D. HUNTER, K. M. A. PERRY, J. D. RICHARDSON and J. B. DE VAN WEIR (1949) *Med. Res. Counc. Memo.* No. 22.

ALLCROFT, R., K. N. BURNS and C. N. HEBERT (1965) *Fluorosis in cattle. II. Development and alleviation: Experimental studies* (Ministry of Agriculture, Fisheries and Food, Animal Disease Surveys Report No. 2, Part II). London: H.M.S.O.

ALLCROFT, R. and J. S. L. JONES (1969) *Vet. Rec.* **84**, 399–402.

ALLCROFT, R., F. J. SALT, R. A. PETERS and M. SHORTHOUSE (1969) *Vet. Rec.* **84**, 403–409.

ALTHER, E. W. (1961) *Chemisch-biologische Untersuchungen zur Fluorose des Rindes* (Dissertation Landwirtschaftliche Hochschule, Hohenheim). Zurich: Conzett & Huber.

BERNSTEIN, S. and P. COHEN (1967) *J. Clin. Endcrinol. & Metab.* **27**, 197–210.

BLAKEMORE, F., T. J. BOSWORTH and H. H. GREEN (1948) *J. Comp. Pathol. & Ther.* **58**, 267–302.

BROWN, W. A. B., P. V. CHRISTOFFERSON, M. MASSLER and M. B. WEISS (1960) *Am. J. Vet. Res.* **21**, 7–34.

BURNS, K. N. and R. ALLCROFT (1964) *Fluorosis in cattle. 1. Occurrence and effects in industrial areas of England and Wales 1954–57* (Ministry of Agriculture, Fisheries and Food, Animal Disease Surveys Report No. 2, Part I). London: H.M.S.O.

CASS, J. S. (1961) *J. Occup. Med.* **3**, 471–477.

GARLICK, N. L. (1955) *Am. J. Vet. Res.* **16**, 38–44.

GREENWOOD, D. A., J. L. SHUPE, G. E. STODDARD, L. E. HARRIS, H. M. NIELSEN and L. E. OLSÓN (1964) *Spec. Rep. Utah Agric. Exp. Stn.* No. 17.

HOBBS, C. S. and G. M. MERRIMAN (1962) *Bull. Tenn. Agric. Exp. Stn.* No. 351.

HODGE, H. C. and F. A. SMITH (1965) In *Fluorine Chemistry*, vol. IV, pp. 2–375, ed. J. H. Simons. New York: Academic Press.

HOOGSTRATTEN, B., N. C. LEONE, J. L. SHUPE, D. A. GRLENWOOD and J. LIEBERMAN (1965) *J. Am. Med. Assoc.* **192**, 26–32.

HUPKA, E. and P. LUY (1929) *Arch. Wiss. Prakt. Tierheilkd.* **60**, 21–39.

JOHNSON, L. C. (1965) In *Fluorine Chemistry*, vol. IV, pp. 424–441, ed. J. H. Simons. New York: Academic Press.

KRUG, O. (1927) *Z. Fleisch- u. Milchhyg.* **37**, 38–39.

LEONE, N. C., E. F. GEEVER and N. C. MORAN (1956) *Public Health Rep. Wash. D.C.* **71**, 459–467.

LEONE, N. C., C. A. STEVENSON, T. F. HILBISH and M. C. SOSMAN (1955) *Am. J. Roentgenol.* **74**, 874–885.

MILLER, G. W. and J. L. SHUPE (1962) *Am. J. Vet. Res.* **23**, 24–31.

MITCHELL, H. H. and M. EDMAN (1952) *Nutr. Abstra. & Rev.* **21**, 787–804.

MORTENSON, F. N., L. G. TRANSTRUM, W. P. PETERSON and W. S. WINTERS (1964) *J. Dairy Sci.* **47**, 186–191.

PHILLIPS, P. H., D. A. GREENWOOD, C. S. HOBBS and C. F. HUFFMAN (1955) *Publ. NAS-NRC.* No. 381.

PHILLIPS, P. H., E. B. HART and G. BOHSTEDT (1934) *Bull. Wis. Agric. Exp. Stn.* No. 123.

PHILLIPS, P. H., J. W. SUTTIE and E. J. ZEBROWSKI (1963) *J. Dairy Sci.* **46**, 513–516.

ROHOLM, K. (1937) *Fluorine Intoxication: a clinical-hygienic study with a review of the literature and some experimental investigations*, pp. 213–253, transl. W. E. Calvert. London: Lewis.

SCHMIDT, J. J., G. W. NEWELL and W. E. RAND (1954) *Am. J. Vet. Res.* **15**, 232–239.

SCHOUR, I. and M. C. SMITH (1934) *Tech. Bull. Ariz. Agric. Exp. Stn.* No. 52, 69–91.

SHAW, J. H. (ed) (1954) *Fluoridation as a Public Health Measure.* Washington D.C.: American Association for the Advancement of Science.

SHUPE, J. L. (1961) *Can. Vet. J.* **2**, 369-376.

SHUPE, J. L. (1963) *Diseases of Cattle*, 2nd. edn, p. 740. Santa Barbara, Calif.: American Veterinary Publications.

SHUPE, J. L. (1966*a*) *Fluorosis, International Encyclopedia of Veterinary Medicine*, vol. 2, pp. 1062–1068. Edinburgh: Green. London: Sweet & Maxwell.

SHUPE, J. L. (1966*b*) *Proc. IV Int. Meet. Dis. Cattle* (Zurich 1966) pp. 1–18.

SHUPE, J. L. (1970*a*) *Am. Ind. Hyg. Assoc. J.* **31**, 240–247.

SHUPE, J. L. (1970*b*) *Fluorosis. Bovine Medicine and Surgery.* pp. 288–301. Santa Barbara, Calif.: American Veterinary Publications.

SHUPE, J. L. and E. W. ALTHER (1966) *Handb. Exp. Pharmakol.* **20**, pt 1, 307–354.

SHUPE, J. L., L. E. HARRIS, D. A. GREENWOOD, J. E. BUTCHER and H. M. NIELSEN (1963) *Am. J. Vet. Res.* **24**, 300–306.
SHUPE, J. L., M. L. MINER and D. A. GREENWOOD (1964) *Ann. N.Y. Acad. Sci.* **111**, 618–637.
SHUPE, J. L., M. L. MINER, D. A. GREENWOOD L. E. HARRIS and G. E. STODDARD (1963) *Am. J. Vet. Res.* **24**, 964–984.
SHUPE, J. L., M. L. MINER, L. E. HARRIS and D. A. GREENWOOD (1962) *Am. J. Vet. Res.* **23**, 777–787.
SHUPE, J. L. and A. OLSON (1971) *J. Am. Vet. Med. Assoc.* **158**, 167–174.
SLAGSVOLD, L. (1934) *Nor. Vet. Tidsskr.* **46**, 2–16, 61–68.
SMITH, F. A. (1962) *J. Am. Dent. Assoc.* **65**, 598–602.
STODDARD, G. E., G. Q. BATEMAN, L. E. HARRIS, J. L. SHUPE and D. A. GREENWOOD (1963) *J. Dairy Sci.* **46**, 720–726.
SUTTIE, J. W. (1969) *J. Air Pollut. Control Assoc.* **19**, 239–242.
SUTTIE, J. W., R. F. MILLER and P. H. PHILLIPS (1957) *J. Dairy Sci.* **40**, 1485–1491.
SUTTIE, J. W. and P. H. PHILLIPS (1959) *J. Dairy Sci.* **42**, 1063–1069.
TAVES, D. R. (1970) *Fed. Proc. Fed. Am. Soc. Exp. Biol.* **29**, 1185–1187.
TAVES, D. R., R. B. FREEMAN, D. E. KAMM, C. P. RAMOS and B. S. SCRIBNER (1968) *Trans. Am. Soc. Artif. Intern. Organs* **14**, 412–414.
UDALL, D. H. and K. P. KELLER (1952) *Cornell Vet.* **42**, 159–184.
VELU, H. (1932) *Arch. Inst. Pasteur Alger.* **10**, 41–118.
WHO (1970) *Monogr. Ser. W.H.O.* No. 59.

Discussion

Peters: Is 30 p.p.m. fluoride safe in a diet for cattle?

Shupe: Yes, if this is based on the amount in the forage and not the drinking water.

Peters: Mrs Shorthouse and I found that our rats were excreting far too much fluoride. We analysed pellets of the rat food, 41B, and to our astonishment they contained between 10 and 20 p.p.m. of inorganic fluoride. I thought this was getting near the limit where one might observe slight changes in the teeth, and sometimes we think we see a little yellowing of the teeth. But the rats grow and breed happily on 20 p.p.m. fluoride.

Shupe: There is slight mottling of the teeth of cattle receiving 25 to 30 p.p.m. of fluoride in forage, but the size and shape of the teeth are normal and they do not wear abnormally. With 40 to 50 p.p.m., mottling is more prominent and abnormal wear may occur with prolonged use.

Armstrong: Is there any evidence from your work that ingestion of increased amounts of fluoride is beneficial in the treatment of osteoporosis? I note that you produced osteoporosis with high doses of fluoride.

Shupe: Osteoporosis was induced by a prolonged high fluoride intake. With a lower but still abnormally high intake, osteosclerosis was induced. We have unpublished data which indicate that small doses of fluoride ingested for a long time influence the physical properties of bone. These changes, up to a point, may have preventive beneficial aspects. In our studies to date, we have no definite evidence that these levels are useful in the treatment of osteoporosis.

Armstrong: Our results with rats do not show any evidence that increases in the intake of fluoride prevent osteoporosis. However, we may have used the wrong model of osteoporosis, because we cut the nerves of the brachial plexus on one side, thus paralysing one limb with the production of 'disuse atrophy' of the humerus. The situation may not be comparable with post-menopausal or senile osteoporosis in the human.

Saunders: Professor Shupe, is there a very high concentration of fluoride in the periosteum of the treated animals?

Shupe: We have never analysed the periosteum itself; however, we have analysed various areas and zones within specific bones and found that in osteofluorosis the sub-periosteal bone had a higher fluoride content than the original cortical bone.

Cain: If the rancher has to take his cattle off polluted pasture his only alternatives are to find non-polluted pastures which could be miles away, or to put his herd on to artificial feeds, which would be expensive. So may I ask: in what form is the polluting fluorine being released from these industrial factories, and once production stops how quickly does the fluorine on pasture leach away in the soil?

Shupe: In one study of comparative toxicity we found that the industrial fluoride effluent was comparable in toxicity to sodium fluoride. The effluent may vary, however, depending on the type of industry, materials processed and other factors. Soil contamination and content are not the major source of fluoride intake in cases of industrially related fluoride toxicity in animals.

Miller: Emission from industrial plants producing phosphate fertilizers or aluminium include fluoride and silicon tetrafluoride. Within a mile's radius of many industrial plants the presence of sub-micron particles containing fluoride is a real problem.

Shupe: When the results of fluoride analyses of vegetation are reported, it is important to state whether the sample was washed or unwashed, because much of the total fluoride may be present as surface contamination.

Cain: But what can the rancher do?

Shupe: Many good management factors may be beneficial in reducing fluoride intake and the degree of fluoride toxicity in animals. Proper utilization of pastures, reduction of total fluoride intake by supplying additional non-contaminated forages, and the use of certain alleviators such as aluminium sulphate, calcium aluminate or calcium carbonate are some beneficial proven practices.

Kent: Are the molecular changes in the collagen structure known in the above conditions? The glycoproteins seem to precede the calcification process. Andrews and Herring (1965) showed that in the resorbing areas of bone there was a glycoprotein which would specifically bind yttrium, americium, thorium and plutonium, even in the presence of calcium. Might there be a fluoride-binding counterpart which would assist in retaining this anion in bone?

Shupe: The exact intracellular effects and precise mechanisms of osteofluorosis and bone remodelling that are associated with fluoride toxicity are not yet fully understood. We do know that there is an induction of an abnormal organic matrix with irregular haphazardly arranged collagenous fibres. Abnormal osteo-blastic activity is thought to cause the formation of abnormal organic matrix with resultant disorderly, defective and irregular mineralization. In cattle, horses and sheep an increase in the fluoride concentration in the bone is associated with a decrease in the citrate concentration in the bone and an increase in crystallinity and bone alkaline phosphatase activity.

Kent: If one decalcifies the fluorotic bone with EDTA is all the fluoride released?

Shupe: Most of the fluoride in fluorotic bone is removed when the bone is decalcified with EDTA; most of the fluoride in the bone is in the inorganic portion which is readily acted upon by decalcifying agents.

Armstrong: Dr Kent, we have seen some unconvincing evidence of the presence of organic fluoride in the collagen and other organic constituents of bone, with respect to teeth, but certainly 99.9% of fluoride in bone is in the mineral phase.

Peters: Dr Ward, Mrs Shorthouse and I had one specimen of fluorotic bone with a very high fluoride content. After a great deal of effort we decided that a small amount of fluorocitrate was present, and this was confirmed by gas chromatography (Peters, Shorthouse and Ward 1969). However, we have been unable to find fluorocitrate in other specimens of fluorotic bone. Miller and Phillips (1955) claimed that when fluoroacetate was fed to rats there was a good deal of fluoride in the bone over and above that found in control rats. The C—F link in fluoroacetate appears to have been split by the rats. I thought this was the first sign of fluoroacetate having fluoride splitting off *in vivo*, but it may have been fluoroacetate going to fluorocitrate. And Ott, Piller and Schmidt (1956) claimed that fluorovalerate improved the teeth, which I do not think has ever been confirmed.

Shorthouse: Professor Shupe, does the fluoride come out of the bone and appear in the urine when the cattle go on a lower fluoride diet, and if it does, what form is it in? Is it acid-labile?

Shupe: The response of animals transferred from a high to a low fluoride diet will vary depending on several factors. Most of the fluoride is eliminated in the urine. After cattle are taken off a high fluoride ration, the urinary fluoride concentration decreases at a slower rate than the plasma concentration. The urine was analysed for total fluoride; the specific form was not determined.

Armstrong: If a dose of fluoride is given, fluoride accumulates in the skeleton fairly quickly, and within a matter of minutes the output of fluoride in the urine begins to increase. We conclude that the skeleton is an important organ in the regulation of the ionic concentrations of body fluids including, and to a special degree, that of fluoride. Thus even though we use our skeletons for rigidity and support of the body and for protection of vital organs, they have a biological role which transcends that of a supporting tissue (Yeh, Singer and Armstrong 1970).

Gál: How much neurotoxicity due to fluoride has been observed? Are there any changes in parathyroid histology?

Taves: There has been some work in sheep where the parathyroid was enlarged after doses of fluoride. We tried to demonstrate an effect of fluoride on the parathyroid of the rat and failed (Raisz and Taves 1967).

Shupe: We have examined the thyroid and parathyroid glands of cattle that were put on controlled diets with various levels of fluoride at $3\frac{1}{2}$ months of age. They were kept on the controlled diets of fluoride for $7\frac{1}{2}$ years. We also examined 655 field cases of animals with fluoride toxicity. No gross or histological abnormalities or changes in the thyroid or parathyroid glands of these animals were detected.

What is your definition of neurotoxicity, Professor Gál?

Gál: Cytological changes in the spinal cord or ganglia or any part of the central nervous system.

Armstrong: Would you regard the osteophyte pressure on the spinal nerves in skeletal fluorosis as neurotoxicity?

Gál: There may be a secondary effect of myelin sheath degeneration.

Armstrong: With large doses of intravenous fluoride the amount of calcium in the blood can be reduced to the concentration that produces convulsions, at least in rabbits.

Peters: As long as the kidneys are regulating properly there will not be an increased concentration of fluoride in the blood, so the bones will not be subjected to a large concentration. In your cattle, Professor Shupe, did the concentration of fluoride in the blood rise much?

Shupe: It rose in the animals on the high fluoride diets. The amount of fluoride in the blood correlated with the various levels of fluoride ingested. There was no evidence of change in blood morphology or detectable adverse effects on the haemopoietic system.

Taves: I do not think the idea that the fluoride concentration of blood is kept constant by the action of the kidney is correct. The clearance of fluoride from the kidney stays fairly constant whether the subject has a low or high fluoride intake. So the concentration of inorganic fluoride in the blood will be roughly proportional to the intake unless there is a change in the clearance in bone. If there is renal disability the clearance goes down, so the retention and the blood concentrations go up. I think the confusion originally arose because total fluoride rather than inorganic fluoride was being measured.

Armstrong: Another factor to be considered here is related to the time of sampling the blood. After a dose of fluoride the concentration of fluoride in plasma reaches a maximum after about an hour (or two); it then starts to decline as a result of clearing of fluoride by bone and excretion in the urine. We sample our patients' plasma at a standard time in order to get meaningful results for plasma fluoride contents.

Bergmann: We have not found any organic fluoride in any plants growing near a phosphate factory which produces relatively large quantities of HF.

Miller: We reported the presence of fluoroacetate in a few samples of crested wheat grass collected near a phosphate factory in Montana (Yu and Miller 1970). I do not understand why it is not found in the forage from all areas within a radius of a mile of the factory. Contamination of the positive samples is a possibility. Also we have analysed many tissues from animals from the same area and so far we have only found indications of fluoroacetate in the kidneys from one horse.

Cain: Mr J. Cooke, at Newcastle, has found no organic fluorine in native wild plants growing in the fluorospar-rich areas of Weardale in Co. Durham where the amount of fluoride in the soil is very high, in the range of 6000–10 000 p.p.m.

Armstrong: Did he find inorganic fluorine?

Cain: Yes, in high concentration.

Miller: In many of our plant samples the inorganic fluoride content exceeded 1000 p.p.m. on a dry weight basis.

Cain: These are the sorts of amounts that Cooke found. In *Agrostic tenuis*, for instance, the leaves contained 1400 p.p.m.; in *Thymus drucei*, the whole shoots plus leaves contained 1500 p.p.m. The shoots of *Campanula rotundifolia* contained 550 p.p.m., those of *Lolus caniculatus* 280 p.p.m., and the leaves of *Deschampsia caespitosa* contained 250 p.p.m. (all figures on a dry weight basis).

Dr A. Davidson, at Newcastle, has recently informed me that occasional values as high as 5000 p.p.m. fluoride have been recorded from plants in this area, and this after vigorous washing of the leaf surfaces to remove adhering fluoride (a solution of EDTA and detergent was found to be the best means of removing adhering surface fluoride). Mr Cooke has even examined the leaf surfaces by stereoscan microscopy to check that no crystals still adhere. We are confident that these fluoride concentrations are internal—though not necessarily soluble.

Saunders: There are many carbon-fluorine compounds that are extremely stable. Sodium fluoroacetate is a very stable and insidious compound, and I personally deprecate its use as a rodenticide, especially as it is used in a rather indiscriminate way.

Peters: I entirely agree with you. In Australia its use is very carefully controlled. But this is not so in the United Kingdom. Fluoroacetate is a marvellous rat killer and is put down in the sewers. But nobody knows where it goes. It may get into the fish, which are not particularly susceptible, and people may eat the fish. I think it is important to find a rat poison that is not nearly as stable as fluoroacetate.

All the symptoms of fluoroacetate poisoning might also be found as a result of other diseases, and it is extraordinarily difficult to convince a pathologist that they are really due to fluoroacetate.

Saunders: I was particularly unhappy when fluoroacetamide was being used as a general systemic insecticide.

Taylor: I think we would all agree that extraordinary care and control must be exercised over any very toxic material that is put in the environment.

References

ANDREWS, A. T. DE B. and G. M. HERRING (1965) *Biochim. Biophys. Acta* **101**, 239.
MILLER, R. F. and P. H. PHILLIPS (1955) *Proc. Soc. Exp. Biol. Med.* **89**, 411–413.
OTT, E., G. PILLER and H. J. SCHMIDT (1956) *Helv. Chim. Acta* **39**, 682–685.
PETERS, R. A., M. SHORTHOUSE and P. V. WARD (1969) *Biochem. J.* **113**, 9P.
RAISZ, L. R. and D. R. TAVES (1967) *Calif. Tiss. Res. 1*, 219–228.
YEH, M. C., L. SINGER and W. D. ARMSTRONG (1970) *Proc. Soc. Exp. Biol. Med.* **135**, 421–425.
YU, M. H. and G. W. MILLER (1970) *Environ. Sci. & Technol.* 492–495.

Chairman's concluding remarks

SIR RUDOLPH PETERS

Department of Biochemistry, University of Cambridge

When this meeting was being prepared, we hoped that three days spent discussing interdisciplinary scientific aspects of monofluoro compounds would present a satisfactorily complete picture of the present state of research in this field. I think that we can be satisfied that we have done this. Furthermore, it is gratifying to note how much new research work has been presented during our discussions. This is a bonus for which we could not have hoped. As our time was limited, we have had perforce to leave out any session devoted to discussions of methods of analysis, important as these may be. But I think that it will be clear to any reader that an evaluation of some of the practical and controversial problems of immediate importance, such as that of fluoridation of water, must rest necessarily upon the scientific background which has been presented here. I hope also that no reader will now think that study of these compounds is solely aimed at 'pollution' problems. One might ask deliberately whether the following can be regarded purely as problems of pollution: the metabolism of fluoride in plants; the various effects of fluoride in combination with sugars and steroids; the presence of fluorine in the antibiotic nucleocidin; the light thrown upon metabolic reactions in the tricarboxylic acid cycle in liver and brain by treatment with fluoroacetate and fluorocitrate, and so upon an important aspect of life; and the clinical use of fluoronucleotides in some forms of cancer and herpes? All of these, and there are more, are enlarging our scientific horizons; and as scientists well know, little progress in practical problems can come without a clarity of the scientific background to inform the prepared mind.

Index of Contributors

Entries in **bold type** *refer to papers; other entries are contributions to discussions*

Subject Index